T0309089

Fermentation: Processes and Applications

Fermentation: Processes and Applications

Edited by Jeffery Goodman

New York

Published by Syrawood Publishing House,
750 Third Avenue, 9th Floor,
New York, NY 10017, USA
www.syrawoodpublishinghouse.com

Fermentation: Processes and Applications
Edited by Jeffery Goodman

International Standard Book Number: 978-1-64740-347-8 (Hardback)

Cataloging-in-publication Data

Fermentation : processes and applications / edited by Jeffery Goodman.
p. cm.
Includes bibliographical references and index.
ISBN 978-1-64740-347-8
1. Fermentation. 2. Biochemical engineering. I. Goodman, Jeffery.

TP156.F4 F47 2023
660.284 49--dc23

TABLE OF CONTENTS

PREFACE

Fermentation is a metabolic process in which organisms anaerobically convert starch or sugar to alcohol or an acid releasing energy. The process of fermentation is catalyzed with the help of enzymes. The science of fermentation is known as zymology. The first process of fermentation is the same as cellular respiration, which involves the formation of pyruvic acid by glycolysis where a total of two ATP molecules are synthesized. In the next process, pyruvate is reduced to lactic acid, ethanol or other products resulting in the formation of NAD+, which is re-utilized back in the glycolysis process. Based on the end-product formed, there are four types of fermentation, namely, lactic acid fermentation, alcohol fermentation, acetic acid fermentation, and butyric acid fermentation. Fermentation process is widely used to make products such as wine, beer, biofuels, yoghurt, pickles, bread, sour foods containing lactic acid, etc. The consumption of fermented food can benefit people by making food nutritious, digestible, and flavorful. It also helps in reducing lactose intolerance and improving the immune system. This book outlines the processes and applications of fermentation in detail. It is appropriate for students as well as for experts seeking detailed information in the area of zymology.

This book is the end result of constructive efforts and intensive research done by experts in this field. The aim of this book is to enlighten the readers with recent information in this area of research. The information provided in this profound book would serve as a valuable reference to students and researchers in this field.

At the end, I would like to thank all the authors for devoting their precious time and providing their valuable contribution to this book. I would also like to express my gratitude to my fellow colleagues who encouraged me throughout the process.

Editor

1

Craft Beers: Current Situation and Future Trends

María Jesús Callejo, Wendu Tesfaye, María Carmen González and Antonio Morata

Abstract

During the twentieth century, the consolidation of large multi-national beer companies and the homogenization of the specified beer types have led to a considerable growth in the beer industry. However, the growing demand by consumers of a single and distinctive product, with a higher quality and better sensory complexity, is allowing for a new resurgence of craft beer segment in recent years. This chapter reviews some different alternatives of innovation in the craft brewing process: from the bottle fermented beers with non-*Saccharomyces* yeast species, to the use of special malts or specific adjuncts, hop varieties, water quality, etc. All of them open a lot of new possibilities to modulate flavor and other sensory properties of beer, reaching also new consumers looking for a specific story in one of the oldest fermented beverages.

Keywords: craft beer, sour beer, non-*Saccharomyces*, new adjuncts, bottle fermentation

1. Introduction

Beer brewing is an established ancient art in different civilization and cultures, but there is no a precise and unanimous agreement on the origin of beer. Recent evidences predominantly based on the archeological and historical evidences explain the origin of brewing across time and space [1]. The timespan for its existence differs over a wide range of geography, from as far back as "The Neolithic Revolution" to the early horizon in South America. It commenced in the agricultural or "Neolithic" revolution period as early as 9000 BC with the advent of the Sumerians in the lowlands of the Mesopotamian alluvial plane [2, 3]. Evidence of rice-based fermented beverage has been found in between 7000 and 5000 BC in China [4–7] and ancient Mesopotamia back to about 6000 BC [8–10]. Similarly, in Northern Africa highlighting Egypt at about 3500 BC [11],
in Europe around 3000 BC [12, 13] and in South America 900-200 BC [14–16], locally fermented alcoholic beverages have been produced. Recent starch [17] and chemical residue studies [18] extend this period as far as 11,000 BC. In broader terms, all these fermented beverages may be considered as a craft beer based on the production scale.

2. Craft beer: as a movement from bottom to top fermentation. The reemergence of craft brewing

Different cultures and different civilization historically produced a number of fermented beverages/beer with different raw materials, which allowed them to have different attributes and different names. Beer is a relatively simple fermented product, mainly water in its composition, which makes easily produced locally; however, for a long time the difficulty to move long distances permits to flourish craft brewers everywhere in the world [19]. However, at present, it is not an impedi-ment due to technological advances and transportation progress.

The craft beer movement or revolution began in the USA after the 13 years of national prohibition of alcohol or "the noble experiment" 1919–1933. In 1965, Fritz Maytag, the man of the craft beer renaissance, bought the Anchor Steam Beer Company of San Francisco with a capacity of 50,000 barrels and developed it as a craft brewery outlet [20, 21]. Regarding the USA, this was the milestone to the expanding innovation and an increasing trend in terms of production and sales of beers with differentiated quality.

Even though this movement marked a shift in several countries recently, to mention some, in 1988 the earliest brewpub lay foundation in Italy [22], while in the Netherlands the craft revolution rouse during the year 1981 [23], in Australia, craft brewing started late 1984 [24]. At the same time, it is very difficult to put a time limit for the begin-ning of craft beer production in some European countries like the UK, Belgium and Germany where these countries were either with a long tradition in "special beers" or the historical existence of small and local producers back to the 1970s [25, 26].

3. Craft beer: statistical viewpoint

One of the indicators for the expansion of craft beer renaissance in different countries is the statistical approach. However, there is no common shared definition of craft beer but different associations and entities of different countries remark based on the size of the firm, production volume raw materials used to produce such drink, degree of independence and way to brew [27]. Even though there is data scarcity on the numbers of microbreweries of different countries, in **Table 1**, those microbreweries actually existing in different countries of the five continents are represented, where beer production is traditional or its consumption is highly relevant at present. The statistical data reflected in this chapter encompasses all beer producers recognized as craft brewery, artisanal brewery, microbrewery, independent brewery, specialty brewery, Brewpub, local brewery, Regional brewery and Contract brewing company in accordance with the regulation rules of different countries without establishing any distinctions among them.

Even though Mergers and acquisitions seem to reduce the number of major brewing firms, the total production in volume is not affected. This tendency provides an opportunity to the merged breweries to take advantage over the microbreweries in terms of economical scale and increased market share. Despite the macrobrewers dominance, worldwide craft beer numbers are increasing at a rapid rate [28].

4. New tendencies: is the glass half-full or half-empty for brewers?

The last two decades brewing landscape continues to rise in number of micro and craft breweries almost everywhere in the world. As it is shown below (**Table 1**), in the last decade, from 2008 to 2017, the number of craft breweries significantly

	2008	2009	2010	2011	2012	2013	2014	2015	2016	2017
Europe	1755	2123	2407	2670	3094	3616	N/A	N/A	N/A	N/A
Czech Republic[1]	57	51	65	90	20	207[8]	238	202	350	402
France[1]	N/A	263	322	373	293	504	566	690	850	1000
Germany[1]	594	628	646	659	665	668	677	723	738	824
The Netherlands	N/A	N/A	115[2]	N/A	N/A	N/A	222	380[2]	434[3]	N/A
Poland	79[5]	89[5]	107[5]	133[5]	164[5]	212[5]	263[5]	308[5]	N/A	N/A
Spain	21	27	46	70	114	203	314	409	465	502
The UK	671	694	778	898	1250	1440	1414	1828	2198	2378
America										
The USA	1321[7]	1596[7]	1754[7]	2016[7]	2420[7]	2898[7]	3739[7]	4544[7]	5424[7]	6266[7]
Canada	N/A	N/A	277[9]	N/A	N/A	N/A	N/A	610[9]	612[3]	N/A
Africa										
South Africa	22[6]	22[6]	27[6]	34[6]	54[6]	63[6]	N/A	N/A	N/A	N/A
Asia										
Japan	N/A	N/A	208[2]	N/A	N/A	N/A	N/A	222[2]	N/A	N/A
Oceania										
Australia	N/A	N/A	172[2]	N/A	N/A	N/A	N/A	358[2]	410[2]	N/A
New Zealand	N/A	N/A	62[4]	59[4]	65[4]	79[4]	98[4]	111[4]	130[4]	N/A

[1] *[29].*
[2] *[24, 26].*
[3] *[29].*
[4] *[30].*
[5] *[31].*
[6] *[32].*
[7] *[27].*
[8] *[29].*
[9] *[33].*

Table 1.
Microbrewery expansion in the last decade sample countries by continent.

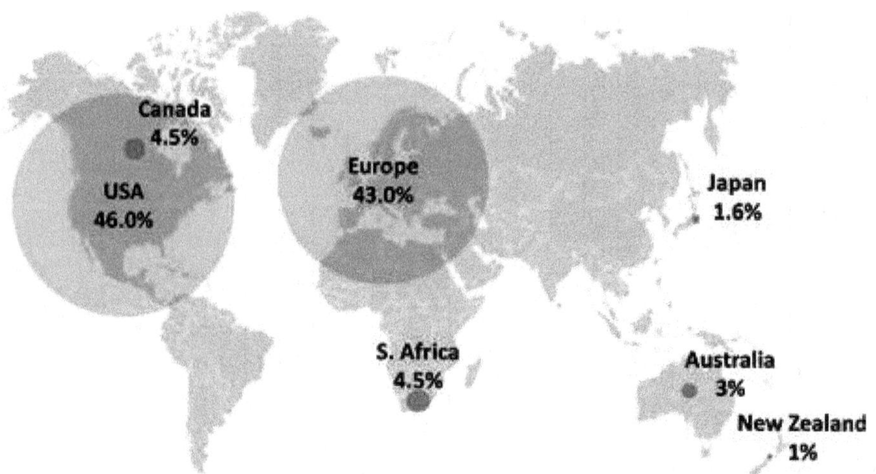

Figure 1.
Percentage of craft beer producers (2013–2017).

increased globally. In fact it passed from 671 to 2378 in the UK (traditionally beer-producing and beer-drinking country), and from 1321 to 6266 in the USA, an increase of 354 and 474% respectively, within the same period (not traditional beer producer country). This increase in number of craft breweries and production volume run up to an increase in compound annual growth rate (CAGR) within the sector. Craft brewing continues to take market share away from the largest brewing companies. According to Brewers Association report (the U.S. beer sales volume growth 2017, National beer sales and production statistical data), the overall U.S. beer volume sales were down 1% in 2017, whereas craft brewer sales continued to grow at a rate of 5% by volume, reaching 12.7% of the U.S. beer market by volume. Craft production grew the most for microbreweries. Retail dollar sales of craft increased 8%, up to $26.0 billion, and now account for more than 23% of the $111.4 billion U.S. beer market [27]. Percentage of craft beer producers (2013–2017) can be seen in **Figure 1**.

There are various factors, which favored this increase in overall craft beer consumption. These factors include per capita income growth, the availability of alternatives toward the production of successful and high levels of quality beers, increased health concerns, and the emergence of new government regulations that affects directly the sustainability issue and consistency and innovation among many others.

5. Craft and special beers: classification

A single beer style, lager beer, has long been the main dominant beer in the world market. However, a worldwide change in trend for the last decade has been registered due to the growing interest in craft and specialized beer [34]. A significant growth in the number of breweries, the variety of styles and the total volume of production had been observed in previous years [35].

But the reasons for the growth are multiple: first, increase in the demand for high more flavorful and stronger beers [34, 36]. This is particularly important in the case of American consumers, often not satisfied with the dominant in the market American pale lagers. An increase in flavors (malted barley, chestnut, honey flavored) and a more readily quality perceived are the main factors to choose craft beer instead of commercial beer between habitual beer drinkers [37]. Second, exclusivity and "unique drinking experiences" are also highly rated by craft beer consumers [34, 38]. Finally, even though traditional brands of beer are closely linked to very specific places [39], craft beer is part of a broader neolocalism movement in which people are demanding goods and services that have a connection with the local community [36].

Taking into account that all beer types evolve from the combination and relationships among ingredients, processing, packaging, marketing and culture, it is therefore necessary to establish some criteria to establish differences between special and craft beers.

This section analyzes the main criteria for classifying beers as special or craft beers (**Figure 2**).

The first element taken into account is the production output of beer per year (**criterion 1**). Craft beer are characterized by small production output and their "small," "independent," and "traditional" character. These characteristics are compatible with others which have been traditionally used to classify beer styles and now they are assuming new importance and making possible to enrich traditional beer brewing: we refer to type of fermentation and yeast strain selection (**criterion 2**). Here, we will look at non-*Saccharomyces* brewing yeasts which require special atten-tion [40, 41]. While malted barley remains the main source of sugars for fermenta-tion in the production of beer, the ingredients can be changed based on the region and preference of the consumer. Innovative ingredients in wort production can be used as a valuable source of variation in craft beer production (**criterion 3**) The two last criteria are relatively recent and novel and are related with the development of special beers in the perspective on health and nutrition (**criterion 4**) and with the use of emerging technologies in brewing (**criterion 5**).

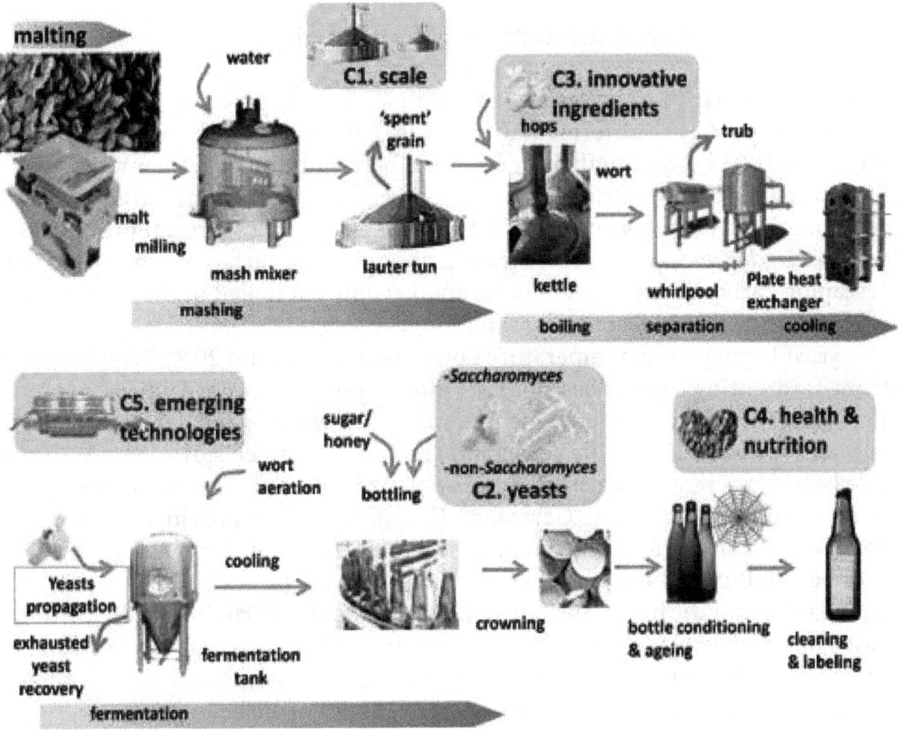

Figure 2.
Main criteria to classify craft and special beers.

5.1 Production output of beer per year (scope of craft beer)

The annual beer production allows distinguishing between *larger breweries mass-producing beer* (annual production capacity of up to 6 million barrels) and *craft beers* or "small" scale breweries (less than 6 million barrels; where 1 BBL = 339 12 oz bottles of beer or 235 half-liter bottles of beer) [26, 42].

According to Kleban and Nickerson [42], small scale beers have different considerations:

- Minimum production quantity: **Nanobreweries.**
- The place of sale of beer: production is sold outside (Microbreweries) or on the same floor of production (**Brewpub**).
- Brewing companies that outsource their production to other already established breweries (**Contract Brewing Company**).
- Over 50% or more of their volume production focuses on all-malt beers and/or their malt flagship (**Regional Craft Brewery**).

The American craft brewing industry assumes that in addition to low volume production, further requirements are expected by the craft beers [36]. They are *independent* in that and not more than 25% of the business is owned by another member of the alcohol industry who is not a craft brewer. *Traditional ingredients* (water, malt, hops, yeast) must also be used in the brewing process although *innovation* in terms of reinterpreting historic beer styles or developing new styles is a hallmark of the industry.

5.2 Selection of the yeast strain and type of fermentation

The main brewing classification criterion particularly relies on the selection of the yeast strain and type of fermentation [35, 41]. Two types of brewing yeast were originally classified based on their flocculation behavior during fermentation.

Beers are classified into two large groups according to the yeast strain and type of fermentation: Ale beers and Lager beers. **Ale yeasts or top-fermenting yeasts,** which are *Saccharomyces cerevisiae* strains, rise up to the surface of the vessel with the escaping carbon dioxide gas bubbles and become entangled in the fermentation head, facilitating their collection by skimming.

Ale yeast fermentation temperature ranges between 15 and 20°C. **Lager yeast or bottom-fermenting yeast** does not rise and becomes entrapped in the foam but settles out at the end of the fermentation. Lager worts often ferment at lower temperatures (8–14°C) than ale yeasts and are therefore much slower.

Ale beers represent only a small percentage of the total beer consumption. They are very common in Britain, Germany, Canada's eastern provinces, the United States and, last but not least, Belgium. Until the sixteenth century, ale was the main type of beer in Europe [43].

Standard/ordinary bitter (Britain), English pale ale (Britain), Mild (Britain), Brown Porter (Britain), Robust Porter (Britain), Dry stout (Ireland), Sweet stout (Britain), Kölsch (Germany, Cologne), Lambic (Belgium), Rauchbier (Germany) and Weizen/Weissbier (Germany) are some examples of ale beer types.

Lager beer is the dominant style in almost all countries and represents more than 90% of the beer produced worldwide [43].

Some principal Lager beer types are: German Pilsner (Pils) (Germany), Bohemian Pilsener (Czech Republic), Classic American Pilsner (United States), Vienna Lager (Austria), Oktoberfest/Märzen (Germany), Dark American Lager (United States), Munich Dunkel (Germany), Schwarzbier (Black Beer) (Germany), Maibock/Helles Bock (Germany), Traditional Bock (Germany), Doppelbock (Germany), Eisbock (Germany).

In all beers cited, the flavor-active compounds such as acids, alcohols, aldehydes, ketones and esters are produced by yeast during fermentation. Although there are many strains of brewing yeast (*Saccharomyces cerevisiae*) for beer production, the choice of suitable yeasts to produce desirable tastes and flavors in beer is very important and significant.

5.2.1 Use of non-Saccharomyces

Several non-*Saccharomyces* yeasts can be used successfully in the making of craft beers with interesting possibilities. Yeasts such as *Lachancea thermotolerans*, *Torulaspora delbrueckii*, *Hanseniaspora vineae* and *Schizosaccharomyces pombe* can help to modulate acidity, aroma, mouthfeel or even color [41, 44]. As the final alcoholic degree in beers is lower than in wines, and normally ranging between 4 and 8% vol, the use of medium fermentative power non-*Saccharomyces* species is possible because most of these yeasts are able to ferment reaching this ethanol level.

Lachancea thermotolerans is trending yeast in fermented beverages because of its ability to ferment until 4–9% vol producing high amounts of lactic acid from sugars. Therefore, it can be used to decrease pH of beverages [45–47]. Moreover, interesting effects in beer aroma can be reached by the production of fruity esters [48]. The use of *L. thermotolerans* has been also described in beer technology [49, 50]. In the brewing of craft beers, *L. thermotolerans* can be used not only in the primary fermenta-tion of the wort but also during the second fermentation in bottle to produce the suitable foam and CO_2 pressure. However, the most interesting application is in the production of sour beers because of the natural biological acidification during wort fermentation [46]. Moreover, even when the early use of *L. thermotolerans* has been proposed in winemaking in which the use of suitable species of these yeasts can produce pH reductions of 0.5 pH units [47] and the use in beer technology is even more effective due to the lower buffer effect in beer compared with wine. In our lab, we reached pH reductions of 1 pH unit [51]. The sensory effect of this acidity is described as a citric acidity without dairy hints because of the low production of acetoin and diacetyl [47]; moreover, the volatile acidity produced by *L. thermotolerans* is very low compared to volatile acidity produced by selected *S. cerevisiae*.

Torulaspora delbrueckii is another versatile yeast suitable for beer production. It has a medium fermentative power and improves the formation of fruity esters in addition to a low production of volatile acidity. These characteristics make it a good yeast for the initial fermentation of the must and the subsequent in bottle [50]. Also it is possible the use of this yeast sequentially or in mixed cultures with *S. cerevisiae* [52] or *S. pombe* [53]. It has been described as yeast able to decrease volatile acid-ity during fermentation. The ability to ferment sugars easily reaching 7–9% vol makes it interesting also for secondary bottle fermentation [52]. The production of 2-phenylethyl acetate, a floral ester with positive floral aroma, is increased during fermentation with *T. delbrueckii*; moreover, high amounts of 3-ethoxy propanol are formed by this species [52]. The release of polysaccharides is also improved by the fermentation with *T. delbrueckii* affecting mouthfeel and structure [54].

Hanseniaspora vineae is an apiculate yeast able to produce fresh and complex fermentation, increasing fruity aroma and producing full bodied structure [55]. It is possible to find strains with fermentative power close to 9% vol, which facilitate its use not only for primary fermentation but also for bottle fermentation. Moreover, it is a persistent yeast that can be found until the end of the alcoholic fermentation in wines and therefore also in beers because of the lower alcoholic degree. During the fermentation with *H. vineae*, an increase in the concentration of acetyl esters, benzenoids, and sesquiterpenes [56, 57], and a decrease in the contents of alcohols and acids occurs. Intense either β-glucosidase or β-xylosidase activities has been described in some strains of *H. vineae* increasing the levels of hotrienol and 2,6-dimethyl-3,7-octadien-2,6-diol during fermentation [58]. It is especially noticeable the production of 2-phenylethyl acetate by *H. vineae* [55], compared with other *Hanseniaspora/Kloeckera* species.

Schizosaccharomyces pombe is a fission yeast able to produce maloalcoholic fermentation, and some strains can reach 13–15% vol of ethanol during fermentation [59, 60]. The peculiar metabolism of *S. pombe* produces an intense degradation of malic acid together with a significant release of pyruvate in the fermentative media [60]. *S. pombe* is especially resistant to some common preservatives such as sulfur dioxide, actidione, benzoic acid, and dimethyl dicarbonate [59, 61]. The main draw-back of this yeast is the high production of volatile acidity. Concerning its structure this species has a peculiar and dense 2-layer cell wall. The autolysis produces the release of high amount of polysaccharides during maturation improving the mouth feel of beers [62]. This property can be especially interesting to produce full-bodied and soft bottle-aged beers. Moreover, we have observed intense bottle fermentation with good foam properties. The aromatic profile in beers is fruity and fresh when this is yeast is used specially in bottle fermentation.

5.3 Innovative ingredients

Raw material in wort production and parameters in production lead to produce an unlimited number of beer types. It might be argued that *beer is a horizontally differentiated product*. [35]. In fact, beers are quite similar in most respects but small differences in their composition can greatly affect both appearance and flavor [63].

We are going to examine each one of the raw materials separately.

5.3.1 Water

Water is quantitatively the main ingredient of beers; it forms more than 90% and often even more than 94% of the final product. The chemical composition of water has a determinant effect on beer properties and contributes significantly to the final beer flavor. The balance of minerals in brewing water will affect the flavor character and flavor perception of malt, hops, and by-products of fermentation. It may also influence the performance of yeast, which in turn influences the flavor, aroma, and mouthfeel of beer.

Chemical composition of water of the localities where famous beer styles were originated are very different in approximate ionic concentrations (in ppm). The chemical composition of water of Pilzen, Munich, Dortmund or Vienna is typi-cal between Lager examples. Burton-on-Trent, Dublin or Edinburgh are typical between ale examples.

5.3.2 Malt

Malted barley is the main source for fermentable sugars used by yeasts in the traditional brewing of beers [64].

Depending on the conditions (time and temperature), pale or amber-colored or even dark malts are obtained; the color being due to caramelization of sugars and to Maillard-type reactions [65]. The variety of barely and the malting process influences the type and quality of beer [66]. **To elaborate craft beer, the right malt** is a **key factor** because craft beers include high proportion of adjuncts and enzymatic activity of malt has to ensure adequate hydrolysis of all the starch present in the wort.

5.3.3 Adjuncts

Malted barley is the main source for fermentable sugars used by yeasts in the traditional brewing, Other grains, malted or not, have been included to provide fermentable carbohydrates to the wort in addition to those from malt [63]. In former times, most cereals were used for malting, emmer, oats, spelt wheat, bread wheat were widely used and, in Estonia, rye was used up until the nineteenth century [67]. Outside Europe, millet, rice, maize and tuber plants have been, and are still, commonly used.

Bogdan and Kordialik-Bogacka [64] estimate that 85–90% of beer worldwide is now produced with adjuncts. Traditionally they had been used because they lead to reduce the cost of raw materials. When adjuncts are selected as unmalted grains, they present the added advantage of improved sustainability, by reducing reliance on the malting process [68] and its associated cost.

Craft brewing is increasing the use of adjuncts [68] because they lead to create a unique beer **flavor/aroma** [69]. **Figure 3** shows the influence of different concen-trations of roasted malt addition on sensory properties of beer.

Appropriately chosen adjuncts can contribute to light or dark colors, improved colloidal or foam stability and prolongation beer shelf-life [64]. The flavor profile can also be changed by altering the sugar and amino acid spectra in wort.

5.3.4 Hops

Hops (*Humulus lupulus* L.) are almost exclusively consumed by the brewing industry. Although hops are only a minority ingredient, they have significant impact on the sensory properties of beer [65]. It contributes not only to bitter flavor but also with the particular character of the selected hop variety [66].

This is mainly due to its particular chemical composition in: the hops resins, the hop oil and hop polyphenols [70].

Figure 3.
Effect of roasted barley addition on beer sensory properties.

In the closing years of the twentieth century, the hop became an icon of the "craft beer revolution" that swept across the United States. The "hopped up" vats created more flavorful and aromatic beers, making them more akin to European specialty varieties than anything seen in United States markets since before prohibition. The hops also became an effective marketing tool [39] from a nutritional and health point of view. It had recently come to light the effect antiviral and anti-HIV of xanthohumol, a phenylated flavonoid isolated from hops [66].

5.4 Perspective on health and nutrition

This section also includes a part on special or craft beers, which meet the **new consumer requirements** related with health and nutrition. In this context, it should include categories such as [66] light or low-calorie beers, low alcohol or non-alcohol beers, gluten free beers and functional beers.

5.4.1 Light beers

Light beer is a relatively new product on the market. Light beers contain at least one-third less calories than conventional beers [71]. However, these products are not widely accepted in Europe compared to North America and Australasia because of their lack of fullness in the taste and low bitterness compared with conventional beer. **Enhanced hop character and addition of a low level of priming syrup** have been proposed to the production of a low-calorie beer with a well-balanced and full beer flavor [38].

From a nutritional point of view [71], light beer contains less carbohydrate than regular beer, low alcohol beer or non-alcoholic beer. Surprisingly, light beer presents more calorie supply than such beers. This may be explained considering that light beer has a significant amount of alcohol (3%) providing a high calorie value.

5.4.2 Low alcohol beers

Low-alcohol beer is a beer with very low- or no-alcohol content. The alcohol by volume (ABV) limits depends on laws in different countries. In recent years, there has been an **increased market share for low alcohol beers**. This is mainly due to health and safety reasons and increasingly strict social regulations [72]. The alcohol-free beers also claim beneficial effects of healthy beer components with a simultaneous effect of the lower energy intake and complete absence of negative impacts of alcohol consumption.

According to Blanco et al. [73], the dealcoholization processes that are commonly used to reduce the alcohol content in beer have negative consequences to beer flavor. Several processes (physical and biological) have been developed for the production of low-alcohol or alcohol-free beer [74]. The physical processes include thermal and membrane processes such as thin-layer evaporation; falling film vacuum evaporation; continuous vacuum rectification; reverse osmosis; and dialysis. The biological processes include cold contact process (CCP); arrested fermentation; and use of special yeasts (*S. ludwigii*).

Overall, the taste defects in alcohol-free beer are mainly attributed to loss of aro-matic esters, insufficient aldehydes, reduction or loss of different alcohols, and an indeterminate change in any of its compounds during the dealcoholization process or as a consequence of incomplete fermentation [73].

5.4.3 Gluten free

The market segment for gluten free (GF) products continues to grow rapidly and gluten free beers are a niche market with **increasing demand** [75, 76].

Beer is considered unsuitable for people suffering from gluten intolerance, but with some modification and removal of proteins which occur during traditional beer processing. The majority of the precipitated protein remains in the spent grain after the lautering process and only a small proportion of gluten passes from malt to sweet wort. A study conducted by [77], in twenty-eight commercial beers, found that 10 of the tested beers contained less than 20 ppm gluten.

There are different alternatives for the reduction of gluten levels below the legislative gluten-free threshold (≤20 ppm) (EC No. 41/2009, 2009), on a daily basis, including precipitation and enzymatic hydrolysis. Deglutinization treatments by enzymatic process were proposed by Fanari et al. [78].

Furthermore, gluten free beers can be produced **using gluten free cereals and pseudocereals**. Currently only sorghum, rice, maize, millet, and buckwheat appear to be successful GF beer ingredients, while others have only shown adjunct possibil-ities. Among cereals, Teff is gaining a lot of popularity in GF beer production. **Teff** grain nutrients are promising and it is also an excellent GF alternative for people with celiac disease and other gluten allergy. Though the α- and β-amylase activities of teff malt are lower than that of barley, it has sufficient level of enzyme activities to be used as a raw material for malting [79] and GF beer production. Mayer et al. [80] has also prepared a GF beer from **all-rice malt** with sufficient endogenous enzyme activity for degradation of the rice components.

A third approach is the production of yeast fermented beverages based on fermentable sugars/syrups [75]. The search for new gluten-free brewing materials is still in its infancy and researchers in this field of study are continuously researching on the malting, mashing, fermentation conditions [78].

5.4.4 Functional beer

There is also scope for positioning low-calorie beers as a source of good carbohydrates, such as the soluble fiber and prebiotics derived from the β-linked glucans and arabinoxylans in the cereal walls [81]. Because these carbohydrates are neither metabolized by the brewing yeast nor they do not contribute toward calorie count but exert health benefits. Prebiotics are dominantly oligosaccharides that are nondi-gestible to human being but selectively stimulate growth and activity of beneficial bacteria (probiotics) in the human gastrointestinal tract.

Further, β-glucans could enhance stress tolerance of intestinal lactobacilli, which may have a positive impact on survival of probiotics. Nonetheless, high molecular weight b-linked glucan materials may have a negative impact on filtration efficiency and optimization of a filtration process will be required.

Probiotics are not limited to bacteria, and there is a well-known probiotic yeast strain of *S. cerevisiae* var. *boulardii*. A novel unfiltered and unpasteurized probiotic beer could be produced by fermenting wort with a probiotic strain of *S. cerevisiae*. A new category of functional beer could be the specialty beer of the future,given the rising consumer recognition and acceptance of probiotics [38].

5.5 Use of new technologies

Emerging technologies as high hydrostatic pressure (HHP) and ultra-high pressure homogenization (UHPH) open new possibilities in beer production. Both technologies are considered as cold techniques allowing the control of microorganisms in beverages [82]. Even when some temperature increasing is produced that can be quantified in 2–3°C/100 MPa in HHP [83] by compression adiabatic heat and until 100°C but just for 0.2 s in UHPH because of intense shear forces and impact [84]. The use of HHP is able to eliminate yeasts at pressures of 400 MPa-10 min but Gram-positive bacteria needs 600 MPa-10 min and spores remain unaffected even with these pressures [85]. Also it has the drawback of being a discontinuous technology. UHPH is now currently highly developed being a fast technology with a good industrial scale-up with equipment that are working at a flow of 10,000 l/h (https://www.ypsicon.com/). Moreover, UHPH is a continuous technology and able to produce sterilization due to the extreme impacts and shear forces produced when the fluid pumped at 300 MPa cross the depressurization valve [84]. In beer produc-tion theoretically is possible to pump the beer at 300 MPa and release the pressure until 4 bar, later is possible to make a sterile iso-barometric bottling. The intense de-polymerization produced by UHPH can also disaggregate colloidal particles improving the beer structure and stability. Potentially it is possible to produce the mechanically lysis of the yeasts formed during fermentation increasing the amount of small size polysaccharides.

Other interesting technology that can be quite useful in beer production and sterilization is pulsed light (PL). This technology produces high energy light during a very short time (few μs) with a strong capacity to inactivate microorganisms and spores allowing sterilization [85]. The light is applied by flash lamps with a range spectra of 160–2600 nm with an intensity 105 folds the sunlight intensity at the seaside level. Power peak can reach 35 MW. PL technology is also a cold technology being a gentle process with sensory quality of beverages. This technique can be applied continuously during beer processing previously to packaging. It is also possible to use this technology to sterilize bottles or packages.

The use of these new technologies opens new possibilities in the processing and preservation of beer. UHPH and PL can be applied in a continuous way being efficient and easily implemented at industrial scale. Both sterilization technologies have a gentle repercussion in sensory quality of beverages.

6. Future trends

The development of new craft and special beers will be focused in the improvement on sensory properties and differentiation. Moreover, health care connotations are essential and should be supported by traditional processes but improved with both new biotechnologies and emerging processes.

Author details

María Jesús Callejo*, Wendu Tesfaye, María Carmen González and Antonio Morata
Universidad Politécnica de Madrid, Spain

*Address all correspondence to: antonio.morata@upm.es

References

[1] Hayden B, Canuel N, Shanse J. What was brewing in the Natufian? An archaeological assessment of brewing technology in the Epipaleolithic. Journal of Archaeological Method and Theory. 2012;**20**(1):102-150. DOI: 10.1007/s10816-011-9127-y

[2] Dineley M, Dineley G. Neolithic ale: Barley as a source of malt sugars for fermentation. In: Fairbairn AS, editor. Plants in Neolithic Britain and Beyond. Oxford: Oxbow; 2000. pp. 137-154

[3] Cabras I, Higgins DM. Beer, brewing, and business history. Business History. 2016;**58**:609-624

[4] McGovern P, Zhang J, Tang J, Zhang Z, Hall G, Moreau R, et al. Fermented beverages of pre- and proto-historic China. Proceedings of the National Academy of Sciences of the United States of America. 2004;**101**(51):17593-17599

[5] Meussdoerffer FG. A comprehensive history of beer brewing. In: Esslinger HM, editor. Handbook of Brewing. Weinheim: Wiley-VCH Verlag GmbH & Co.; 2009. pp. 1-42

[6] Bai J, Huang J, Rozelle S, Boswell M. Beer battles in China: The struggle over the World's largest beer market. In: Swinnen JFM, editor. The Economics of Beer. Oxford: Oxford University Press; 2011. pp. 267-286

[7] Jiajing W, Li L, Terry B, Linjie Y, Yuanqing L, Fulai X. Revealing a 5,000-y-old beer recipe in China. Proceedings of the National Academy of Sciences of the United States of America. 2016;**113**(23):6444-6448

[8] Hardwick WA. History and antecedents of brewing. In: Hardwick WA, editor. Handbook of Brewing. New York: Marcel Dekker; 1994. pp. 37-52

[9] Cortacero-Ramirez S, De Castro MHB, Segura-Carretero A, Cruces-Blanco C, Fernandez-Gutierrez A. Analysis of beer components by capillary electrophoretic methods. Trends in Analytical Chemistry. 2003;**22**(7):440-455

[10] Michel C. L'alimentation au Proche-Orient ancien: Les sources et leur exploitation. Dialogues d'Histoire Ancienne. 2012;**7**:17-45

[11] Maksoud SA, Hadidi MN, Amer WN. Beer from the early dynasties (3500-3400 cal. B.C.) of Upper Egypt, detected by archaeochemical methods. Vegetation History and Archaeobotany. 1994;**3**(4):219-224

[12] Nelson M. The barbarian's Beverage: A History of Beer in Ancient Europe. London/New York: Routledge; 2005. DOI: 10.4324/ 9780203309124. Available from: https://scholar.uwindsor.ca/llcpub/26

[13] Poelmans E, Swinnen JFM. From monasteries to multinationals (and back): A historical review of the beer economy. The Journal of Wine Economics. 2011;**6**(2):196-216

[14] Moore J. Pre-Hispanic beer in coastal Peru: Technology and social context of prehistoric production. American Anthropologist. 1989;**91**(3):682-695

[15] Burger RL, Van Der Merwe NJ. Maize and the origin of Highland Chavín civilization: An isotopic perspective. American Anthropologist. 1990;**92**(1):85-95

[16] Hastorf CA, Johannessen S. Pre-Hispanic political change and the role of maize in the Central Andes of Peru. American Anthropologist. 1993;**95**(1):115-138

[17] Liua L, Wanga J, Rosenbergb D, Zhaoc H, Lengyeld G, Nadel D. Fermented beverage and food storage in 13,000 y-old stone mortars at Raqefet cave, Israel: Investigating Natufian ritual feasting. Journal of Archaeological Science: Reports. 2018;**21**:783-793

[18] Perruchinia E, Glatza C, Haldb MM, Casanac J, Toneyd JL. Revealing invisible brews: A new approach to the chemical identification of ancient beer. Journal of Archaeological Science. 2018;**100**:176-190

[19] Howard PH. Too big to ale? Globalization and consolidation in the beer industry. In: Patterson MW, Pullen NH, editors. The Geography of Beer: Regions, Environment, and Society. Dordrecht, The Netherlands: Springer; 2014; pp. 155-165

[20] Sewell SL. The spatial diffusion of beer from its Sumerian origins to today. In: Patterson M, Hoalst-Pullen N, editors. The Geography of Beer: Regions, Environment, and Society. Dordrecht: Springer; 2014. pp. 23-29. https://doi.org/10.1007/978-94-007-7787-3_3

[21] Elzinga K, Tremblay C, Tremblay V. Craft beer in the United States: History, numbers, and geography. The Journal of Wine Economics. Dordrecht, Springer; 2015;**10**(3):242-274. DOI: 10.1017/jwe.2015.22

[22] Garavaglia C. Birth and Diffusion of Craft Breweries in Italy. In: Garavaglia C, Swinnen J, editors. Economic Perspectives on Craft Beer: A Revolution in the Global Beer Industry. London: Palgrave Macmillan; 2017

[23] van Dijk M, Kroezen J, Slob B. From Pilsner Desert to craft beer oasis: The rise of craft brewing in the Netherlands. In: Economic Perspectives on Craft Beer. Cham, Switzerland: Palgrave McMillan; 2017. pp. 259-293. DOI: 10.1007/978-3-319-58235-1_10. ISBN: 978-3-319-58235-1

[24] Sammartino A. Craft brewing in Australia, 1979-2015. In: Garavaglia C, Swinnen J, editors. Economic Perspectives on Craft Beer: A Revolution in the Global Beer Industry. London/New York: Palgrave Macmillan; 2018. pp. 397-423

[25] Depenbusch L, Ehrich M, Pfizenmaier U. Craft Beer in Germany– New Entries in a Challenging Beer Market. In: Garavaglia C, Swinnen J, editors. Economic Perspectives on Craft Beer: A Revolution in the Global Beer Industry. London, New York: Palgrave Macmillan; 2018

[26] Garavaglia C, Swinnen J. The craft beer revolution: An international perspective. Choices. 2017;**32**(3):1-8. Available from: http://www.choicesmagazine.org/choices-magazine/theme-articles/global-craft-beer-renaissance/the-craft-beer-revolution-an-international-perspective

[27] Brewers Association. 2019. Retrieved from: https://www.brewersassociation.org/statistics/ [Accessed: April 1, 2019]

[28] Bamforth C, Cabras I. Interesting times: Changes for brewing. In: Cabras I, Higgins D, Preece D, editors. Brewing, Beer and Pubs: A Global Perspective. London: Palgrave Macmillan; 2016. pp. 13-33

[29] The Brewers of Europe. Beer Statistics—2018 and Previous Years. 2018. Retrieved from: https://brewersofeurope.org/site/index.php [Accessed: April 1, 2019]

[30] Australia and New Zealand Banking Group Limited (ANZ). New Zealand craft beer industry insights: ANZ industry reports. 2017. Available from: https://comms.anz.co.nz/businsights/

article/report.html?industry=Craft% 20 Beer [accessed: April 1, 2018]

[31] Tripes S, Dvořák J. Strategic forces in the Czech brewing industry from 1990-2015. Acta Oeconomica Pragensia. 2017;**3**:3-38

[32] Rogerson CM, Collins KGE. Developing beer tourism in South Africa: International perspectives. African Journal of Hospitality, Tourism and Leisure. 2015;**4**(1):1-15

[33] Beer Canada. *2015 Industry Trends*. 2016. Available from: http://www.beercanada.com/sites/default/files/2015_industry_trends_final.pdf

[34] Gómez-Corona C, Lelievre-Desmas M, Buendía HBE, Chollet S, Valentin D. Craft beer representation amongst men in two different cultures. Food Quality and Preference. 2016;**53**:19-28

[35] Clemons EK, Gao GG, Hitt LM. When online reviews meet hyperdifferentiation: A study of the craft beer industry. Journal of Management Information Systems. 2006;**23**(2):149-171

[36] Reid N, McLaughlin RB, Moore MS. From yellow fizz to big biz: American craft beer comes of age. Focus on Geography. 2014;**57**(3):114-125

[37] Smith S, Farrish J, McCarroll M, Huseman E. Examining the craft brew industry: Identifying research needs. International Journal of Hospitality Beverage Management. 2017;**1**(1):3

[38] Yeo HQ , Liu SQ. An overview of selected specialty beers: Developments, challenges and prospects. International Journal of Food Science & Technology. 2014;**49**(7):1607-1618

[39] Kopp P. The global hop: An agricultural overview of the brewer's gold. In: Patterson M, Hoalst-Pullen N, editors. The Geography of Beer. Dordrecht: Springer; 2014

[40] Tataridis P, Kanelis A, Logotetis S, Nerancis E. Use of non-*Saccharomyces Torulaspora delbrueckii* yeast strains in winemaking and brewing. Zbornik Matice Srpske za Prirodne Nauke. 2013;**124**:415-426

[41] Callejo MJ, González C, Morata A. Use of non-*Saccharomyces* yeasts in bottle fermentation of aged beers. In: Kanauchi M, editor. Brewing Technology. Rijeka, Croatia: IntechOpen; 2017. DOI: 10.5772/intechopen.68793. Available from: https://www.intechopen.com/books/brewing-technology/use-of-non-saccharomyces-yeasts-in-bottle-fermentation-of-aged-beers

[42] Kleban J, Nickerson I. To brew, or not to brew-that is the question: An analysis of competitive forces in the craft brew industry. Journal of the International Academy for Case Studies. 2012;**18**(3):59

[43] Pavsler A, Buiatti S. Non-lager beer. In: Beer in Health and Disease Prevention. London, United Kingdom: Academic Press; 2009. pp. 17-30

[44] Budroni M, Zara G, Ciani M, Comitini F. *Saccharomyces* and non-*Saccharomyces* starter yeasts. In: Kanauchi M, editor. Brewing Technology. Rijeka, Croatia: IntechOpen; 2017. DOI: 10.5772/intechopen.68792. Available from: https://www.intechopen.com/books/brewing-technology/saccharomyces-and-non-saccharomyces-starter-yeasts

[45] Gobbi M, Comitini F, Domizio P, Romani C, Lencioni L, Mannazzu I, et al. *Lachancea thermotolerans* and *Saccharomyces cerevisiae* in simultaneous and sequential co-fermentation: A strategy to enhance acidity and improve the overall quality of wine. Food Microbiology. 2013;**33**:271-281. DOI: 10.1016/j.fm.2012.10.004

[46] Morata A, Loira I, Tesfaye W, Bañuelos MA, González C, Suárez Lepe JA. *Lachancea thermotolerans* applications in wine technology. Fermentation. 2018;**4**:53. DOI: 10.3390/fermentation4030053

[47] Morata A, Bañuelos MA, Vaquero C, Loira I, Cuerda R, Palomero F, et al. *Lachancea thermotolerans* as a tool to improve pH in red wines from warm regions. European Food Research and Technology. 2019;**245**:885-894. DOI: 10.1007/s00217-019-03229-9

[48] Escott C, Morata A, Ricardo-da-Silva JM, Callejo MJ, González MC, Suarez-Lepe JA. Effect of *Lachancea thermotolerans* on the formation of polymeric pigments during sequential fermentation with *Schizosaccharomyces pombe* and *Saccharomyces cerevisiae*. Molecules. 2018;**23**:2353. DOI: 10.3390/molecules23092353

[49] Domizio P, House JF, Joseph CML, Bisson LF, Bamforth CW. *Lachancea thermotolerans* as an alternative yeast for the production of beer. Journal of the Institute of Brewing. 2016;**122**:599-604. DOI: 10.1002/jib.362

[50] Callejo MJ, García Navas JJ, Alba R, Escott C, Loira I, González MC, et al. Wort fermentation and beer conditioning with selected non-*Saccharomyces* yeasts in craft beers. European Food Research and Technology. 2019;**245**:1229-1238. DOI: 10.1007/s00217-019-03244-w

[51] Vanooteghem M. Impact of non-*Saccharomyces* fermentations on the flavour profile of craft beer [MS thesis]. Madrid, Spain: Technical University of Madrid; 2019

[52] Loira I, Vejarano R, Bañuelos MA, Morata A, Tesfaye W, Uthurry C, et al. Influence of sequential fermentation with *Torulaspora delbrueckii* and *Saccharomyces cerevisiae* on wine quality. LWT-Food Science and Technology. 2014;**59**:915-922. DOI: 10.1016/j.lwt.2014.06.019

[53] Loira I, Morata A, Comuzzo P, Callejo MJ, González C, Calderón F, et al. Use of *Schizosaccharomyces pombe* and *Torulaspora delbrueckii* strains in mixed and sequential fermentations to improve red wine sensory quality. Food Research International. 2015;**76**:325-333. DOI: 10.1016/j.foodres.2015.06.030

[54] Comitini F, Gobbi M, Domizio P, Romani C, Lencioni L, Mannazzu I, et al. Selected non-*Saccharomyces* wine yeasts in controlled multistarter fermentations with *Saccharomyces cerevisiae*. Food Microbiology. 2011;**28**:873-882. DOI: 10.1016/j.fm.2010.12.001

[55] Martin V, Valera MJ, Medina K, Boido E, Carrau F. Oenological impact of the *Hanseniaspora/Kloeckera* yeast genus on wines—A review. Fermentation. 2018;**4**:76. DOI: 10.3390/fermentation4030076

[56] Martin V, Giorello F, Fariña L, Minteguiaga M, Salzman V, Boido E, et al. De novo synthesis of benzenoid compounds by the yeast *Hanseniaspora vineae* increases the flavor diversity of wines. Journal of Agricultural and Food Chemistry. 2016;**64**:4574-4583

[57] Martin V, Boido E, Giorello F, Mas A, Dellacassa E, Carrau F. Effect of yeast assimilable nitrogen on the synthesis of phenolic aroma compounds by *Hanseniaspora vineae* strains. Yeast. 2016;**33**:323-328

[58] López S, Mateo JJ, Maicas S. Characterisation of *Hanseniaspora* isolates with potential aroma enhancing properties in Muscat wines. South African Journal of Enology and Viticulture. 2014;**35**:292-303

[59] Suárez-Lepe JA, Palomero F, Benito S, Calderón F, Morata A.

Oenological versatility of *Schizosaccharomyces* spp. European Food Research and Technology. 2012;**235**:375-383

[60] Loira I, Morata A, Palomero F, González C, Suárez-Lepe JA. *Schizosaccharomyces pombe*: A promising biotechnology for modulating wine composition. Fermentation. 2018;**4**:70. DOI: 10.3390/fermentation4030070

[61] Escott C, Loira I, Morata A, Bañuelos MA, Suárez-Lepe JA. Wine spoilage yeasts: Control strategy. In: Morata A, Loira I, editors. Yeast-Industrial Applications. London, UK: InTech; 2017. pp. 89-116

[62] Palomero F, Morata A, Benito S, Calderón F, Suárez-Lepe JA. New genera of yeasts for over-lees aging of red wine. Food Chemistry. 2009;**112**:432-441

[63] Buiatti S. Beer composition: An overview. In: Beer in Health and Disease Prevention. London, United Kingdom: Academic Press; 2009. pp. 213-225

[64] Bogdan P, Kordialik-Bogacka E. Alternatives to malt in brewing. Trends in Food Science & Technology. 2017;**65**:1-9

[65] De Keukeleire D. Fundamentals of beer and hop chemistry. Quimica Nova. 2000;**23**(1):108-112

[66] Sohrabvandi S, Mortazavian AM, Rezaei K. Health-related aspects of beer: A review. International Journal of Food Properties. 2012;**15**(2):350-373

[67] Behre KE. The history of beer additives in Europe—A review. Vegetation History and Archaeobotany. 1999;**8**(1-2):35-48

[68] Kok YJ, Ye L, Muller J, Ow DSW, Bi X. Brewing with malted barley or raw barley: What makes the difference in the processes? Applied Microbiology and Biotechnology. 2019;**103**(3):1059-1067

[69] Schnitzenbaumer B, Arendt EK. Brewing with up to 40% unmalted oats (*Avena sativa*) and sorghum (*Sorghum bicolor*): A review. Journal of the Institute of Brewing. 2014;**120**(4):315-330

[70] Steenackers B, De Cooman L, De Vos D. Chemical transformations of characteristic hop secondary metabolites in relation to beer properties and the brewing process: A review. Food Chemistry. 2015;**172**:742-756

[71] Blanco CA, Caballero I, Barrios R, Rojas A. Innovations in the brewing industry: Light beer. International Journal of Food Sciences and Nutrition. 2014;**65**(6):655-660

[72] Brányik T, Silva DP, Baszczyňski M, Lehnert R, e Silva JBA. A review of methods of low alcohol and alcohol-free beer production. Journal of Food Engineering. 2012;**108**(4):493-506

[73] Blanco CA, Andrés-Iglesias C, Montero O. Low-alcohol beers: Flavor compounds, defects, and improvement strategies. Critical Reviews in Food Science and Nutrition. 2016;**56**(8):1379-1388

[74] Montanari L, Marconi O, Mayer H, Fantozzi P. Production of alcohol-free beer. In: Beer in Health and Disease Prevention. London, United Kingdom: Academic Press; 2009. pp. 61-75

[75] Hager AS, Taylor JP, Waters DM, Arendt EK. Gluten free beer—A review. Trends in Food Science & Technology. 2014;**36**(1):44-54

[76] Watson HG, Vanderputten D, Van Landschoot A, Decloedt AI. Applicability of different brewhouse technologies and gluten-minimization treatments for the production of gluten-free (barley) malt beers: Pilot-to

industrial-scale. Journal of Food Engineering. 2019;**245**:33-42

[77] Guerdrum LJ, Bamforth CW. Prolamin levels through brewing and the impact of prolyl endoproteinase. Journal of the American Society of Brewing Chemists. 2012;**70**:35-38

[78] Fanari M, Forteschi M, Sanna M, Zinellu M, Porcu MC, Pretti L. Comparison of enzymatic and precipitation treatments for gluten-free craft beers production. Innovative Food Science & Emerging Technologies. 2018;**49**:76-81

[79] Gebremariam MM, Zarnkow M, Becker T. Teff (*Eragrostis tef*) as a raw material for malting, brewing and manufacturing of gluten-free foods and beverages: A review. Journal of Food Science and Technology. 2014;**51**(11):2881-2895

[80] Mayer H, Ceccaroni D, Marconi O, Sileoni V, Perretti G, Fantozzi P. Development of an all rice malt beer: A gluten free alternative. LWT- Food Science and Technology. 2016;**67**:67-73

[81] Bamforth CW. Beer, carbohydrates and diet. Journal of the Institute of Brewing. 2005;**111**(3):259-264

[82] Morata A, Loira I, Vejarano R, González C, Callejo MJ, Suárez- Lepe JA. Emerging preservation technologies in grapes for winemaking. Trends in Food Science & Technology. 2017;**67**:36-43. DOI: 10.1016/j.tifs.2017.06.014

[83] Bañuelos MA, Loira I, Escott C, Del Fresno JM, Morata A, Sanz PD, et al. Grape processing by high hydrostatic pressure: Effect on use of non-*Saccharomyces* in must fermentation. Food and Bioprocess Technology. 2016;**9**:1769-1778. DOI: 10.1007/s11947-016-1760-8

[84] Loira I, Morata A, Bañuelos MA, Puig-Pujol A, Guamis B, González C, et al. Use of ultra-high pressure homogenization processing in winemaking: Control of microbial populations in grape musts and effects in sensory quality. Innovative Food Science and Emerging Technologies. 2018;**50**:50-56

[85] Morata A, Loira I, Vejarano R, Bañuelos MA, Sanz PD, Otero L, et al. Grape processing by high hydrostatic pressure: Effect on microbial populations, phenol extraction and wine quality. Food and Bioprocess Technology. 2015;**8**:277-286. DOI: 10.1007/s11947-014-1405-8

2

Lactic Acid Bacteria as Microbial Silage Additives: Current Status and Future Outlook

Pascal Drouin, Lucas J. Mari and Renato J. Schmidt

Abstract

Silage making is not a novel technique. However, the agricultural industry has made great strides in improving our understanding of—and efficiency in—producing high-quality silage for livestock. Silage microbiology research has been using the newest molecular techniques to study microbial diversity and metabolic changes. This chapter reviews important research that has laid the foundation for field-based utilization of silage inoculants. We also outline areas of current, and future, research that will improve global livestock production through the use of silage.

Keywords: silage, forage, inoculants, additives

1. Introduction

Fermentation of forage is harder to control than other fermentation processes such as industrial fermentation of food. Whole plants cannot be manipulated to remove contaminating microorganisms, and this can lead to important variations in the quality of the forage. Harvesting machinery can also contribute to the inclusion of soil or manure particles as contaminants. Other factors have an impact on silage quality, which include harvesting management, packing rate, weather events during harvest, selection of the ensiling structure, and selection of a microbial or chemical additive to preserve the crops. **Figure 1** provides an overview of the interactions between the main parameters involved in the produc-tion of high-quality silage.

This chapter will evaluate the recent published literature and will expand on the current knowledge in the study of the microbiota, search for silage inoculants, issues with aerobic instability, and understanding nonusers of forage inoculants. We will also review important research areas of microbial inoculants: fiber digestibility, analyzing "big data" functional studies, co-ensiling with by-products or food-processing wastes, and how lactic acid bacteria (LAB) used as forage additives influence animal performance.

2. Microbiota diversity during ensiling

Characterization of the different microbial species observed throughout the different phases of the ensiling process was traditionally performed using

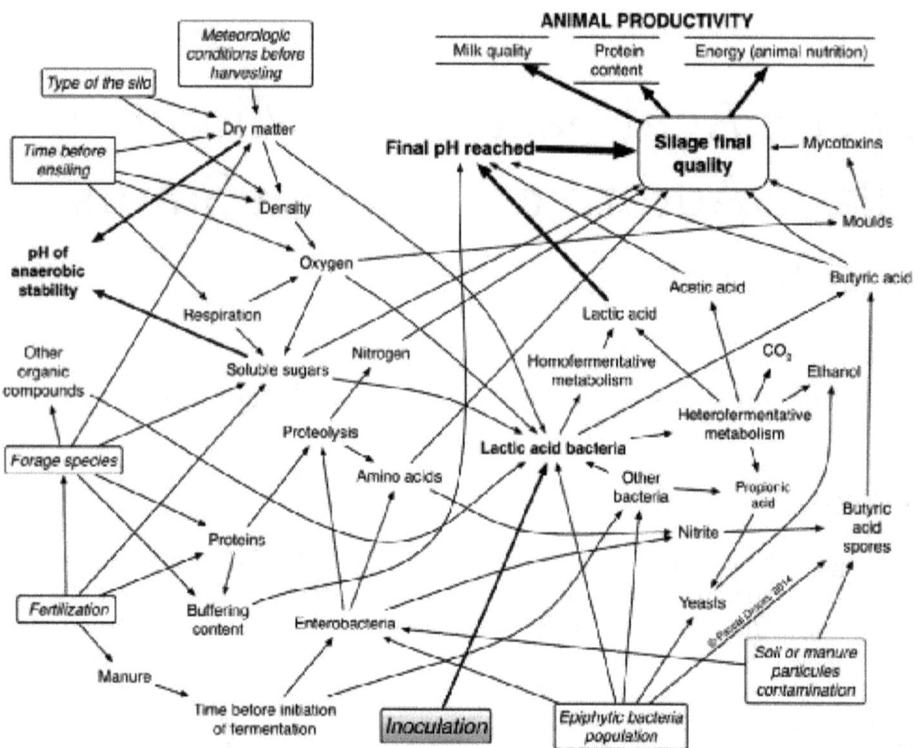

Figure1.
Ensiling involves several biochemical and microbiological descriptors that are influencing silage quality and could be controlled by different management criteria (boxed elements), which are directly influencing the main fermentation parameters of forage as well as animal productivity.

culture-dependent methods, following the isolation of strains and the determination of their taxonomic classification. The use of selective media has several shortcomings, including limited knowledge on how composition of the different defined culture media influences the growth of organisms within the targeted species range. Dormant or inactive cells (viable but nonculturable) may not have been accurately measured [1].

New techniques based on DNA profiling have helped understanding the microbial diversity of silage within specific families or genera [2]. These techniques were diverse and included denaturating gel electrophoresis [3] or metabolic fingerprinting by Fourier transform infrared spectroscopy [4].

Next generation sequencing (NGS) technologies provide more complete details on microbiota diversity. The first application of NGS in silage was performed on ensiled grass to help understand how inoculation would influence the microbial communities [5]. Three years passed before a second paper would be published using NGS studying spatial and temporal microbial variations in commercial bunkers [6]. Several more papers or communications were performed afterward (see **Table 1**).

One of the complexities facing ensiling of forage is that several factors will influence the size and diversity of the microbial community at harvest. Microbial diversity will change according to the plant species, weather conditions during growth and prior to harvesting, fertilization management, physiological state of the forage, and so on. As an example of the potential variation, important differences in the composition of the epiphytic bacterial population were observed from different organs of whole plant corn in the weeks prior to harvesting (**Figure 2**). Leaves, silk,

Forage	DM (g kg)	Time of fermentation	Temperature	Inoculation and rate	Abundance of *Lactobacillus* (max)	16S rDNA amplicons	ITS amplicons	Reference
Time-related dynamic								
Alfalfa-grass	395	7 periods, up to 64 days	20°C	*L. buchneri* and *L. hilgardii* (4 × 10^5 CFU g FM)	61%	V3–V4	ITS1-4	[7]
Alfalfa	421	4 periods, up to 90 days	22–25°C	*L. plantarum* or *L. buchneri* (1 × 10^6 CFU g FM)	93%	Full 16S—PacBio	No	[13]
Corn	381	9 periods, up to 90 days	n.a.	*L. plantarum* MTD1 (10^6 CFU g FM)	97%	V3–V4	ITS1-2	[9]
Corn	352	8 periods, up to 64 days	20°C	*L. buchneri* and *L. hilgardii* (4 × 10^5 CFU g FM)	95%	V3–V4	ITS1-4	[7]
Manyflower	410	6 periods, up to 30 days	Ambient	No	75%	V4–V5	No	[81]
Oat	456	6 periods, up to 90 days	n.a.	*L. plantarum* (1 × 10^6 CFU g FM)	97%	V4–V5	No	[82]
Commercial silos								
Corn (bunker)	n.a.	Vary	n.a.	n.a.	96%	V1–V3	No	[83]
Corn (bunker)	212–373	60 days	n.a.	n.a.	8–90%	V4	No	[84]
Corn-Sorghum (bunker)	320–510	Vary	n.a.	n.a.	>90%	V4	No	[6]
Corn (bag silo)	383	150	n.a.	*L. buchneri* and *L. hilgardii* (3 × 10^5 CFU g FM)		V3–V4	ITS1-4	Unpublished
Corn (bunker)	360	150	n.a.	*L. hilgardii* (1.5 × 10^5 CFU g FM)		V3–V4	ITS1-4	Unpublished
Experimental silos								
Alfalfa and sweet corn	187–222	65 days	25 °C	No	91–96%	V3–V4	No	[85]

Forage	DM (g kg)	Time of fermentation	Temperature	Inoculation and rate	Abundance of *Lactobacillus* (max)	16S rDNA amplicons	ITS amplicons	Reference
Corn (whole)	380	100 days	23°C	*L. buchneri* 40788 (4 × 10^5 CFU g FM) and *P. pentosaceus* (1 × 10^5 CFU g FM)	34 (con)–99%	V4	ITS1	[86]
Corn	234	90 days	22–25°C	*L. plantarum* or *L. buchneri* (1 × 10^6 CFU g FM)	>98%	Full 16S—PacBio	No	[32]
Grass (not further defined)	368	14 and 58 days	n.a.	*L. buchneri* CD034 (10^6 CFU g FM)	35–67% (inoculated)	V3–V4	No	[5]
High moisture corn	751	10, 30 and 90 days	20–22°C	*L. buchneri* and/or *L. hilgardii* (4 × 10^5 CFU g FM)	95%	V3–V4	ITS1–4	Unpublished
Moringa oleifera	n.a.	60 days	15 and 30°C	4 species of LAB (individual) (10^5 CFU g FM)	61–97%	V3–V4	No	[11]
Moringa oleifera	233	60 days	25–32°C	No (vacuum bags)	86%	V3–V4	No	[87]
Purple prairie clover	300	76 days	22°C	No	30%	V3–V4	V4-V5	[88]
Small grain (mix)	385	90 days	22°C	No	82%	V3–V4	V4-V5	[89]
Oat	450	217 days	23°C	*L. buchneri* 40788 (4 × 10^5 CFU g FM) and *P. pentosaceus* (1 × 10^5 CFU g FM)	57%	V4	ITS1	[10]
Soybean + corn	340	60 days	15–30°C	No	60–80%	V3–V4	No	[90]
Sugar cane	n.a.	90 days	20–35°C	No	50%	V4	No	[91]

Table 1.
Characteristics of the silage and experimental design from publications using amplicon-based metagenomic to study the microbiome.

and tassels harbored different proportions of the main epiphytic bacterial families even though the variation in microbiota composition was small between the sampling periods. *Cytophagaceae* and *Methylobacteriaceae* were mainly observed on the leaves, while *Enterobacteriaceae* and *Pseudomonadaceae* were observed on silk, cob, and tassel [7].

Published results of microbiome analysis were performed from varied forages from temperature and tropical regions, including pure strands of legumes or grasses and mixed forages. Several studies performed time-based samplings to describe changes in the microbial communities in relation to the fermentation periods [7–9] (**Table 1**). Generally, the relationship between the time of fermentation and the microbial composition was similar to the general succession pattern previously reported by culture-dependent microbiological techniques. For example, with corn silage inoculated with either *Lactobacillus plantarum* or *Lactobacillus buchneri* and/or *Lactobacillus hilgardii*, it was possible to observe that the succession to Firmicutes was rapid, in a matter of hours after sealing the experimental mini-silos. A second observation was that *Leuconostocaceae* (mainly *Weissella* sp.) was the dominant operational taxonomic unit (OTU) during early fermentation. In both studies, there were important changes in bacteria richness during the fermentation, with either values below 50 OTUs after incubation of 30 days [9] or decreasing throughout fermentation to a similar level of OTUs [8]. In both studies, fungal richness dropped throughout fermentation.

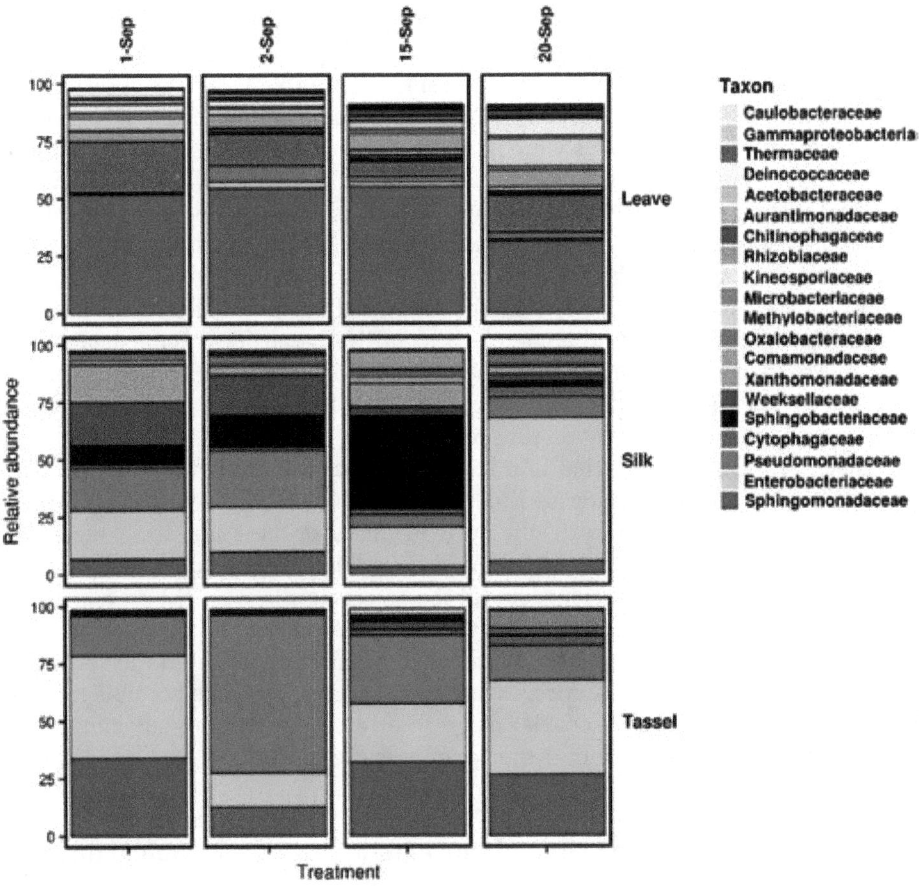

Figure 2.
Bacterial microbiome from different corn organs (leave, silk, and tassel) at four time points prior to harvesting.

These changes in microbial population were also observed in samples collected on farms. Under commercial conditions, comparing silage made from the same forage between sites is difficult since differences in dry matter (DM), packing density, and other physical parameters will influence efficiency of the fermentation and the microflora. Associating those parameters to NGS studies could improve the understanding of this process. It will then be possible to comprehend how other physical variables may contribute, e.g., the impact of high or low temperature on microbial succession, the impact of length of storage, length of time at a high temperature, and the impact of DM variations within the same forage.

To date, most of the data collected from experimental silos was performed with incubation periods shorter than 100 days and at a temperature around 20–25° C. These conditions offer an initial set of parameters but must be expanded to simulate real-life conditions in silos, which could include variances of more than 20°C above ambient temperatures during fermentation and long fermentation periods [10].

Most of the published studies included a comparison between control and a microbial silage additive or between different strains of LAB. The general trend on microbial diversity is that inoculation with LAB reduces the microbial diversity, but the impact differs in relation to the forage and the species of LAB. As observed by Wang et al. [11], microbial diversity was influenced by the inoculation of *Moringa oleifera* differentially for each of the four LAB species inoculated as well as from the temperature of incubation.

Comparisons between studies tend toward similar changes in microbial composition. To facilitate comparisons, it will be necessary to standardize DNA isolation and preparation of the amplicons prior to sequencing. By summarizing the main methodology information from different trials (**Table 1**), it was observed that some studies did not include fungal diversity, and the amplified DNA region differed. Most bacterial studies were performed following the amplification of the V3–V4 region, but there was a trend toward using the V4 or V4–V5 region, which offers potential for longer DNA strand and improves comparison scores against the database. Using a good quality database is also a critical step that is often overlooked during analysis [12]. The drawback of the current methodology for amplicon-based metagenomic is that the amplified region is short and does not provide enough coverage of the complete 16S rRNA gene. Two published studies were able to gather near complete fragments by sequencing the 16S rRNA gene on a PacBio sequencer instead of the Illumina model [12, 13]. This expanded the analysis of diversity to the species, or even subspecies, level.

Currently, no study has tried to mix the potential offered by polymerase chain reaction (PCR)-based profiling technology—like PCR-DGGE—with NGS capacities. Instead of amplifying with universal primers, primers targeting regions of lower variations within ribosomal DNA, or in other genes, provide more precise results allowing higher similarity scores at the species level.

Microbial communities continuously evolve during the storage period, even during the anaerobic stable phase. By improving our knowledge on the succession between communities, genus, species, and even strains, it will be possible to refine how strains are selected as microbial silage additives. This could easily allow selection of strains for particular forage species or climatic conditions.

3. Searching for new forage inoculants in temperate and tropical forages

The fermentation capability—or the acidification potential—depends directly on the DM content, at the level of water-soluble carbohydrates (WSC), and, inversely, on the buffering capacity of a given forage [14]. Due to their compositions, the ensiling potential is completely different among the different families of forages: tropical (C4), temperate (C3) grasses, and legumes.

Studies conducted by Wilkinson [15] with C3 grasses have concluded that the minimal concentration of the WSC should be at least 2.5–3.0% of the fresh forage. Below 2% of WSC of fresh crop weight, forages are prone to undesirable fermentations. The average level of WSC found by Zopollatto et al. [16] in a review of micro-bial additives in Brazil for tropical grasses was only 1.6%, far from the minimum for a good fermentation.

Tropical grasses provide large quantities of DM, which can reach up to 30 tons of DM per hectare. This great yield, however, comes at the optimal stage of maturity in terms of nutrients with other types of challenges: wilting is an issue, and the excess moisture can lead to important losses of nutrients through effluent production [17]; additionally, its nutritive value sharply declines as maturity advances.

The microflora existing on the vegetative parts of plants consist mainly of microorganisms considered undesirable from the point of view of the fermentation process. These include anaerobic bacilli of the genus *Clostridium*; aerobic bacteria of the genus *Bacillus* ; coliform bacilli, including *Escherichia coli* , *Enterobacter* spp., *Citrobacter* spp., and *Klebsiella* spp.; as well as bacteria of the genus *Listeria*, *Salmonella*, and *Enterococcus* (*E. faecium*, *E. faecalis*, *E. mundtii*, *E. casseliflavus*, *E. avium*, and *E. hirae*); and the occurrence of actinomycetes. Species of *Clostridium* are responsible for large losses because they produce CO_2 and butyric acid instead of lactic acid. Yeast and molds also form a large group [18].

Concerning the presence of LAB, Pahlow et al. [19] found in grasses that *L. plantarum*, *L. casei*, *E. faecium* and *Pediococcus acidilactici* were the most frequently observed species. However, with the development and the use of DNA sequenc-ing profile techniques, it is possible to identify hundreds of species as mentioned earlier. Most of the studies done by scientific groups were based on the efforts to find any microorganisms, especially bacteria, able to drive a good fermentation and inhibit undesirable and detrimental microorganisms.

Zielińska et al. [20] demonstrated that microbial inoculants altered many parameters of silages, but the strength of the effects on fermentation depends on specific characteristic of an individual strain. Several research teams have been searching for new strains able to perform better than the ones currently on the market. For example, Agarussi et al. [21] searched for new promising strains for alfalfa silage inoculants and isolated *Lactobacillus pentosus* 14.7SE, *L. plantarum* 3.7E, *Pediococcus pentosaceus* 14.15SE, and a mixture of *L. plantarum* 3.7E and *P. pentosaceus* 14.15SE. The authors concluded that all of the tested strains had a positive effect on at least one chemical feature of the silage during the fermentation process, although the most promising strain found in that trial was the *P. pentosaceus* 14.15SE.

Moreover, Saarisalo et al. [22] searched for LAB capable of lowering the pH of grass silages with low proteolytic activity. The researchers found a potential strain of *L. plantarum*, which was effective in reducing the deamination in silages.

Besides aiming to enhance silage fermentation, aerobic stability has been an important topic in the last 20 years. During silage feedout, accelerated growth of spoilage organisms (yeasts) results in high temperatures and nutrients and DM losses, leading to increased silage deterioration [23]. According to McDonald et al. [24], even though yeasts can grow from 5 to 50°C, the optimum growth of most species occurs at 30°C. Other spoiling microorganisms, such as molds and *Clostridium* bacteria, grow between 25 and 37°C, respectively. Considering the specific temperature and humidity ranges of different microbes for growth, it is possible to see that tropical climates are more prone to spoilage than temperate ones.

4. Improving aerobic stability using forage inoculants

Silage feedout is the final phase of the ensiling process. At that moment, oxygen can slowly diffuse inside the silage mass. Diffusion speed will be influenced by different factors, including the level of humidity, porosity, and temperature of the silage [25].

The process of aerobic deterioration of silage involves a shift to aerobic metabolism in some microorganisms and the reactivation of strict aerobes that were dormant. Reduce nutritional value due to oxidation of the fermentation products, of carbohydrates, amino acids, and lipids to H_2O, CO_2, and heat. Simultaneously, the higher metabolic activity will increase the silage temperature, accelerating microbial growth. Several microorganisms are involved, but yeast and acetic acid bacteria are adapted to tolerate the initially low pH conditions and thus able to exploit this niche before pH increases following the catabolism of the organic acids. Crops with higher levels of easily accessible carbohydrates are more prone to aerobic deterioration, i.e., corn, sorghum, and sugarcane, since these sugars can be readily fermented by spoilage microorganisms in the presence of oxygen.

Following the isolation of a *L. buchneri* strain [26], researchers described its unique metabolic pathway, which consisted of converting moderate amounts of lactate under low pH to equal parts of acetate and 1,2-propanediol [27]. The latter chemical is an intermediate in the potential synthesis of propionic acid. *L. buchneri* does not have the gene to complete the reaction, so another species of LAB has to be involved to convert 1,2-propanediol to an equimolar amount of 1-propanol and propionic acid [28]. This conversion was initially observed in silage by *Lactobacillus diolivorans* [29], but other members of the buchneri group also possess the genetic system [30], like *Lactobacillus reuteri* [28].

Compared to lactic acid, the key feature of acetic and propionic acids in improving aerobic stability of silage is based on the difference in pK_a between these weak acids and lactic acid, which is a stronger acid, with a pK_a of 3.86. At higher pK_a, 4.76 for acetic acid and 4.86 for propionic acid, these weak organic acids will have a low dissociation level under most ensiling conditions, thus allowing for passive diffusion inside the yeast or other microorganism cytoplasm. Once inside the cytoplasm, propionic acid will dissociate to the corresponding salt since internal pH is above pK_a value. The same process is also possible for acetic acid. Constant pumping of the protons released inside the cytoplasm causes physiological stresses impacting several metabolic pathways in yeast cells [31].

Length of fermentation and establishment of heterofermentative LAB population are now considered critical toward the establishment of a good aerobic stability level. The facultative, or obligate heterofermentative, strains of LAB have lower growth rates than homofermentative strains, including rods like *L. plantarum* or coccids of the genera *Leuconostoc*, *Enterococcus*, or *Lactococcus*. The growth conditions after several days of ensiling are also more restrictive for physiologi-cal activities considering the low pH usually encountered. The strains succeeding the earlier colonizer need to be more tolerant to both acidity and osmotic stresses, simultaneously. Observation of the succession of different species of LAB during the anaerobic stability phase often leads to high abundance of LAB belonging to the *L. buchneri* taxonomical group [32], leading to specific adaptation to this ecological niche by these strains. Although few physiological studies on *L. buchneri* strains had been published, Heinl and Grabherr recently published a complete analysis of the genetic potential of the strain CD034 compared to other genomes from public databases [33]. One of the comparisons performed aimed to describe how the genetic system of this species can cope with high concentration of organic acids, including lactic acid. The anaerobic conversion system of lactic acid to 1,2-propanediol (to acetic acid and CO_2 under aerobic condition) represents one of those properties. It is possible to extend these observations to the results gathered from transcriptomic analysis on the strain *L. buchneri* CD034 [34] following the aeration of culture grown under anaerobic conditions. The team described the functions of 283 genes induced by the presence of oxygen. They also observed physiological adaptation related to changing oxygen concentration. Genes required by lactic acid fermenta-tion systems were hardly affected.

Co-inoculation with different heterofermentative strains has recently been tested in the field or in commercialized conditions. This was the case for *L. buchneri* and *L. diolivorans*, tested on the fermentation of sourdough [35]. The authors showed an increase in the accumulation of propionic acid following the inoculation with both strains together. Co-inoculation of *L. buchneri* and *L. hilgardii* was tested in different ensiling trials [36, 37] inducing better fermentation and higher aerobic stability level. *L. hilgardii*, an obligate heterofermentative strain, was not only previ-ously observed as a contaminant of wine but also represents one of the dominant LAB strains in water kefir [38]. Strains of this species are often observed in sugar cane silage [39, 40] and provide increased aerobic stability levels for this challeng-ing crop. Improvement in fermentation and aerobic stability of sugarcane silage allowed increasing DM intake and milk yield [41].

Two recent meta-analyses [42, 43] provided a complete overview of the impact of inoculation of LAB and described the importance of fermentation and aerobic stability in relation to the specificities of the forages and the activity of homofermentative, facultative heterofermentative, and obligate heterofermentative strains. In particular, the meta-analysis of Blajman et al. [42] analyzed the role of inoculation on reducing the amount of yeast in silage.

Improving aerobic stability to reduce overall losses during the storage and feed-out is one of the main reasons to apply microbial inoculants on the forage at the time of ensiling. The value of silage inoculants is important, but optimal management of silos at all steps of the ensiling process is critical.

5. Improving adoption of forage inoculant use by increasing awareness of the economic value of forage inoculants

According to the 2017 National Agricultural Statistics Survey [44] census report, approximately 120,000,000 tons of whole-plant corn alone was harvested for silage in the United States. Even with this huge quantity of silage, there is little reliable survey data about the use of forage inoculants.

Based on an independent market survey of U.S. beef and dairy producers, two thirds of respondents indicated that forage additives used on their operations are microbial based. The main reason for their use is to minimize mold and spoilage in silage. Other reasons cited include preventing heat damage and increasing herd productivity [45]. Most inoculant users plan on continuous using and investing in this technology each year (personal communication).

Product performance, ease of use, and cost are the main influencers on the purchasing decision of inoculants. In addition, nutritionists and consultants are important sources for providing information on forage inoculants and the most involved outside sources in the purchase decision (personal communication).

Most producers do not have a detailed understanding of the different types of inoculant products, but they instead recognize the value and return on investment (ROI) that these technologies can bring to their operation. Value-added services and education offered by inoculant companies are also reasons to purchase, especially for larger producers.

Producers may often choose not to purchase forage inoculants due to the cost of the products. Other top reasons that influence purchase decisions are (1) not believing inoculants work, (2) lack of knowledge, or (3) lack of specific equipment for inoculating the forage. With all these factors in mind, there is a strong need for proper education on the application and showing the cost-to-benefit calculation of these forage additives (personal communication).

Even though some producers are nonusers, they believe that inoculants have the potential to improve consistency of silage quality, enhance ration quality, and increase feedout stability. In the same question, just 40% answered that improving ROI is one of the most important benefits of purchasing inoculants. Even though some producers do not associate inoculants with contributing to overall herd ROI and profitability, they positively associate the word "fresh" to silage having a good smell and high palatability (personal communication).

During typical field and harvest management conditions, silage losses are easily reported between 15 and 20%. If inoculant use can reduce DM losses by 5 percentile points, there would be savings of $2000 (US$) per thousand tons of silage, assuming the silage is valued at $40.00 (US$) per ton FM. Moreover, silage with high degree of deterioration not only has less overall tonnage to be fed, but the feed is also of lower nutritional quality.

6. Optimizing fiber and carbohydrate digestibility

The main metabolic activity of LAB during the ensiling process consists of reducing soluble carbohydrates to organic acids to acidify and preserve the for-age for long-term storage. It has been observed that animal performance has been increased following the use of microbial inoculants, even if no or small changes in silage fermentation parameters were observed [2]. Future research is needed to explain why these improvements are observed. Yet, past research has made several important advancements.

As discussed previously, inoculation with LAB contributes to important modifications of the silage microbiota, for both the bacterial and the fungal communities. Some of these modifications could partly explain the contribution of the inoculant to one or more nutritional characteristics of silage. This could also support the theory of an indirect positive impact of these nutritional characteristics to the rumen microbial population and functions.

The rumen environment may also be affected by LAB forage inoculants. Some strains of LAB used as inoculants were shown to survive in the rumen fluid [46] and shift gas production toward other products or microbial cells [47]. Weinberg et al. [48] observed that LAB inoculants applied at ensiling, or into the rumen, had the potential to increase DM and fiber digestibility.

Studies using different inoculants showed increases in animal performance and milk production [49]. Mohammed et al. were also able to quantify elevated levels of *L. plantarum* in the rumen of cows eating the treated silage [50].

To help explain this improved animal performance, results from the studies of LAB used as a human probiotic may offer some clues. In a review of the metabolism of oligosaccharides and starch by lactobacilli, Gänzle and Follador [51] described limitations of the conversion of oligosaccharides since most related enzymes in LAB are active intracellularly and their substrates must be transported inside the cells to hydrolyze (**Figure 3**). By studying the genome of several LAB species, they report that most lactobacilli could generally metabolize α-glucans. They would require contribution of a trans-membrane transporter in order to hydrolyze small

Figure 3.
Starch granules of corn after several months of ensiling. Rod shape bacteria, putatively LAB, were thriving on fiber particles surrounding the starch granule but not on the granules. Micrograph provided by Lallemand Specialties Inc.

oligosaccharides. Like some other lactobacilli, *L. plantarum* genome includes a gene encoding for an extracellular amylase with endoamylases activity. The presence of this amylase in the genome is strain specific as reported by Hattingh [52] for strains of *L. plantarum* isolated from barley.

Selecting strains with a functional trait, for example, fiber- or starch-degrading functions, represents the initial step in the development of a new inoculant. The strain has to cope with the different stresses of silage and also compete against epiphytic LAB and other microorganisms. The function has also to be expressed under the targeted microbial niche. The extracellular enzymes then have to be optimized for the acidic conditions and cope with the specific nature of polysaccharide substrates.

Access by fibrolytic enzymes to cellulose is difficult due to steric hindrance of the lignin-hemicellulose-homocellulose matrix. Improving cellulose degradation was targeted by selecting a LAB strain producing ferulate esterase [53]. This enzyme releases ferulic acid from arabinoxylans, improving access to other fibrolytic enzymes of the lignin-cellulose layer within cell walls.

More research is needed in this area. The complexity and dynamic of the microbial communities following the inoculation provide an important challenge in understanding the impact and role of the key players involved in this beneficial effect of microbial silage additive [54].

7. Improving animal performance with LAB forage inoculants

The expected effects of using a LAB forage additive are improved fermentation and enhanced feedout stability, which in turn lead to better recovery of nutrients and DM. However, expectations from producers are often beyond better silage characteristics, such as improvements in feed efficiency and, subsequently, animal performance.

Scientific evidence shows positive impact from the use of microbial inoculants on increases animal performance and production, in addition to enhancing the fermentation. However, these improvements are difficult to quantify.

Some of the existing theories are that these bacteria may have a beneficial influence in the rumen environment, including altering the fermentation profile and interacting with the animal's existing digestive microbiota [48] and inhibiting undesirable microorganisms, which subsequently help reduce the potential for toxin production [55].

Oliveira et al. [43] analyzed 31 studies—including animal performance results. This meta-analysis showed that microbial inoculation at a rate of at least 10^5 colony-forming units (CFU) of LAB per gram of forage significantly increased milk production by 0.37 kg/d, increased DM intake, and had no effect on feed efficiency and total tract DM digestibility. Furthermore, the contents of milk fat and milk protein tended to be higher for cows fed inoculated silage. The effects on increased milk production due to LAB inoculation happened regardless of the type of forage and diet, inoculant bacterial species and application rate (10^5 vs. 10^6 CFU/g of forage), and level of milk production.

Among the animal performance trials, there are cases when the inoculant had no effect on the silage fermentation compared to untreated silage, although animal productivity was increased [56]. Therefore, this indicates that some LAB strains are positively affecting the rumen microbial community and the digestive tract environment, resulting in improved effects on animal performance.

Recent research has described these effects by evaluating the impact of inoculated silages in the populations of the rumen microbial community, but no significant changes were observed [51]. However, nitrogen efficiency seemed to be improved due to lower levels of milk urea nitrogen in cows fed inoculated silage and greater ruminal DM digestibility on the inoculated silage ration [57]. Since LAB were shown to attach to the fiber inside the rumen [58], isolation methodology needs to be adapted to target the correct ecological niche.

Changes in nitrogen compounds during ensiling are expected. For example, over half of the true protein in alfalfa is degraded to soluble nonprotein compounds initially by the plant's own proteases, and then later by microbial activity within the cow, resulting in inefficient nitrogen use to the cow [59].

Specifically, in the corn kernel or other cereal grain, a protein matrix (prolamins) around the starch granules partially prevents ruminal starch digestion. It has been reported that a slow and continuous breakdown of the prolamins during the storage phase makes the starch more digestible with longer storage time [60]. The authors explained that this effect is due to natural proteolytic mechanisms. This event, however, requires months of storage for the optimum level of starch digestibility in the rumen, in which it is not always feasible in commercial operations. One alternative solution would be to shorten the time necessary for storage to help enhance starch digestibility by inoculation with bacteria that possess high proteolytic activity, but, to date, limited research has been reported and results are inconsistent.

Improvement of fiber digestibility has to be considered in relation to the activity of silage inoculants. Some strains of LAB have been reported to produce the enzyme ferulic acid esterease, which breaks the esterease bond between the lignin and the hemicellulose fraction, leading to more digestible fiber portions for the rumen microorganisms [61]. However, data from animal performance or production studies did not show consistencies in the improvements [61, 62]. While *in vitro* and *in situ* effects may be conceivable, the expression of this phenomenon within *in vivo* environments needs additional research to be better understood.

There is still a need to better understand how the microbial additives for ensiling positively affect animal performance, so this should be used as criteria for a new generation of this type of additive.

8. Understanding the impact of ensiling on a global scale

Silage represents an important part of animal diets. Challenges in production, reducing losses, and the impact on agricultural practices are often overlooked compared to other nutritive benefits provided. Microbial activity during fermentation produces several compounds besides the desirable organic acids. Some of those compounds were identified as negatively influencing air quality around farms. They are classified as alcohols, esters, and aldehydes [63, 64]. Production and volatiliza-tion of these compounds contribute to a reduction in quality of the stored feed, inducing ground-level ozone, and influence emission of greenhouse gases by the agricultural sector [65].

Forage characteristics and yield potential are influenced by several factors, including geographic and meteorological conditions. New analytical technologies and statistical methodologies now allow more comprehensive understanding of ensiling techniques and analyze productivity and nutritional quality on a broader scale.

Comparison between farms is always challenging, even between neighboring farms, since they could differ on animal husbandry, genetics of the herd, field management, harvesting periods, type and size of silos, management of the silos, and so on. On a broader geographic area, these differences will be minimized by the inclusion of higher numbers of farms, up to a point that patterns of variations could be analyzed. This type of analysis was performed by Gallo et al. in two recent studies [65, 66]. The team used a multivariate analysis technique, Principal Component Analysis (PCA), to evaluate ensiling of corn silage on 68 dairy farms [66] and generated a fermentation quality index to rank the silage [67]. Using 36 variables measured on every individual samples, they were able to group the silage according to quality parameters in relation to silo management techniques to discriminate between well-preserved and poorly preserved forages.

At the farm level, quality parameters from silage and feed analysis reports could be analyzed to identify trends in animal health and performance. Different types of data could be collected and analyzed to understand the main variations in milk quality and yield on a yearly or multi-year basis. Linking milk quality parameters to farm management practices was performed following the analysis of milk constituent using Fourier transformed mid-infrared spectroscopy results gathered from 33 farms [68]. The difference between observed high and low *de novo* fatty acid composition of milk allowed characterizing differences in feeding management (one or two feeding periods—fresher silage) and higher animal management scores (freestall stocking—lower housing density).

Up to now, few data analysis included data specific to silage fermentation beside the main fermentation acids. This is truer for other parameters related to silage production and management, including yield from the field, management of the silos, losses during fermentation, or type of silage additive used. This needs to be addressed considering important changes to the microbiota following the inoculation discussed previously and to differentiate in other fermenta-tion chemicals or their relationship with the nature of the additives applied, as observed by Daniel et al. [69].

9. Increasing the understanding of the fermentation process

Compared to other research domains in agricultural and environmental sciences, using new sequencing technologies to understand the dynamics of the microbial communities in silage is recent. McAllister et al. [12] published a review providing a technological and methodological overview. Currently, the number of trials performed using this technique is small enough that repetitions between geographical regions and over time are nonexistent.

Amplicon-based metasequencing represents the entry level of the -omic techniques. For silage research, the industry could also consider metagenomic, proteomic, transcriptomic, or epigenomic as a potential area of study. A review of the possibilities offered by metabolomics in agriculture was recently published [70].

Since ensiling is based on the fermentation of forage crops, knowledge of the metabolic activity of the forage prior to ensiling would be useful. A review by Rasmussen et al. [71] provides an insight into how plants are coping with physiological changes due to breeding strategies, associations with endophytes or rhizobia, responses to nutrients, and, more interestingly, on the metabolic responses to the osmotic stress. Harvesting and wilting will directly influence plant cell activities and nutrient cycling. The authors reported that amino acids, fatty acids, and phytosterols generally decrease following the water stress, while sugars and organic acids increased. Since the fermentation process requires fermentable sugars for optimal acidification of the forage, wilted plants may respond positively toward ensiling. We need to consider the speed of those changes in concentration of metabolites during wilting compared in order to propose a model of the response to an osmotic stress. Ould-Ahmed et al. [72] provided some knowledge on this response to wilting while studying changes in fructan, sucrose, and some associated hydrolytic enzymes, concluding there is a positive effect toward ensiling requirements from the different metabolites.

Metabolomic profiling of silage was performed in a study aiming to understand the role of inoculation with *L. plantarum* or *L. buchneri* in alfalfa silage against a noninoculated control [13]. The authors were able to distinguish all three inoculation treatments by a PCA of the 102 metabolites surveyed. The major metabolites observed were related to amino acids, organic acids, polyhydric alcohols, and some derivatives. One of the main observations was an increase in free amino acids and 4-aminobutyric acid following the inoculation with *L. buchneri* and a decrease in cadaverine and succinic acid following the inoculation with *L. plantarum*.

Testing the same two LAB strains on whole plant corn silage instead of alfalfa, Xu et al. [32] observed a total of 979 chemical substances, from which 316 were identified and quantified. The PCA allowed separating the three inoculation treatments along the first axis, representing nearly 80% of the variations between samples. The second axis was able to further distinguish how inoculation with *L. buchneri* influenced the fermentation. Inoculation with either *L. plantarum* or *L. buchneri* contributes to increase the concentration of amino acids and phenolic acids, 4-hydroxycinnamic acid, 3,4-dihydroxycinnamic acid, glycolic acids, and other organic acids. Inoculation with *L. buchneri* also induces higher concentration of 2-hydroxybutanoic acid, saccharic acid, mannose, and alpha-D-glucosamine- 1-phosphate, among others. Other substances were increased by ensiling without specific impact of the inoculants, such as catechol and ferulic acid that could have antioxidant functions.

Metabolomic studies can also be used in defining a metabolomic signature specific of different forage and silage on feed efficiency of ruminants. With the aim of identifying feed efficiency traits in beef cattle, Novais et al. [73] investigated how serum metabolomic profiles could be used to predict feed intake and catabolism. They identified different molecules having feed efficiency role. Two molecules from the retinol pathway, vitamin A synthesis, were significantly associ-ated with feed efficiency (higher concentration of retinal and lower concentration of retinoate).

Besides the studies of Guo et al. [13] and Xu et al. [32], one other study combined different -omic techniques in understanding the ensiling process. The first glimpse of that study was presented at the International Silage Conference in 2018 [8] with data on microbiota dynamic between 1 and 64 days of fermentation of corn silage. Analysis of the amplicon-based metasequences, metagenomic, and metabo-lomic data set is currently underway.

The potential of transcriptomic was also shortly covered by the *in vitro* trial of Eikmeyer et al. [34], which aimed to understand induction of genes in *L. buchneri* CD034 under different incubation settings. It is expected that additional studies performed directly under ensiling conditions may be published in the next few years.

Metabolomic data have shown how inoculation of LAB strains induces changes to the ensiled forage that goes beyond the simple production of lactic and acetic acids from the fermentation of sugars under anaerobic conditions. Increases in a whole array of molecules were observed, but the change also extends to the fibers and is either a direct or an indirect effect of the inoculant. Inoculation of alfalfa by *L. plantarum* or *Pediococcus pentosaceus* strains increased the release of different hemicellulose polysaccharides, including homogalacturonan, rhamnogalacturonan, and arabinogalactan from the cell walls [74].

These new technologies will allow greater understanding of the impact of bacterial inoculants on improvements of the silage and their contribution in the induction of specific genes and proteins by other members of the microbial community at different stages of the ensiling process.

10. Co-ensiling forage with food processing waste and TMR conservation

Food processing residues represent high-energy organic material already used in some way that could include either food-processing residues from food industries or distiller's grains from the ethanol production. These residues could easily be used by farms closely located to the production site, but their relatively high humidity content renders them prone to a rapid deterioration. New ensiling techniques allow mixing them with low moisture forage or grain in order to perform a fermentation that is enclosed in a kind of total mixed ration (TMR) acidic conservation.

Aiming to use a bakery co-product waste, Rezende et al. [75] tested possibilities of re-hydration, treating it with acid whey or water and levels of urea. The authors found that the resulting silages had reduced populations of molds and yeast by acidification process. However, the initial population of these microorganisms was high, mainly accounting of *Penicillium* and *Aspergillus* spp. Inoculating with a bacteria that could produce antifungal chemicals, including acetic and propionic acids, might be considered for this kind of co-product.

TMR silage is an important source of ruminant feed. This practice has been more common in some places, where companies or producers mix wet co-products with dry feeds to prepare TMR that is then preserved as silage. Based on conventional criteria, aerobic deterioration could occur easily in TMR silage, because lactic acid prevails during fermentation and any sugars remaining unfermented can serve as substrates for the growth of yeasts. However, some trials [76, 77] have been shown that when added concentrate, the brewer's grains or soybean curd residue, the main co-products used in TMR preserved do not show heating in the TMR. For the trial with brewers' grain-based TMR, the main bacteria found in the stable silages were *L. buchneri*, but for the soybean curd-based TMR, the main LAB found were *P. acidilactici* and *L. brevis* [78], showing potential association of those bacteria to preserve TMR silages. A similar trial was performed by Ferraretto et al. [79] to test how the process influenced luminal *in vitro* starch digestibility. They used dry ground corn to adjust the humidity level of wet brewers' grain and observed an increase in digestibility of the starch from the combined feed.

Nishino and Hattori [80] evaluated two bacterium-based additives in wet brewer's grains stored as a TMR in laboratory silos with lucerne hay, cracked maize, sugar beet pulp, soya bean meal, and molasses. The additives tested were the homofermentative LAB, *L. casei*, and the heterofermentative LAB *L. buchneri*. This last one was responsible for controlling yeast growth and the homolactic one helped in the fermentative profile of the ensiled TMR.

11. Final comments

General microbiology techniques have helped to understand the basic dynamic of microbial communities, the diversity of species, the biochemical pathways involved at each phase of the fermentation process, and the metabolic functions of the main spoiling agents involved in degrading the nutritional quality of the silage. NGS helped observe microbial communities, and metabolic profiling does not cease to evolve. This fact directly influences the nutritional characteristics of the silage.

In this chapter, the authors reviewed the main research activities that helped the agricultural industry understand silage, as it is known today and also pointed to experimental techniques that will continue to improve the understanding of metabolic pathways and functional aspects of the ensiling process. It is clear that these techniques will allow the scientific community to discover new inoculants that will combine our knowledge of silage fermentation, understand nutritional quality, improve rumen function, and contribute to better animal health. We are looking forward to the third generation of forage inoculants and seeing their positive impact.

Acronyms and abbreviations

AS	aerobic stability
CFU	colony-forming units
DM	dry matter
LAB	lactic acid bacteria
NGS	next generation sequencing
OTU	operational taxonomic unit
PCR	polymerase chain reaction
PCA	principal component analysis
TMR	total mixed ration
WSC	water-soluble carbohydrate

Author details

Pascal Drouin[1*], Lucas J. Mari[2] and Renato J. Schmidt[1]

1 Lallemand Animal Nutrition, Milwaukee, Wisconsin, United States

2 Lallemand Animal Nutrition, Aparecida de Goiânia, Goiás, Brazil

*Address all correspondence to: pdrouin@lallemand.com

References

[1] Pinto D, Santos MA, Chambel L. Thirty years of viable but nonculturable state research: Unsolved molecular mechanisms. Critical Reviews in Microbiology. 2015;**41**:61-76

[2] Muck RE. Recent advances in silage microbiology. Agricultural and Food Science. 2013;**22**:3-15

[3] Julien MC, Dion P, Lafrenière C, Antoun H, Drouin P. Sources of clostridia in raw milk on farms. Applied and Environmental Microbiology. 2008;**74**:6348-6357

[4] Johnson HE, Broadhurst D, Kell DB, Theodorou MK, Merry RJ, Griffith GW. High-throughput metabolic fingerprinting of legume silage fermentations via Fourier transform infrared spectroscopy and chemometrics. Applied and Environmental Microbiology. 2004;**70**:1583-1592

[5] Eikmeyer F, Köfinger P, Poschenel A, et al. Metagenome analyses reveal the influence of the inoculant *Lactobacillus buchneri* CD034 on the microbial community involved in grass ensiling. Journal of Biotechnology. 2013;**167**:334-343

[6] Kraut-Cohen J, Tripathi V, Gatica J, et al. Temporal and spatial assessment of microbial communities in commercial silages from bunker silos. Applied Microbiology and Biotechnology. 2016;**100**:6827-6835

[7] Drouin P, Chaucheyras F. How do time of fermentation and lactic acid bacteria inoculation influence microbial succession during ensiling? In: Gerlach K, Südekum KH, editors. XVIII International Silage Conference; 24-26 July 2018; Bonn, Germany. Germany: OundZ GmbH; 2018. pp. 32-33. ISBN 978-3-86972-044-9

[8] Gerlach K, Südekum K-H. How Do Time of Fermentation and Lactic Acid Bacteria Inoculation Influence Microbial Succession during Ensiling? Bonn, Germany: University of Bonn; 2018

[9] Keshri J, Chen Y, Pinto R, Kroupitski Y, Weinberg ZG, Sela S. Microbiome dynamics during ensiling of corn with and without *Lactobacillus plantarum* inoculant. Applied Microbiology and Biotechnology. 2018;**102**:4025-4037

[10] Romero JJ, Zhao Y, Balseca-Paredes MA, Tiezzi F, Gutierrez-Rodrigues E, Castillo MS. Laboratory silo type and inoculation effects on nutritional composition, fermentation, and bacterial and fungal communities of oat silage. Journal of Dairy Science. 2017;**100**:1812-1822

[11] Wang Y, He L, Xing Y, et al. Bacterial diversity and fermentation quality of *Moringa oleifera* leaves silage prepared with lactic acid bacteria inoculants and stored at different temperatures. Bioresource Technology. 2019;**284**:349-358

[12] McAllister TA, Dunière L, Drouin P, et al. Silage review: Using molecular approaches to define the microbial ecology of silage. Journal of Dairy Science. 2018;**101**:4060-4074

[13] Guo XS, Ke WC, Ding WR, et al. Profiling of metabolome and bacterial community dynamics in ensiled *Medicago sativa* inoculated without or with *Lactobacillus plantarum* or *Lactobacillus buchneri*. Nature Scientific Reports. 2018;**8**:357

[14] Oude Elferink SJW. Silage fermentation processes and their manipulation. In: FAO Electronic conference on tropical silage. Food and Agriculture Organization of the United Nation. pp. 17-30

[15] Wilkinson JM. Silage. Shedfield, UK: Chalcombe Publications; 1990

[16] Zopollatto M, Daniel JLP, Nussio LG. Aditivos microbiológicos em silagens no Brasil: Revisão dos aspectos da ensilagem e do desempenho de animais. Revista Brasileira de Zootecnia. 2009;**38**:170-189

[17] Ferrari EJ, Lavezzo W. Qualidade da silagem de capim-elefante (*Pennisetum purpureum*, Schum) emurchecido ou acrescido e de farelo de mandioca. Revista Brasileira de Zootecnia. 2001;**30**:1424-1431

[18] O'Brien M, O'Kiely P, Forristal PD, Fuller HT. Quantification and identification of fungal propagules in well-managed baled grass silage and in normal on-farm produced bales. Animal Feed Science and Technology. 2007;**132**:283-297

[19] Pahlow G, Muck RE, Driehuis F, Oude Elferink SJWH, Spoelstra SF. Microbiology of ensiling. In: Buxton DR, Muck RE, Harrison JH, editors. Silage Science and Technology. Madison, Wisconsin, USA: American Society of Agronomy; 2003. pp. 31-93

[20] Zielińska K, Fabiszewska A, Stefańska I. Different aspects of *Lactobacillus* inoculants on the improvement of quality and safety of alfalfa silage. Chiliean Journal of Agricultural Research. 2015;**75**:298-306

[21] Agarussi MCN, Pereira OG, da Silva VP, Leandro ES, Ribeiro KG, Santos SA. Fermentative profile and lactic acid bacterial dynamics in non-wilted and wilted alfalfa silage in tropical conditions. Molecular Biology Reports. 2019;**46**:451-460

[22] Saarisalo E, Skyttä E, Haikara A, Jalava T, Jaakkola S. Screening and selection of lactic acid bacteria strains suitable for ensiling grass. Journal of Applied Microbiology. 2007;**102**:327-336

[23] Bernardes TF, Daniel JLP, Adesogan AT, et al. Silage review: Unique challenges of silages made in hot and cold regions. Journal of Dairy Science. 2018;**101**:4001-4019

[24] McDonald P, Henderson N, Heron S. The Biochemistry of Silage. Vol. 340. Marlow Bottom: Chalcombe Publications; 1991

[25] Wilkinson JM, Davies DR. The aerobic stability of silage: Key findings and recent development. Grass and Forage Science. 2013;**68**:1-19

[26] Driehuis F, Elferink SJWHO, Spoelstra SF. Anaerobic lactic acid degradation during ensilage of whole crop maize inoculated with *Lactobacillus buchneri* inhibits yeast growth and improves aerobic stability. Journal of Applied Microbiology. 1999;**87**:583-594

[27] Oude Elferink SJWH, Krooneman J, Gottschal JC, Spoelstra SF, Faber F, Driehuis F. Anaerobic conversion of lactic acid to acetic acid and 1,2-propanediol by *Lactobacillus buchneri*. Applied and Environmental Microbiology. 2001;**67**:125-132

[28] Sriramulu DD, Liang M, Hernandez-Romero D, et al. *Lactobacillus reuteri* DSM 20016 produces cobalamin-dependent diol dehydratase in metabososomes and metabolizes 1,2-propanediol by disproportionation. Journal of Bacteriology. 2008;**190**:4559-4567

[29] Krooneman J, Faber F, Alderkamp AC, et al. *Lactobacillus diolivorans* sp. nov., a 1,2-propanediol-degrading bacterium isolated from aerobically stable maize silage. International Journal of Systematic and Evolutionary Microbiology. 2002;**52**:639-646

[30] Zielińska K, Fabiszewska A, Świątek M, Szymanowska-Powalowska D.

Evaluation of the ability to metabolize 1,2-propanediol by heterofermentative bacteria of the genus *Lactobacillus*. Electronic Journal of Biotechnology. 2017;**26**:60-63

[31] Lourenco AB, Ascenso JR, Sá-Correia I. Metabolic insights into the yeast response to propionic acid based on high resolution 1H NMR spectroscopy. Metabolomics. 2011;7:457-468

[32] Xu D, Ding W, Ke W, Li F, Zhang P, Guo X. Modulation of metabolome and bacterial community in whole crop corn silage by inoculating homofermentative *Lactobacillus plantarum* and *Lactobacillus buchneri*. Frontiers in Microbiology. 2019;**9**:3299

[33] Heinl S, Grabherr R. Systems biology of robustness and flexibility: *Lactobacillus buchneri*—A show case. Journal of Biotechnology. 2017;**257**:61-69

[34] Eikmeyer FG, Heinl S, Marx H, Pühler A, Grabherr R, Schlüter A. Identification of oxygen-responsive transcripts in the silage inoculant *Lactobacillus buchneri* CD034 by RNA sequencing. PLOS ONE. 2015;**10**(7):e0134149

[35] Zhang C, Brandt MJ, Schwab C, Gänzle MG. Propionic acid production by cofermentation of *Lactobacillus buchneri* and *Lactobacillus diolivorans* in sourdough. Food Microbiology. 2010;**27**:390-395

[36] Ferrero F, Piano S, Tabacco E, Borreani G. Effects of conservation period and *Lactobacillus hilgardii* inoculum on the fermentation profile and aerobic stability of whole corn and sorghum silages. Journal of the Science of Food and Agriculture. 2019;**99**:2530-2540

[37] Reis CB, de Oliveira dos Santos A, Carvalho BF, Schwan RF, da Silva Ávila CL. Wild *Lactobacillus hilgardii* (CCMA 0170) strain modifies the fermentation profile and aerobic stability of corn silage. Journal of Applied Animal Research. 2018;**46**:632-638

[38] Waldherr FW, Doll VM, Meißner D, Vogel RF. Identification and characterization of a glucan-producing enzyme from *Lactobacillus hilgardii* TMW 1.828 involved in granule formation of water kefir. Food Microbiology. 2010;**27**:672-678

[39] Carvalho BF, Ávila CLS, Miguel MGCP, Pinto JC, Santos MC, Schwan RF. Aerobic stability of sugar-cane silage inoculated with tropical strains of lactic acid bacteria. Grass and Forage Science. 2014;**70**:308-323

[40] Ávila CLS, Carvalho BF, Pinto JC, Duarte WF, Schwan RF. The use of *Lactobacillus* species as starter cultures for enhancing the quality of sugar cane silage. Journal of Dairy Science. 2014;**97**:940-951

[41] Santos WP, Ávila CLS, Pereira MN, Schwan RF, Lopes NM, Pinto JC. Effect of the inoculation of sugarcane silage with *Lactobacillus hilgardii* and *Lactobacillus buchneri* on feeding behavior and milk yield of dairy cows. Journal of Animal Science. 2017;**95**:4613-4622

[42] Blajman JE, Páez RB, Vinderola CG, Lingua MS, Signorini ML. A meta-analysis on the effectiveness of homofermentative and heterofermentative lactic acid bacteria for corn silage. Journal of Applied Microbiology. 2018;**125**:1655-1669

[43] Oliveira AS, Weinberg ZG, Ogunade IM, et al. Meta-analysis of effects of inoculation with homofermentative and facultative heterofermentative lactic acid bacteria on silage fermentation, aerobic

stability, and the performance of dairy cows. Journal of Dairy Science. 2017;**100**:4587-4603

[44] Service USDANAS. Geographic Area Series—Part 51. Washington: United States Department of Agriculture; 2019. p. 820

[45] Kung LJ, Shaver RD, Grant RJ, Schmidt RJ. Silage review: Interpretation of chemical, microbial, and organoleptic components of silages. Journal of Dairy Science. 2018;**101**:4020-4033

[46] Weinberg ZG, Muck RE, Weimer PJ. The survival of silage inoculant lactic acid bacteria in rumen fluid. Journal of Applied Microbiology 2003;**94**:1066-1071

[47] Muck RE, Filya I, Contreras- Govea FE. Inoculant effects on alfalfa silage: In vitro gas and volatile fatty acid production. Journal of Dairy Science 2007;**90**:5115-5125

[48] Weinberg ZG, Shatz O, Chen Y, et al. Effect of lactic acid bacteria inoculants on in vitro digestibility of wheat and corn silages. Journal of Dairy Science. 2007;**90**:4754-4762

[49] Contreras-Govea FE, Muck RE, Mertens DR, Weimer PJ. Microbial inoculant effects on silage and *in vitro* ruminal fermentation, and microbial biomass estimation for alfalfa, bmr corn, and corn silages. Animal Feed Science and Technology. 2011;**163**:2-10

[50] Mohammed R, Stevenson DM, Beauchemin KA, Muck RE, Weimer PJ. Changes in ruminal bacterial community composition following feeding of alfalfa ensiled with a lactic acid bacterial inoculant. Journal of Dairy Science. 2012;**95**:328-339

[51] Gänzle MG, Follador R. Metabolism of oligosaccharides and starch in lactobacilli: A review. Frontiers in Microbiology. 2012;**3**:340

[52] Hattingh M, Alexander A, Meijering I, Van RCA, Dicks LMT. Amylolytic strains of *Lactobacillus plantarum* isolated from barley. African Journal of Biotechnology. 2015;**14**:310-318

[53] Nsereko VL, Smiley BK, Rutherford WM, et al. Influence of inoculating forage with lactic acid bacterial strains that produce ferulate esterase on ensilage and ruminal degradation of fiber. Animal Feed Science and Technology. 2008;**145**:122-135

[54] Jin L, Dunière L, Lynch JP, McAllister TA, Baah J, Wang Y. Impact of ferulic acid esterase producing lactobacilli and fibrolytic enzymes on conservation characteristics, aerobic stability and fiber degradability of barley silage. Animal Feed Science and Technology. 2015;**207**:62-74

[55] Ellis JL, Hindrichsen IK, Klop G, et al. Effects of lactic acid bacteria silage inoculation on methane emission and productivity of Holstein Friesian dairy cattle. Journal of Dairy Science. 2016;**99**:7159-7174

[56] Daniel JLP, Morais G, Junges D, Nussio LG, editors. Silage Additives: Where Are we Going? Piracicaba, Brazil: University of Sao Paulo; 2015

[57] Kuoppala K, Rinne M, Vanhatalo A. Lactating Cow Response to Luerne Silage Inoculated with *Lactobacillus plantarum*. Hameenlinna, Finland: MTT Agrifood Research Finland, University of Helsinki, Helsinki, Finland; 2012

[58] Yang HE, Zotti CA, McKinnon JJ, McAllister TA. Lactobacilli are prominent members of the microbiota involved in the ruminal digestion of barley and corn. Frontiers in Microbiology. 2018;**9**:718

[59] Muck RE, Nadeau EMG, McAllister TA, Contreras-Govea FE, Santos MC, Kung LJ. Silage review: Recent advances and future uses of silage additives. Journal of Dairy Science. 2018;**101**:3980-4000

[60] Hoffman PC, Esser NM, Shaver RD, et al. Influence of ensiling time and inoculation on alteration of the starch-protein matrix in high-moisture corn. Journal of Dairy Science. 2011;**94**:2465-2474

[61] Lynch JP, Jin L, Lara EC, Baah J, Beauchemin KA. The effect of exogenous fibrolytic enzymes and a ferulic acid esterase-producing inoculant on the fibre degradability, chemical composition and conservation characteristics of alfalfa silage. Animal Feed Science and Technology. 2014;**193**:21-31

[62] Lynch JP, Baah J, Beauchemin KA. Conservation, fiber digestiblity, and nutritive value of corn harvested at 2 cutting heights and ensiled with fibrolytic enzymes, either alone or with a ferulic acid esterase-producing inoculant. Journal of Dairy Science. 2015;**98**:1214-1224

[63] Hafner SD, Howard C, Muck RE, et al. Emission of volatile organic compounds from silage: Compounds, sources, and implications. Atmospheric Environment. 2013;**77**:827-839

[64] Weiss K. Volatile organic compounds in silages—Effects of management factors on their formation: A review. Slovak Journal of Animal Science. 2017;**50**:55-67

[65] Åby BA, Randby ÅT, Bonesmo H, Aass L. Impact of grass silage quality on greenhouse gas emissions from dairy and beef production. Grass and Forage Science. 2019;**74**(3):525-534

[66] Gallo A, Bertuzzi T, Giuberti G, et al. New assessment base on th use of principal factor analysis to investigate corn silage quality from nutritional traits, fermentation end products and mycotoxins. Journal of the Science of Food and Agriculture. 2015;**96**:437-448

[67] Gallo A, Giuberti G, Bruschi S, Fortunati P, Masoero F. Use of principal factor analysis to generate a corn silage fermentative quality index to rank well- or poorly preserved forages. Journal of the Science of Food and Agriculture. 2015;**96**:1686-1696

[68] Woolpert ME, Dann HM, Cotanch KW, et al. Management practices, physically effective fiber, and ether extract are related to bulk tank milk de novo fatty acid concentration on Holstein dairy farms. Journal of Dairy Science. 2017;**100**:5097-5106

[69] Daniel JLP, Weiß K, Custódio L, et al. Occurence of volatile organic compounds in sugarcane silages. Animal Feed Science and Technology. 2013;**185**:101-105

[70] do Prado RM, Porto C, Nunes E, de Aguiar CL, Pilau EJ. Metabolomics and agriculture: What can be done. mSystems. 2018;**3**:e00156-17

[71] Rasmussen S, Parsons AJ, Jones CS. Metabolomics of forage plants: A review. Annals of Botany. 2012;**110**:1281-1290

[72] Ould-Ahmed M, Decau M-L, Bertrand A, Prud'homme M-P, Lafrenière C, Drouin P. Fructan, sucrose and related enzyme activities are preserved in timothy (*Phleum pratense* L.) during wilting. Grass and Forage Science. 2015;**72**:64-79

[73] Novais FJ, Pieres PRL, Alexandre PA, et al. Identification of a metabolomic signature associated with feed efficiency in beef cattle. BMC Genomics. 2019;**20**:8

[74] Drouin P, Ordaz S, Verrastro L, Sivakumar P. Impacts of silage bacterial additives on forage neutral detergent-soluble fiber. In: American Dairy Science Association Annual Meeting, Knoxville, TN. 2018

[75] Rezende AVD, Rabelo CHS, Sampaio LDM, et al. Ensiling a dry bakery by-product: Effect of hydration using acid whey or water associated or not at urea. Revista Brasileaira de Saúde e Produção Animal. 2016;**17**:626-641

[76] Nishino N, Harada H, Sakaguchi E. Evaluation of fermentation and aerobic stability of wet brewers' grains ensiled alone or in combination of various feeds as a total mixed ration. Journal of the Science of Food and Agriculture. 2003;**83**:557-563

[77] Wang F, Nishino N. Ensiling of soybean curd residue and wet brewers grains with or without other feeds as a total mixed ration. Journal of Dairy Science. 2008;**91**:2380-2387

[78] Li Y, Wang F, Nishino N. Lactic acid bacteria in total mixed ration silage containing soybean curd residue: Their isolation, identification and ability to inhibit aerobic deterioration. Asian Australasia Journal of Animal Science. 2016;**29**:516-522

[79] Ferraretto LF, Silva Filho WI, Fernandes T, Kim DH, Sultana H. Effect of ensiling time on fermentation profile and ruminal in vitro starch digestiblity in rehydrated corn with or without varied concentrations of wet brewers grains. Journal of Dairy Science. 2018;**101**:4643-4649

[80] Nishino N, Hattori N. Resistance to aerobic deterioration of total mixed ration silage inoculated with and without homofermentative or heterofermentative lactic acid bacteria. Journal of the Science of Food and Agriculture. 2007;**87**:2420-2426

[81] Lianhua L, Yongming S, Zhenhong Y, et al. Effect of microalgae supplementation on the silage quality and anaerobic digestion performance of Manyflower silvergrass. Bioresource Technology. 2015;**189**:334-340

[82] Kreshri J, Chen Y, Pinto R, Kroupitski Y, Weinberg ZG, Saldinger SS. Bacterial dynamics of wheat silage. Frontiers in Microbiology. 2019;**10**:1532

[83] Gharechahi J, Kharazian ZA, Sarikhan S, Jouzani GS, Aghdasi M, Salekdeh GH. The dynamics of the bacterial communities developed in maize silage. Microbial Biotechnology. 2017;**10**:1663-1676

[84] Guan H, Yan Y, Li X, et al. Microbial communities and natural fermentation of corn silages prepared with farm bunker-silo in Southwest China. Bioresource Technology. 2018;**265**:282-290

[85] Wang M, Wang L, Yu Z. Fermentation dynamics and bacterial diversity of mixed lucerne and sweet corn stalk silage ensiled at six ratios. Grass and Forage Science. 2019;**74**(2):264-273

[86] Romero JJ, Joo Y, Park J, Tiezzi F, Gutierrez-Rodrigues E, Castillo MS. Bacterial and fungal communities, fermentation, and aerobic stability of conventionnal hybrids and brown midrib hybrids ensiled at low moisture with or without a homo- and heterofermentative inoculant. Journal of Dairy Science. 2018;**101**:3057-3076

[87] Wang C, He L, Xing Y, et al. Fermentation quality and microbial community of alfalfa and stylo silage mixed with *Moringa oleifera* leaves. Bioresource Technology. 2019;**284**:240-247

[88] Peng K, Jin L, Niu YD, et al. Condensed tannins affect bacterial and

fungal microbiomes and mycotoxin production during ensiling and upon aerobic exposure. Applied and Environmental Microbiology. 2018;**84**:e02274-e02217

[89] Dunière L, Xu S, Long J, et al. Bacterial and fungal core microbiomes associated with small grain silages during ensiling and aerobic spoilage. BMC Microbiology. 2017;**17**:50

[90] Ni K, Zhao J, Zhu B, et al. Assessing the fermentation quality and microbial community of the mixed silage of forage soybean with crop corn or sorghum. Bioresource Technology. 2018;**265**:563-567

[91] Ren F, He R, Zhou X, et al. Dynamic changes in fermentation profiles and bacterial commjnity composition during sugarcane top silage fermentation: A preliminary study. Bioresource Technology. 2019;**285**:121315

3

Streamlining the Fermentation Process Using Mixed Cultures

Keukeu Kaniawati Rosada

Abstract

Fermentation technology is still being developed in all aspects, with the aim of improving the yields and qualities of products and reducing the costs of production. Increasing the yields of fermentation products can be accomplished by optimizing the factors that influence the process, including both the microbe itself and the environment. For example, the acetic acid production process from raw materials can be performed simultaneously with submerged batch fermentation using mixed cultures of anaerobic and facultative anaerobic *S. cerevisiae* and obligate aerobic *A. aceti*. This system is very simple because it only has one stage. In this system, efforts can be made to enhance the yields of acetic acid production, including evalu-ating the availability of nutrients in the medium and determining the optimum proportion of microbial abundance and agitation speed. Under optimal conditions, the resulting increases in acetic acid yields occur with high conversion efficiency. These results can then be applied on an industrial scale by integrating these findings with advanced technologies in the operating system.

Keywords: acetic acid production, *Acetobacter aceti*, aerobic submerged fermentation, mixed culture, *Saccharomyces cerevisiae*

1. Introduction

The fermentation industry has developed rapidly, especially as bioreactors have become the center of the process, as previously described [1]. The factors that have been the focus of development include the feeding of the bioreactor (batch, fed-batch, and continuous mode of operation), the use of microbial cultures (single strain or mixed culture processes), the availability of oxygen (aerobic, microaerobic, and anaerobic processes), and the mixing of the bioreactor during the process, particularly in the production of acetic acid. Acetic acid is produced from alcohol, and alcohol is produced from sugar. These two processes require different types of microorganisms. The microorganisms most commonly used in the fermentation of alcohol are yeasts, such as *Saccharomyces cerevisiae*, and bacteria, such as *Zymomonas mobilis*. However, for industrial fermentation, *Z. mobilis* appears to be inferior to *S. cerevisiae*, due to the reduced biomass production of the bacterium when pH decreases [2]. Commonly used acetic acid bacteria (AAB) include *Acetobacter* and *Gluconacetobacter*, two AAB genera that oxidize ethanol more easily than sugars [3], and exhibit resistance to high acetic acid concentrations and low pH [4]. For large-scale industries, the efficiency of the fermentation process design and operation continues to be developed, with the aim of improving the yields and qualities of the products and reducing the costs of production.

2. The development of fermentation technology in the production of acetic acid

In principle, the production of acetic acid from raw material is performed in two phases: the acetic acid fermentation process occurs under aerobic conditions, while alcoholic fermentation occurs under anaerobic conditions. Traditionally, the two processes are performed separately, under static and uncontrolled conditions [5, 6]. However, in its development, the production of acetic acid tends to occur in two or more stages, using either batch, fed-batch, or continuous types of operations. Many modifications have been made to the process, some of which are listed in **Table 1**. These modifications include the identification of alternative raw materials, the use

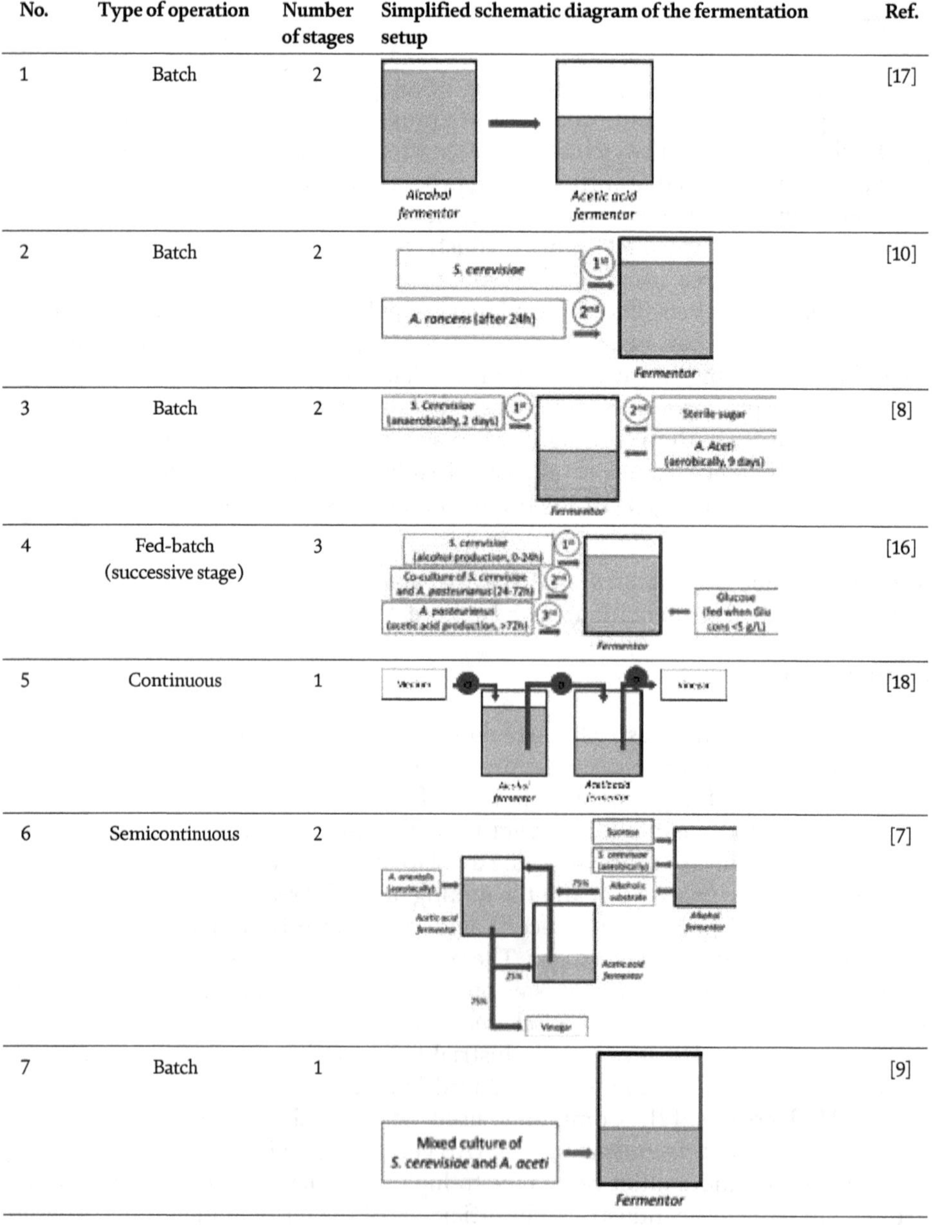

No.	Type of operation	Number of stages	Simplified schematic diagram of the fermentation setup	Ref.
1	Batch	2		[17]
2	Batch	2		[10]
3	Batch	2		[8]
4	Fed-batch (successive stage)	3		[16]
5	Continuous	1		[18]
6	Semicontinuous	2		[7]
7	Batch	1		[9]

Table 1.
The various fermentation processes used during acetic acid production from raw materials.

of different types of microorganisms, the implementation of different fermentation process operations and stages, and the manipulation of environmental conditions.

The identification of alternative raw materials is performed not only to identify new sources of material but also to address economic problems, such as the existence of surplus agricultural products, including onions [7], palm [8], or apples [9], and the utilization of sugar-containing waste materials, including pineapple peels [10]. The fermentation processes use different types of microorganisms. The identifica-tion and use of new types of microorganisms is performed with the aim of obtaining new strains with superior properties and abilities to produce high-quality and high-yield products [11–15]. Furthermore, various types of fermentation operations and modifications to the stages within these operations have been tested to determine the most simple and efficient methods capable of producing high yields because the operational procedures of acetic acid fermentation can be complicated and require a long time when multiple processes are required. Finally, the manipulation of envi-ronmental conditions, such as altering temperatures and aeration/agitation rates, is performed to obtain the optimal fermentation conditions [7, 9, 10, 16].

3. Enhanced acetic acid production from raw materials using mixed cultures during batch-type fermentation

Acetic acid fermentation has been studied using apples as a substrate. The fermentation was performed using submerged batch cultivation with mixed cultures of *S. cerevisiae* and *A. aceti*, which were inoculated simultaneously at the beginning of the process. These two microorganisms have different physiological properties: *S. cerevisiae* is a facultative anaerobe and requires anaerobic conditions to produce alcohol, and *A. aceti* is strictly aerobic. Because these microorganisms have opposing characteristics, using both cultures simultaneously is challenging, especially because acetic acid is a strong inhibitor of yeast, whereas yeast makes the medium anaerobic and unsuitable for AAB growth [5].

During this fermentation process, the first thing to be considered is the availability of sugar in the substrate, which represents a carbon source for the growth of the two microorganisms and the production of acetic acid. We must determine whether the sugar requirements are met by the substrate or whether sugar must be added, as described in previous studies [7, 8, 16]. Another factor that must be considered during this process is the inoculum ratio between the two microbes, as the strong competition between the two microbial groups must be anticipated and balanced to allow the production of acetic acid. Furthermore, because the different stages of acetic acid fermentation demand different oxygen requirements, the appropriate agitation speed is also important to consider.

3.1 Availability of sugar for microbial growth and acetic acid production

The microbial requirements for sugar during acetic acid fermentation and the availability of sugar in the substrate can be observed using different experiments, such as those shown in **Figure 1**. The system uses one-third of the working volume. Because this type of fermentation uses a batch culture with only one stage, all materials are added simultaneously at the beginning of the process, including sugar, at concentra-tion of 0, 10, and 20% ($^{w}/_{v}$). Changes in the sugar, alcohol, and acetic acid contents and changes in the pH values during this process were evaluated, as previously described [9]. The results demonstrated that the conversions of sugar into alcohol and of alcohol into acetic acid were accompanied by decreases in the pH of the medium. This result indicates that the fermentation process has been successfully performed.

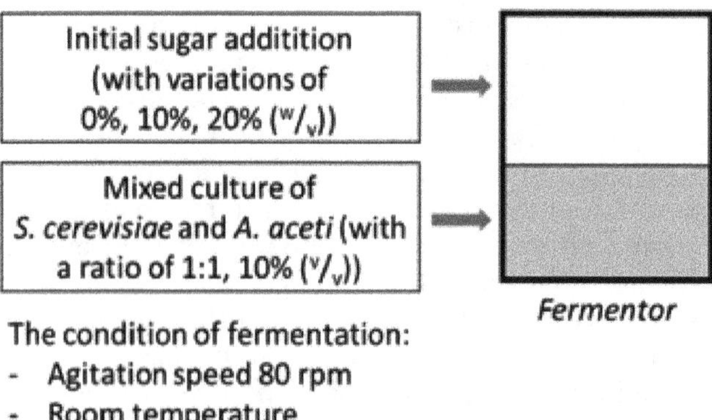

Figure 1.
Schematic diagram of the submerged batch fermentation process to evaluate the availability of sugar in the medium.

The percentage of initial sugar added to the medium ($^{w}/_{v}$)	0%	10%	20%
Conversion efficiency	233%	46.60%	6.40%

Table 2.
The conversion efficiency from sugar to acetic acid during 10 days of the fermentation process.

The sugar availability requirements during fermentation were determined by calculating the efficiency of the conversion of sugars into acetic acid (**Table 2**). According to **Table 2**, the conversion of sugar to acetic acid occurred with the highest efficiency when no sugar was added to the medium. Therefore, there is no need to add sugar to the medium because the substrate itself is sufficient to meet the needs of the microbes involved in the fermentation process and still produce a high acetic acid yield. The conversion efficiency of sugar into acetic acid during the fermentation process without the addition of sugar was greater than 100%. This result can be the result of the hydrolysis of starches contained in the medium, either chemically due to the decrease in pH [19, 20] or by *S. cerevisiae* to support growth and metabolic activity [21]. In addition to using glucose as its primary substrate, *S. cerevisiae* is able to grow on a wide range of carbon compounds, is able to metabolize some carbohydrates after they have undergone extracellular hydrolysis, and is able to ensure the efficient metabolism of those hydrolyzed carbohydrates [22]. Therefore, the addition of sugar to the fermentation medium is not required because *S. cerevisiae* is able to decompose and utilize the sugars that already exist in apples, which is evident from the relatively stable fermentative sugar content found in the fermentation medium during the fermentation process [9]. These results suggest that the nutrients contained in the apples were sufficient to support the maximum activity levels of the microbes.

In general, a higher sugar concentration in the medium results in the formation of a greater acetic acid content. However, excess sugar in the fermentation medium will not increase the microbial activity above its maximum threshold, and high sugar concentrations can limit the production of yeast biomass [23]. In addition, high levels of sugar can create anaerobic or microaerobic environmental conditions, which can inhibit the growth and activity of aerobic obligate bacteria, such as *A. aceti*, which is not optimal for acetic acid production. Thus, the availability of complex forms of sugar within the natural medium presents the advantage of providing a gradual carbon source to meet the needs of microbes.

3.2 Inoculum ratio of *S. cerevisiae* and *A. aceti*

In addition to the availability of sugar in the medium, the other factor that must be considered when performing acetic acid fermentations using mixed cultures is the optimal inoculum ratio of all cultures involved; in this case, *S. cerevisiae* and *A. aceti* were used. Because the two groups of microbes have different physiological properties, especially in terms of oxygen requirements, they also have different needs for carbon, different metabolic properties, and different growth rates. As mentioned above, *S. cerevisiae* is a facultative anaerobe that is able to grow on a wide range of carbon compounds and is able to produce alcohol under anaerobic condi-tions, whereas *A. aceti* is an obligate aerobe that is able to use ethanol, glycerol, and glucose as carbon sources for growth but is unable to hydrolyze lactose and starch and can oxidize ethanol to acetic acid and acetate to CO_2 and H_2O [4, 24]. Moreover, *S. cerevisiae* has a longer growth rate than *A. aceti* [9]. The metabolism and physiol-ogy of these two microbes have been described previously, in detail [4, 22, 24–26]. With these differences, the regulation of species dominance in mixed cultures by adjusting the inoculum ratios is expected to result in a syntrophic state that maxi-mizes the production of acetic acid.

An example of an experimental design to determine the best ratio of the cultures used during acetic acid fermentation is shown in **Figure 2**. The ratios of *S. cerevisiae* and *A. aceti* cultures used were 3:7, 1:1, and 7:3. The performances of these microbes when used at different ratios during acetic acid production can be observed by measuring the changes in acetic acid contents and pH values during the process (**Figure 3**). The results showed that the highest acetic acid concentration with the lowest pH value was achieved on day 8 using mixed cultures of *S. cerevisiae* and *A. aceti* at a 7:3 ratio.

According to **Figure 3**, the acetic acid levels produced by the ratio of the 3:7 of *S. cerevisiae* to *A. aceti* are higher at the beginning of the process than those produced by the other ratios. In this period, the dominance of *A. aceti* over *S. cerevisiae* results in *A. aceti* rapidly utilizing glucose to convert the ethanol produced by *S. cerevisiae* into acetic acid. According to Maier [26], the initial inoculum size controls the length of the lag phase. However, during the next stage, the resulting acetic acid contents decreased. The larger ratio of *A. aceti* causes this microbe to require more nutrients, which the smaller ratio of *S. cerevisiae* cannot provide. The limited nutri-ents available to *A. aceti* result in suboptimal cell growth and enzymatic activity, causing the metabolic processes of *A. aceti* to not work properly and the resulting

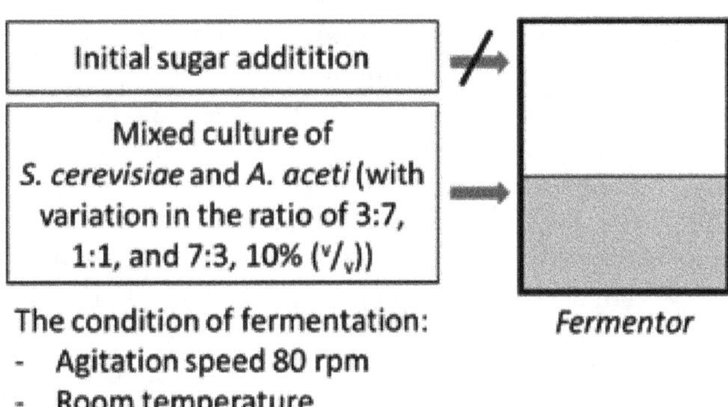

Figure 2.
Schematic diagram of the submerged batch fermentation process to determine the optimum inoculum ratio for the cultures used.

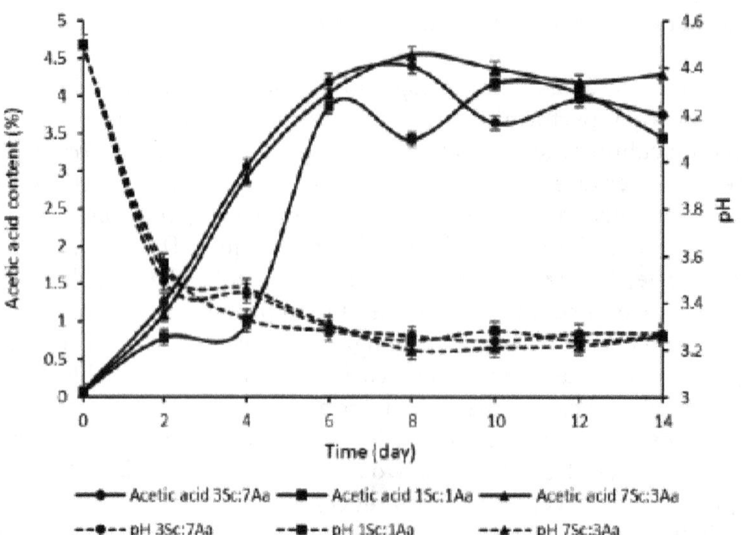

Figure 3.
Changes in the acetic acid contents and pH values with variations in the inoculum ratios between S. cerevisiae and A. aceti during the fermentation process.

acetic acid levels to decrease. Moreover, the acetic acid that is already produced undergoes overoxidation by *A. aceti* via the tricarboxylic acid (TCA) cycle [27].

The highest level of acetic acid was achieved on day 8, using the 7:3 inoculum ratio of *S. cerevisiae* to *A. aceti*. At the beginning of the fermentation process, the acetic acid concentration for this ratio was lower than for the 3:7 inoculum ratio, due to the dominance of *S. cerevisiae*. However, under aerobic conditions, *S. cerevisiae* is still able to produce alcohol in small amounts, and the large population of *S. cerevisiae* cells can produce enough alcohol to meet the nutrient requirements of *A. aceti*. During the later stages, the low levels of oxygen consumption by *S. cerevisiae* during alcohol production cause the availability of oxygen in the medium to become sufficient for *A. aceti* growth, and the resulting acetic acid contents increase.

3.3 Agitation speed for optimal mixing

As explained above, under aerobic conditions, the fermentation process using mixed cultures can work well, as indicated by the greater than 100% conversion efficiency from sugar to acetic acid. These results were achieved using an agitation speed of 80 rpm. Agitation plays an important role in fermentation processes, causing surface renewal; aiding in the dissolution of oxygen found at the top of the fermentor; improving the transfer of oxygen, heat, and mass through the system; and maintaining homogeneous physical and chemical conditions within the medium [28, 29]. Thus, the effect of agitation speed on the production of acetic acid in this system was evaluated by examining agitation speeds of 80 and 160 rpm (**Figure 4**). The percentage of acetic acid produced from both treatments can be seen in **Figure 5**.

The results showed that faster agitation speeds consistently resulted in higher acetic acid contents. The highest acetic acid level, 6.47%, was achieved on day 10 using an agitation speed of 160 rpm. Agitation is an important parameter for all aerobic processes [29]. The purpose of agitation during a submerged fermentation process is to homogeneously increase the availability and solubility of oxygen in the medium. Increased dissolved oxygen concentrations, generated by increased

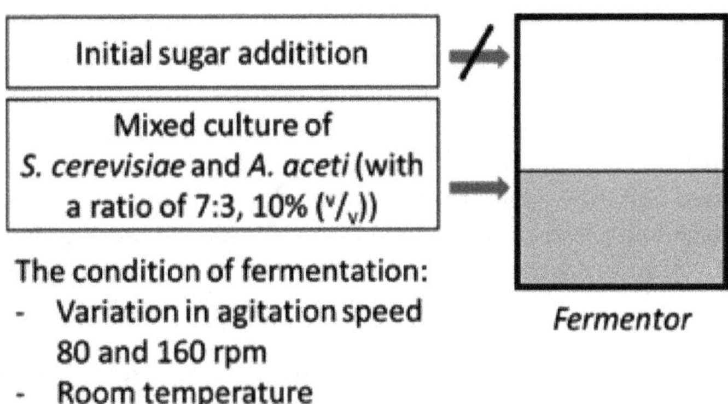

Figure 4.
Schematic diagram of the submerged batch fermentation process to determine the optimum agitation speed.

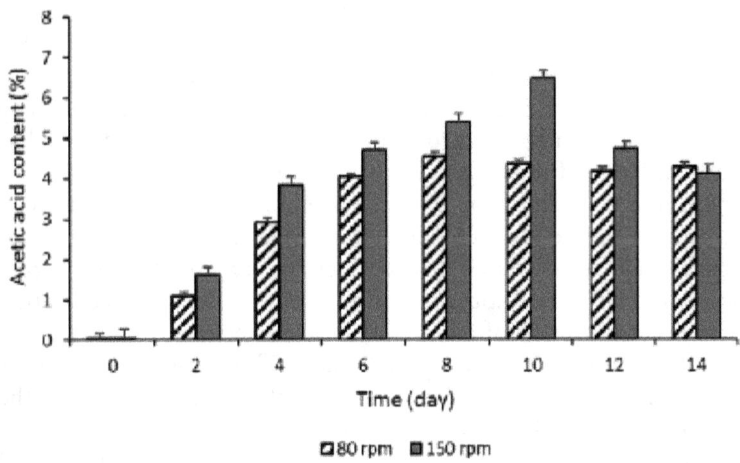

Figure 5.
The percentage of acetic acid produced from fermentation at different agitation speeds.

agitation speeds, resulted in a shortened lag time for cell growth and increased biomass formation [28]. Oxygen is needed not only by *A. aceti* but also by *S. cerevisiae* for growth [30, 31]. Massive oxygen consumption by both microbes simultaneously can create anaerobic conditions, causing *S. cerevisiae* to shift its metabolism from respiratory to fermentative and to produce alcohol. According to Navarro and Durand [32], during fermentation, yeast growth is rapidly stopped when the concentration of alcohol in the medium increases; however, fermentative activity is not entirely inhibited until high alcohol concentrations are reached. However, alcohol consumption by *A. aceti* prevents the concentration of alcohol in the medium from reaching the maximum value, preventing the inhibition of *S. cerevisiae* growth and activity, as indicated by the increase of glucose and alcohol contents in the medium. However, oxygen remains available in the medium, due to rapid agitation, allowing the growth and the activity of *A. aceti* to remain at high levels. *A. aceti* can directly use dissolved oxygen to grow and to produce acetic acid, and, simultaneously, the environment becomes anaerobic or microaerobic, allowing *S. cerevisiae* to produce alcohol, which is then used by *A. aceti* as a substrate for the production of acetic acid.

According to Zhou et al. [29], agitation can cause shear forces that can influence changes in cell morphology, variations in the growth and formation of products, and damages to the cell structure. However, increasing the speed of agitation results in stronger mixing processes, more rapid contacts between nutrients and microbes, and higher oxygen transfer rates (OTR) and oxygen uptake rates (OUR); therefore, aerobic and anaerobic environmental conditions are created simultaneously over a short period of time. Therefore, increasing agitation speed, up to a certain level, can lead to the production of larger amounts of acetic acid over shorter periods of time. In addition, the high dissolved oxygen content caused by the increased agitation speed in this system does not appear to cause oxidative stress or damage to proteins in cells, which could inhibit *A. aceti* growth [33]. As a whole, under conditions using an optimal inoculum ratio and an optimal agitation speed, the conversion efficiency from sugar to acetic acid increased to 362%.

3.4 The dynamics of changes in the sugar, alcohol, and acetic acid contents and in the pH value during fermentation under optimal conditions

The dynamics of changes in the sugar, alcohol, and acetic acid contents and in the pH values during the fermentation of acetic acid from apple juice under optimal con-ditions can be observed in **Figure 6**. In the beginning, when the sugar level is high, *S. cerevisiae* works to produce alcohol, increasing the alcohol contents. In conjunction with the production of alcohol, *A. aceti* began to produce acetic acid, causing the acetic acid level to increase. As *S. cerevisiae* produces alcohol, *A. aceti* simultaneously grows until the alcohol contents produced by *S. cerevisiae* are sufficient for *A. aceti* to produce acetic acid. In the mixed culture fermentation, *A. aceti* which is an obligate aerobic microbe uses dissolved oxygen for growth and for the oxidation of alcohol into acetic acid. However, the medium also undergoes an anaerobic state due to a lack of oxygen, allowing *S. cerevisiae* to convert sugar into alcohol. Another advantage of the use of mixed cultures is that the continuous consumption of oxygen by *A. aceti* appears to cause *S. cerevisiae* to grow without the multiplication of cell mass. Thus, the sugar present in the substrate can maximally be converted into alcohol by *S. cerevisiae*, and the alcohol can subsequently maximally be converted into acetic acid by *A. aceti*. However, at the end of the process, a decrease in the resulting acetic acid levels was observed. This decrease may be due to the unfavorable pH of the medium, which could inhibit the microbes from metabolizing substrates and producing acetic acid, or may be due to acetic acid overoxidation due to the limited

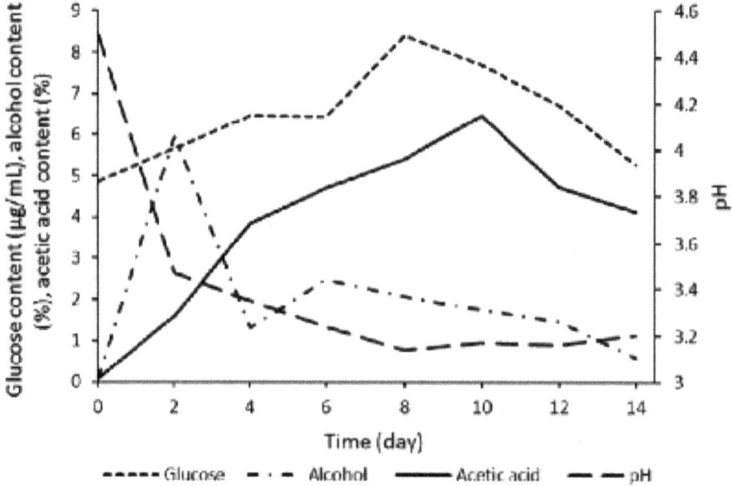

Figure 6.
The dynamics of the changes in glucose, alcohol, and acetic acid contents and pH values under optimal fermentation conditions.

availability of nutrients. For the purposes of harvesting the products, the fermentation process can be stopped when the highest yield is achieved.

3.5 Future outlook

Under the right conditions, the production of acetic acid can be maximized by using a simple system, such as submerged batch fermentation using a mixed culture that acts synchronously. Optimization can be performed by considering the character and needs of all microbes involved, which are the nutritional adequacy of the medium, the microbial proportions in the inoculum, and the agitation speed. The use of a mixed culture could shorten the fermentation time, reduce fermentation losses, and increase the acetic acid yields [16]. Some other advantages of this system compared with a gradual system are the relatively simple operation and easy handling of this system, which no particular control is required during the fermentation process, and the low risk of contamination. Thus, the application of this system for industrial purposes can be considered. However, the future scaling up of this process should consider other factors, including automation systems and the use of cutting-edge technologies in both the production and monitoring processes, to further improve the productivity and product quality without increasing produc-tion costs.

4. Conclusion

The efficiency of the acetic acid fermentation process can be assessed using a simplified system with mixed cultures. Some of the aspects evaluated in this system were the availability of sugar in the medium, the inoculum ratio of the cultures used, and the speed of agitation. By optimizing this system, the resulting acetic acid levels can be increased.

Author details

Keukeu Kaniawati Rosada
Department of Biology, Padjadjaran University, Sumedang, Indonesia

*Address all correspondence to: keukeu@unpad.ac.id

References

[1] Paulová L, Patáková P, Brányik T. Advanced fermentation processes. In: Teixeira JA, Vicente AA, editors. Engineering Aspects of Food Biotechnology. 1st ed. Boca Raton: CRC Press; 2013. pp. 89-105. DOI: 10.1201/b15426

[2] Karsch T, Stahl U, Esser K. Ethanol production by *Zymomonas* and *Saccharomyces*, advantages and disadvantages. European Journal of Applied Microbiology and Biotechnology. 1983;**18**:387-391. DOI: 10.1007/BF00504750

[3] Gullo M, Giudici P. Acetic acid bacteria in traditional balsamic vinegar: Phenotypic traits relevant for starter cultures selection. International Journal of Food Microbiology. 2008;**125**(1):46-53. DOI: 10.1016/j.ijfoodmicro.2007.11.076

[4] Raspor P, Goranovič D. Biotechnological applications of acetic acid bacteria. Critical Reviews in Biotechnology. 2008;**28**:101-124. DOI: 10.1080/07388550802046749

[5] Giudici P, Lemmetti F, Mazza S. Balsamic Vinegars: Tradition, Technology, Trade. Cham: Springer International Publishing; 2015. DOI: 10.1007/978-3-319-13758-2. 171p

[6] Li S, Li P, Feng F, Luo LX. Microbial diversity and their roles in the vinegar fermentation process. Applied Microbiology and Biotechnology. 2015;**99**:4997-5024. DOI: 10.1007/s00253-015-6659-1

[7] Lee S, Lee JA, Park GG, Jang JK, Park YS. Semi-continuous fermentation of onion vinegar and its functional properties. Molecules. 2017;**22**:1313. DOI: 10.3390/molecules22081313

[8] Ghosh S, Chakraborty R, Chatterjee G, Raychaudhuri U. Study on fermentation conditions of palm juice vinegar by response surface methodology and development of a kinetic model. Brazilian Journal of Chemical Engineering. 2012;**29**:461-472. DOI: 10.1590/S0104-66322012000300003

[9] Rosada KK. Enhanced acetic acid production from manalagi apple (*Malus sylvestris* Mill) by mixed cultures of *Saccharomyces cerevisiae* and *Acetobacter aceti* in submerged fermentation. Journal of Physics: Conference Series. 2018;**1013**:012171. DOI: 10.1088/1742-6596/1013/1/012171

[10] Singh R, Singh S. Design and development of batch type acetifier for wine-vinegar production. Indian Journal of Microbiology. 2007;**47**:153-159. DOI: 10.1007/s12088-007-0029-3

[11] Díaz C, Molina AM, Nähring J, Fischer R. Characterization and dynamic behavior of wild yeast during spontaneous wine fermentation in steel tanks and amphorae. BioMed Research International. 2013;**2013**:540465. DOI: 10.1155/2013/540465

[12] Viktor MJ, Rose SH, Van Zyl WH, Viljoen-Bloom M. Raw starch conversion by *Saccharomyces cerevisiae* expressing *Aspergillus tubingensis* amylases. Biotechnology for Biofuels. 2013;**6**:167. DOI: 10.1186/1754-6834-6-167

[13] Štornik A, Skok B, Trček J. Comparison of cultivable acetic acid bacterial microbiota in organic and conventional apple cider vinegar. Food Technology and Biotechnology. 2016;**54**:113-119. DOI: 10.17113/ftb.54.01.16.4082

[14] Wu X, Yao H, Cao L, Zheng Z, Chen X, Zhang M, et al. Improving acetic acid production by over-expressing PQQ-ADH in *Acetobacter pasteurianus*.

Frontiers in Microbiology. 2017;**8**:1713. DOI: 10.3389/fmicb.2017.01713

[15] Wang J, Hao C, Huang H, Tang W, Zhang J, Wang C. Acetic acid production by the newly isolated *Pseudomonas* sp. Csj-3. Brazilian Journal of Chemical Engineering. 2018;**35**:1-9. DOI: 10.1590/0104-6632.20180351s20160500

[16] Wang Z, Yan M, Chen X, Li D, Qin L, Li Z, et al. Mixed culture of *Saccharomyces cerevisiae* and *Acetobacter pasteurianus* for acetic acid production. Biochemical Engineering Journal. 2013;**79**:41-45. DOI: 10.1016/j.bej.2013.06.019

[17] Giudici P, Altieri C, Cavalli R. Aceto balsamico tradizionale, preparazione del fermentato di base. Industrie delle Bevande. 1992;**21**:478-483

[18] Saeki A. Continuous vinegar production using twin bioreactor made from ethanol fermentor and acetic acid fermentor. Nippon Shokuhin Kogyo Gakkaishi. 1991;**38**:891-896. DOI: 10.3136/nskkk1962.38.891

[19] Woiciechowski AL, Nitsche S, Pandey A, Soccol SR. Acid and enzymatic hydrolysis to recover reducing sugars from *Cassava bagasse*: An economic study. Brazilian Archives of Biology and Technology. 2002;**45**:393-400. DOI: 10.1590/S1516-89132002000300018

[20] Spets JP, Kuosa M, Granström T, Kiros Y, Rantanen J, Lampinen MJ, et al. Production of glucose by starch and cellulose acid hydrolysis and its use as a fuel in low-temperature direct-mode fuel cells. Materials Science Forum. 2010;**638-642**:1164-1169. DOI: 10.4028/www.scientific.net/MSF.638-642.1164

[21] Walker G, Stewart G. *Saccharomyces cerevisiae* in the production of fermented beverages. Beverages. 2016;**2**:30. DOI: 10.3390/beverages2040030

[22] Kruckeberg AL, Dickinson RJ. Carbon metabolism. In: Dickinson JR, Schweizer M, editors. The Metabolism and Molecular Physiology of *Saccharomyces cerevisiae*. 2nd ed. London: CRC Press; 2004. pp. 42-103

[23] Bokulich NA, Bamforth CW. The microbiology of malting and brewing. Microbiology and Molecular Biology Reviews. 2013;**77**:157-172. DOI: 10.1128/MMBR.00060-12

[24] Sievers M, Swings J. *Acetobacter*. In: Bergey's Manual of Systematics of Archaea and Bacteria. Hoboken, NJ: John Wiley & Sons, Inc., in association with Bergey's Manual Trust; 2015. pp. 1-7. DOI: 10.1002/9781118960608.gbm00876

[25] Saichana N, Matsushita K, Adachi O, Frébort I, Frebortova J. Acetic acid bacteria: A group of bacteria with versatile biotechnological applications. Biotechnology Advances. 2015;**33**:1260-1271. DOI: 10.1016/j.biotechadv.2014.12.001

[26] Maier RM. Bacterial growth. In: Maier RM, Pepper IL, Gerba CP, editors. Environmental Microbiology. 2nd ed. San Diego: Academic Press; 2009. pp. 37-54. DOI: 10.1007/978-94-017-8908-0

[27] Matsushita K, Toyama H, Adachi O. Respiratory chains in acetic acid bacteria: Membrane bound periplasmic sugar and alcohol respirations. In: Zannoni D, editor. Respiration in Archaea and Bacteria, Advances in Photosynthesis and Respiration. Dordrecht: Springer; 2004. pp. 81-99. DOI: 10.1007/978-1-4020-3163-2

[28] Rodmui A, Kongkiattikajorn J, Dandusitapun Y. Optimization of agitation conditions for maximum ethanol production by coculture. Kasetsart Journal (Natural Science). 2008;**42**:285-293. DOI: 10.1590/S0101-98802008000100019

[29] Zhou Y, Han LR, He HW, Sang B, Yu DL, Feng JT, et al. Effects of agitation, aeration and temperature on production of a novel glycoprotein GP-1 by *Streptomyces kanasenisi* ZX01 and scale-up based on volumetric oxygen transfer coefficient. Molecules. 2018;**23**:125. DOI: 10.3390/molecules23010125

[30] Aceituno FF, Orellana M, Torres J, Mendoza S, Slater AW, Melo F, et al. Oxygen response of the wine yeast *Saccharomyces cerevisiae* EC1118 grown under carbon-sufficient, nitrogen-limited enological conditions. Applied and Environmental Microbiology. 2012;**78**:8340-8352. DOI: 10.1128/aem.02305-12

[31] Salari R, Salari R. Investigation of the best *Saccharomyces cerevisiae* growth condition. Electronic Physician. 2017;**9**:3592-3597. DOI: 10.19082/3592

[32] Navarro JM, Durand G. Alcohol fermentation: Effect of temperature on ethanol accumulation within yeast cells. Annals of Microbiology. 1978;**129B**:215-224

[33] Cabiscol E, Tamarit J, Ros J. Oxidative stress in bacteria and protein damage by reactive oxygen species. International Microbiology. 2000;**3**:3-8. DOI: 10.2436/im.v3i1.9235

4

Current Status of Alkaline Fermented Foods and Seasoning Agents of Africa

Jerry O. Ugwuanyi and Augustina N. Okpara

Abstract

Fermented foods and seasoning agents play central roles in the food and nutrition security of nations across the world, but particularly so in Africa, Asia, South America and Oceania. As several people across the world gravitate back to "eating natural," there is a new emphasis on these fermented foods and seasoning agents which are also critical cultural foods in countries and societies where they are important. The result is the growth in demand for these products beyond what the traditional kitchen technologies is able to cope with. In Africa, many of the seasoning agents are products of alkaline fermentation of legume seeds, pulses and in some cases animal proteins and sea foods. There is an upswing in the popularity of these seasoning agents and around them, new cottage industries are growing, as against the kitchen technology that sustained them through the ages. This chapter will explore the state of biotechnological developments around these foods and seasoning agents and point the way to good manufacturing practice and industrial development and the need to grow this value chain that has helped to sustain societ-ies through ages.

Keywords: alkaline fermentation, African seasoning agents, fermented foods, *okpeye*, *dawadawa*, *ugba*, *ogiri*, *soumbala*

1. Introduction

Fermented foods are products of edible or inedible raw materials that have undergone desirable physic-chemical and biochemical modifications through the activities of microorganisms and/ or their metabolites, but in which the weight of the microorganism (relative to substrate) in the food is small [1]. A distinct group of fermented foods is the traditional alkaline fermented products often used as food condiments/ seasoning agents [2]. Fermented foods and seasoning agents play central roles in the food and nutrition security of many nations, but particularly so in Africa, Asia, South America and Oceania [3]. As several people across the world gravitate back to "eating natural", there is a new emphasis on fermented foods and seasoning agents which are also critical cultural foods in countries and societies where they are important. In Africa, many of the seasoning agents are products of fermentation of legume seeds, a process that causes an increase, to alkaline regions, in pH of the product. This results from microbial degradation of seed proteins to peptides and amino acids and finally to ammonia [3, 4]. Fermentation of raw materials such as fish, legumes and plant oil seeds for the production of food condiments with desirable organoleptic properties and enhanced nutritional values has historically been a popular practice in Africa, particularly in West and Central Africa. Currently, there is an upswing in the popularity of these seasoning agents, and around them new cottage industries are growing, as against the kitchen technology that sustained them through the ages [3]. This resurgence in alkaline fermented foods are results of a better understanding of fermentation processes, as well as increased knowledge of the nutritional, and health-promoting benefits of fermented foods [5]. This chapter will explore the state of biotechnological developments around these foods and seasoning agents and point the way to good manufacturing practice and industrial and market development.

1.1 The beginning of fermented foods

The art of food fermentation dates back to prehistoric times and are the oldest methods for producing new foods from existing substrates, and of prolonging the shelf life of foods [3, 6]. Historically, fermentation has been used to modify the composition of foods without any scientific knowledge of the processes or benefits, and this art has been practiced for thousands of years [7–9]. As at 2000–4000 BC, the Egyptians were producing alcoholic beverages [6]. According to records [10, 11], fermentation has been in practice also in Sudan, (1500 BC) and Mexico, (2000 BC). Despite advances in biotechnology and efforts towards industrialization of the traditional fermentations, uncontrolled traditional techniques/ kitchen technologies are still predominantly used for the processing of alkaline African fermented foods and seasonings.

Modern food technology practices such as the use of good manufacturing practice (GMP) protocols, as well as new innovations like the use of starter cultures in controlled fermentations continue to play little or no role in the developing countries. The disposition to understanding traditional food processing is now beginning to gain some ground in developing countries. It is essential to recognize the significance of biotechnology-based innovations and applications in food pro-cessing in order to ensure quality and safety of products [12]. More recently, process techniques used in traditional fermented foods are being redefined and diversified through the use of molecular biology-based tools, enabling fermentation technology around these processes to evolve towards sustainable commercialization and industrialization. This lift from artisanal production has stimulated new interests in food research, such that today a lot of scientific works [13–26] have been devoted to these fermented foods. With modern biotechnology new and better methods for processing foods under GMP are developing.

2. Fermentation processes

Across cultures, a variety of traditional techniques are used for producing fermented foods and seasoning agents. The techniques differ based on microorganisms, raw material and fermentation conditions [27, 28]. Basically, processes involved in food product development by fermentation are of four types (**Figure 1**), viz.: alcoholic, lactic acid, acetic acid and alkaline fermentation [6, 27, 29].

Alcoholic fermentation is mainly performed by yeasts leading to the production of ethanol. Products include wine, beer, other alcoholic beverages and bread. Lactic acid fermentation is driven by lactic acid bacteria (LAB), which produce organic acid and other compounds in various foods. Acetic acid fermentation is carried out by the acetic acid bacteria which convert alcohol to acetic acid under aerobic process as in vinegar. Alkaline fermentation usually takes place during the fermentation of

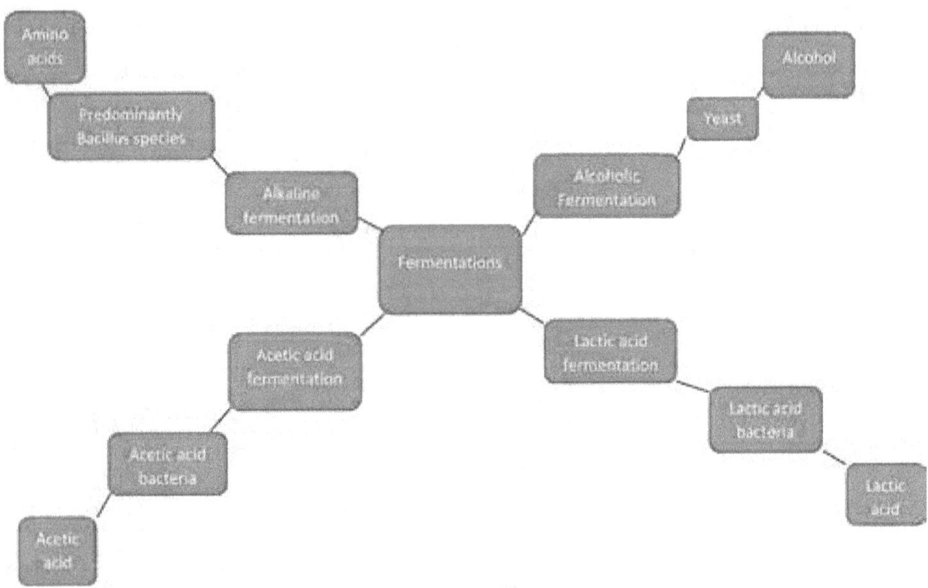

Figure 1.
An illustration of the various types of fermentations, based on microorganisms, fermentation condition and end product. Source: Anal [27].

Food group/class	Substrate	Derived product	Country
Starchy foods	Root tubers		
	Cassava	*Garri, akpu, loi-loi*	Nigeria
		lafun/aribo	Nigeria
			Nigeria
		Kokobele	Nigeria
	Cereal-based		
	Maize	*Ogi/akamu*	Nigeria
		Kito	Tanzania
		Mawe	Nigeria (Benin)
		Njera	Ethiopia
		Mahewu	South Africa, Kenya
		Uji	Uganda, Tanzania
	Millet	*Kenkey/banku*	Ghana
		Ogi/akamu	Nigeria
		Uji	Uganda, Tanzania
	Sorghum	*Busa*	Egypt
		Nasha	Sudan
		Kisra	Sudan
		Bogobe	Botswana
		Uji	Uganda, Tanzania

Food group/class	Substrate	Derived product	Country
Alcoholic and non-alcoholic beverages	Cereal	*Burukutu/pito/otika*	Nigeria
	Palm sap	*Ngwo/nkwuenu*	Nigeria
	Grape	*Pique*	Mexico
	Cane sugar	*Sake*	Japan
		Tape	Indonesia
Animal-based product	Milk	*Nono*	Nigeria
	Beef tripe	*Afo-nama*	Nigeria
	Milk	*Warankasi*	Nigeria
Fish/sea food product	Fish	*Azu-okpo*	Nigeria
	Crab	*Nshiko*	Nigeria
	Cray fish/shrimp	*Uponi/oporo*	Nigeria
	Fish	*Garum*	Europe
	Fish	*Suan yu*	China
	Fish (herring)	*Surstromming*	Sweden
Plant based alkaline product (seasoning agents)	Refer to (**Table 2**)	Refer to (**Table 2**)	Refer to (**Table 2**)

Sources: [4, 30–37].

Table 1.
The main classes of fermented foods based on the substrate from which they were derived.

fish, legumes and other plant seeds (raw materials with high protein content, and in which principal metabolic processes center around protein degradation) to produce seasoning agents including *dawadawa* from Locust bean and *ugba* from oil bean and several related products.

Fermented foods can be classified in different ways based on the type of substrate, microorganisms involved in the fermentation and even the processing methods. Based on the substrate or raw material from which they are manufactured [30], foods derived by fermentation can be classified into five main categories namely:

1. Starchy foods such as root tubers (cassava), examples: *garri, akpu, lafun*, cereals (maize, sorghum, millet). 2. Alcoholic and non-alcoholic beverages (palm wine) *ngwo, nkwu enu, kunu-zaki*, 3. Animal protein based products (milk) example *nono*, warankasi 4. Fish/ sea food based foods examples *azu-okpo*, garum 5. Plant-seed and legume based products (seasoning agents), including *daddawa, iru* and *netetu* among others. **Table 1** presents the main classes of fermented foods based on the substrate from which they were derived.

3. Diversity of alkaline fermented foods and seasoning agents

Alkaline fermentation mainly relate to the fermentation of legumes (soy bean), protein rich oil seeds (African oil bean) and fish to produce condiments. During these processes, there is always an increase in pH up to 8 and above. The increase in pH has been attributed to the metabolic activities of the microbes that break-down the protein of the raw material into peptides, amino acids and ammonia [2]. A diversity of alkaline fermented foods including seasoning agents are available world-wide, particularly in countries in Africa and Asia where these products are an integral part of the cultural diets of the native communities [2, 3, 12, 31]. They are prepared from a wide range of raw materials including soybean, African locust bean and various species of fish [32]. Most of these are highly priced seasoning agents prepared by solid state fermentations in which *Bacillus* species are the key organisms. **Table 2** shows some common alkaline seasoning agents that are key players in traditional food systems across the world, particularly in Africa and Asia. Some

Raw material	Product local name	Distribution/ country	State of development	References
Soy-bean (*Glycine max*)	*Daddawa*	West Africa, Nigeria	A, B	[15, 19]
	Soy-dawadawa	Ghana	A, B	[3, 16, 20]
	Kinema	India	A, B	[14, 41, 42]
	Hawaijar	India	A, B	[43–45]
	Aakhune	India	A, B	[27]
	Bekang	India	A, B	[27]
	Peruyaan	India	A, B	[27]
	Tungrymbai	India	A, B	[27]
	Thua-nao	Asia, Thailand	A, B, C	[27, 45–47]
	Natto	Asia, Japan	A–D	[27, 48]
	Douchi	China, Taiwan	A, B	[75, 76]
	Chungkokjang or jeonkukjang or cheonggukjang	Korea	A, B	[27, 50, 51]
	Meju	Korea	A, B	[52]
	Miso	Japan	A–C	[45, 53, 54]
	Shoyu	Japan, Korea, China	A, B	[54]
	Tauco	Indonesia	A, B	[55]
	Tempe	Indonesia (origin), The Netherlands, Japan, USA	A–D	[45, 56, 57]
	Yandou	China	A, B	[45, 58, 59]
African locust bean (*Parkia biglobosa*)	*Soumbala*	Burkina-Faso	A, B, C	[3, 60]
	Afitin/sonru/	Mali, Côte d'Ivoire and Guinea, Nigeria (Benin)	A, B	[24, 27, 38, 61]
	Netetu	Senegal	A, B	[62]
	Kinda	Sierra Leone	A, B	[63]
	Dawadawa/iru	West Africa (Nigeria)	A, B	[3, 24, 60, 61]
Mesquite (*Prosopis africana*)	*Okpehe/okpeye/okpiye*	West Africa/middle belt and southern Nigeria	A–C	[3, 22]
	Kpaye/afiyo	Northern Nigeria	A, B	[64, 65]
Castor oil/fluted pumpkin/melon	*Ogiri*	West Africa/ Eastern Nigeria	A, B	[4, 15, 24, 66]
African oil bean (*Pentaclethra macrophylla*)	*Ugba/Ukpaka*	West Africa/ Southern Nigeria	A–C	[4, 67]

Raw material	Product local name	Distribution/ country	State of development	References
Roselle (*Hibiscus sabdariffa*)	*Bikalga*	Burkina-Faso	A–C	[68, 79]
	Daton	Mali		[68]
	Furandu	Sudan		[2]
	Mbuja	Cameroon		[2]
Cathormion altissimum	*Oso*	West Africa/ Nigeria	A, B	[69, 70]
Saman tree (*Albizia saman*)	*Aisa*	Nigeria	A, B	[21]
African yam bean	*Owoh*	Nigeria	A, B	[21, 71, 72]
Cotton seed (*Gossypium hirsitium*)	*Owoh*	Nigeria	A, B	[2, 73]
Leaves of Cassia	*Kawal*	Sudan	A, B	[45, 74]

State of development is based on published information. Product may not be in the market. Key: A: Microorganisms involved are known, B: Roles of organisms known or inferred, C: Starter Cultures have been developed or suggested, D: Pilot or improved technologies or industrial plant (s) available.

Table 2.
Some of the plant-based alkaline fermented seasoning agents.

of these are described briefly to show state of the art and current application of modern technology (biotechnology) to the production processes.

Fermented condiments are cherished by consumers due to their peculiar organoleptic properties, nutritional and health significance as well as durability. The seasoning agents are characteristically used in small quantities to flavor traditional dishes, but their unique aroma eventually become central to the properties of those foods [9]. The quality of fermented foods is influenced by the starting raw material, microbiota as well as the processing methods [6]. The starting raw material for producing a particular seasoning can vary. For instance, *ogiri* is traditionally produced from any of three substrates namely: castor oil seeds, fluted pumpkin seeds (*Telfairia vulgaris*) and melon seeds. Given the diversity of raw materials from which comparable products are obtained, it is clear that the basis for uniformity in flavor characteristics relate considerably to the biochemical and physiological features of the microorganisms that drive the process. In different parts of the world, and even within the same region or country, the same fermented product may be known by different local names. For instance, in Nigeria *dawadawa /iru* is the traditional name for a fermented seasoning from African locust bean [33] while *netetu* is used in Senegal to refer to a food condiment from the same substrate [34] and *soumbala* is the traditional name used in Burkina-Faso [35].

3.1 Traditional plant-based alkaline fermented seasoning agents

Among the various substrates used for preparation of traditional fermented seasonings, soy beans is the most popular because of its wide spread distribution across the globe and importance as a rich source of plant protein [36]. In East and Southeast Asia and in West Africa, a wide range of alkaline fermented seasonings are produced from soy bean [27]. These include West African *dawadawa,* Japanese *natto* and Thai *thua nao*. Soy bean can be fermented by either bacteria or fungi. For bacteria-based soy bean products, *Bacillus* species (predominantly *B. subtilis*) are the predominant microorganisms. Fungi-based soy bean products are produced using filamentous mold (Mostly *Aspergillus, Mucor, Rhizopus*) [27]. The other common substrates used to prepare alkaline fermented products in Africa include African locust bean (*dawadawa, iru, kinda, soumbala*), African mesquite seeds (*Okpeye/ okpehe/ kpaye/ afiyo*) and African oil bean (*ugba*).

Apart from these popular legumes and oil seeds, other less popular and less utilized legumes and vegetables are also used for production of alkaline condiments in Africa. These include *Albizia saman* seeds for production of aisa [22], cotton seeds (*Gossypium hirsutum*) for used production of *owoh* [2] and *Hibiscus sabdariffa* for production of bikalga [37].

3.1.1 Dawadawa

Dawadawa is probably the most popular and commercially successful traditional seasoning agent in West and Central African Savannah where it is known by different ethnic names [38, 39]. *Dawadawa* is processed from the solid substrate fermentation of cotyledons of locust bean (*Parkia biglobosa*). It is widely consumed as a food seasoning in the Northern and some part of Southern Nigeria [3, 15]. The traditional process may vary slightly depending on the processor and locality. In perhaps the most popular process (**Figure 2**) the basic steps include boiling of

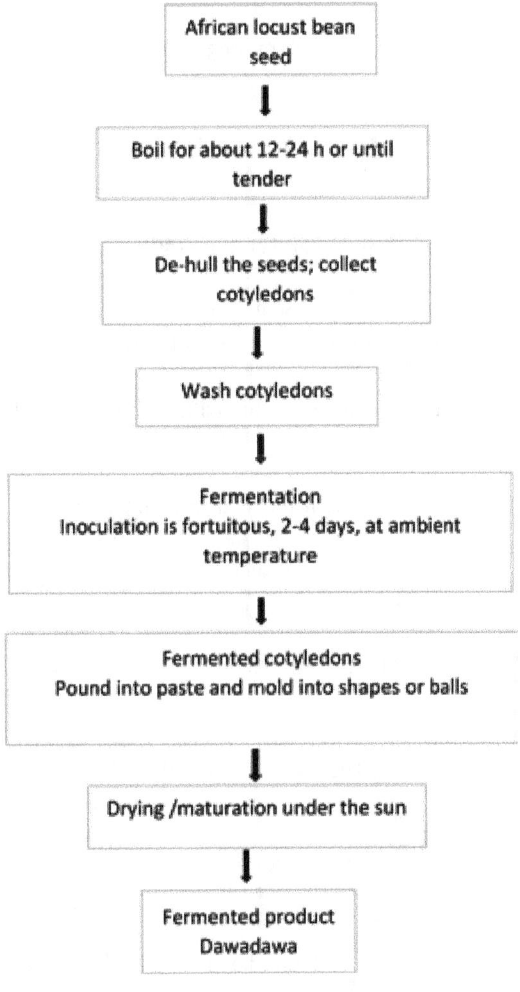

Figure 2.
Flow chart for the traditional method of producing dawadawa.

the locust bean seeds for 12–24 h, followed by manual de-hulling to remove the seed coat. The de-hulled cotyledons are collected and washed thoroughly and then boiled again for 1 h. The cotyledons are placed in jute bags or wrapped with banana leaves and allowed to ferment for 2–4 days at ambient temperature. During fermen-tation, the pH increases from near neutral to 8.1 or higher due to the breakdown of protein to amino acids and ammonia. As with other traditional processes, inocula-tion is usually fortuitous from production environment and equipment used or (rarely by back slopping). Microorganisms that drive the process are predominantly species of *Bacillus* [14, 40]. Other associated organisms include *Staphylococcus* [41], but the roles of these minor populations are contentious. At the end the product is sticky, with pungent odor and covered with mucilaginous grayish layer.

Dawadawa is the most scientifically studied traditional seasoning in West Africa. Several studies designed to improve the traditional process of have been published [41, 42]. **Figure 3** shows a modified procedure for *dawadawa* production [43]. Although, many African alkaline seasoning agents are still produced by the old-aged traditional cottage and kitchen processes, the preparation of some condiments like *dawadawa* has achieved pilot commercial status and is now considerably carried out on large scale by

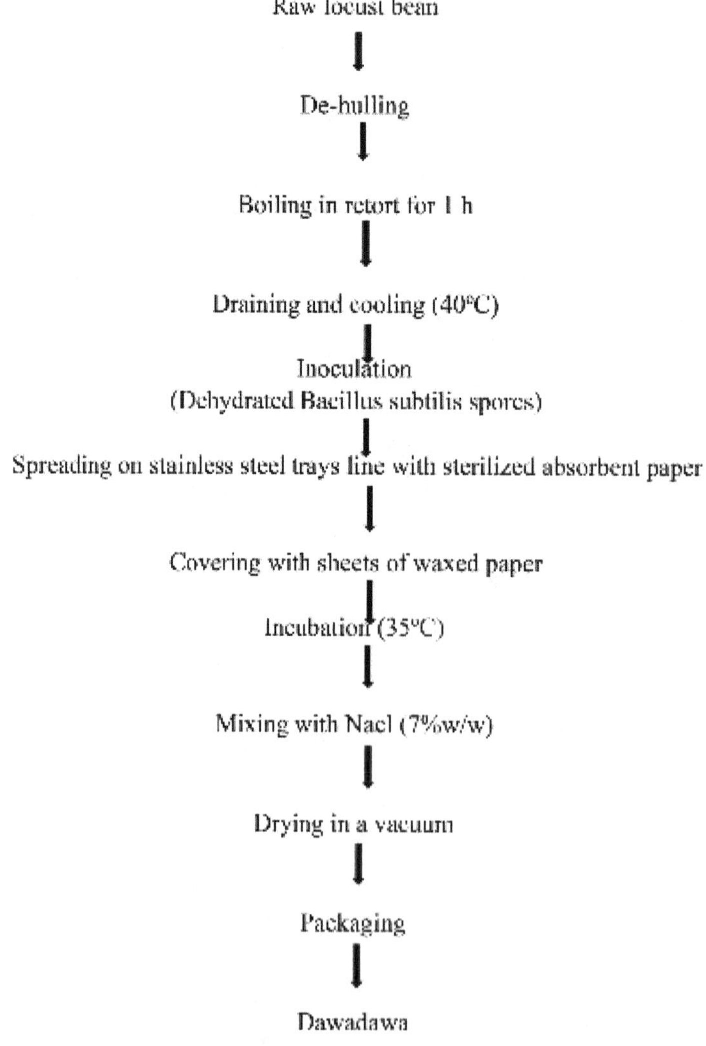

Figure 3.
Flow chart for the modern production of dawadawa. Source: [15, 31, 38].

entrepreneurs [29]. In Nigeria and some other parts of West Africa, *dawadawa* cubes are available in the markets [43] with improved quality, durability and packaging. These are produced using reasonably reproducible protocol and are marketed under various brand names. However, the most significant achievement towards improving the traditional fermentation technology is that the mechanism of flavor production during the fermentation, as well as flavor components generated in *dawadawa* have been studied by international food manufacturers and used as basis for the development of flavors from legumes hydrolysis for bouillon-type products [44].

3.1.2 *Soumbala*

Soumbala is a traditional condiment popular in Burkina Faso and other countries in West and Central Africa. It is also known by different names by different local communities [45]. Similar to *dawadawa*, it is prepared by solid state alkaline fer-mentation of African locust beans. It contributes significantly to protein nutrition of the consumers. The traditional process for preparing *soumbala* is uncontrolled and similar to *dawadawa* with minor variations based on ethnic preferences. The principal microorganisms involved in the fermentation of *soumbala* are *Bacillus* spp. [34]. The ability of *Bacillus* species involved in *soumbala* fermentation to inhibit undesirable bacteria including *Bacillus cereus* and *E. coli* has been reported [46–48].

3.1.3 *Okpeye*

Okpeye, much like *dawadawa* is a traditional seasoning produced by solid substrate alkaline fermentation of *Prosopis africana* (African mesquite) seeds. *P. africana* grows across the African Savannah and rain forest regions, but is mostly used as source of seasoning in the middle belt and parts of the Southeastern Nigeria [3]. Like *dawadawa*, the household technology used for producing *okpeye* can also vary between cultures. Perhaps, the most common procedure for preparation of *okpeye*, as practiced in parts of Southeastern Nigeria is as described (**Figure 4**) [9]. The process involves boiling of the mesquite seeds for 12–24 h to cook the seeds, soften the seed coat and ease the de-hulling process. This is followed by de-hulling in a very laborious manual process. The cotyledons are washed thoroughly, drained and reheated (dry heat) in a pot lined with the leaves of *Alchornea cordifolia* popu-larly known as (*akwukwo okpeye*) by the native people. Other leaves such as banana leaves may be used when the conventional leaves are unavailable. The cotyledons are spread to a few cm depth in a shallow raffia basket already lined with leaves of *Alchornea cordifolia*, covered with more leaves and weighted with pebbles. This solid substrate fermentation arrangement is then placed outside under the sun in the day time and inside the house in the night (avoiding precipitation and moisture for the duration of the process). Fermentation proceeds for 4 days at uncontrolled tem-perature which varies from less than 30°C at night to over 37°C in the afternoons during very sunny days. At the end of this stage the fermented cotyledons now dark brown in color with strong ammonia-like smell are ground into a smooth paste and molded into different shapes and sizes. At this moment the product may be used but for more desirable quality, it is usually sun dried for a variable length of time at the end of which the product becomes hard and black with a more mellow and prefera-ble aroma. The dried condiment is resistant to spoilage by microorganisms and has a very long shelf life with occasional re-drying under the sun [3]. **Figure 5** shows the stages in the traditional process of *okpeye* production. During the natural process the pH increases from an initial of 6.0–6.2 in the boiled unfermented substrate to 8.0–8.8 (sometimes pH of 9.0–9.2 may be achieved) in the fermented product [49]. In our laboratory a diversity of microorganisms were established to be involved in

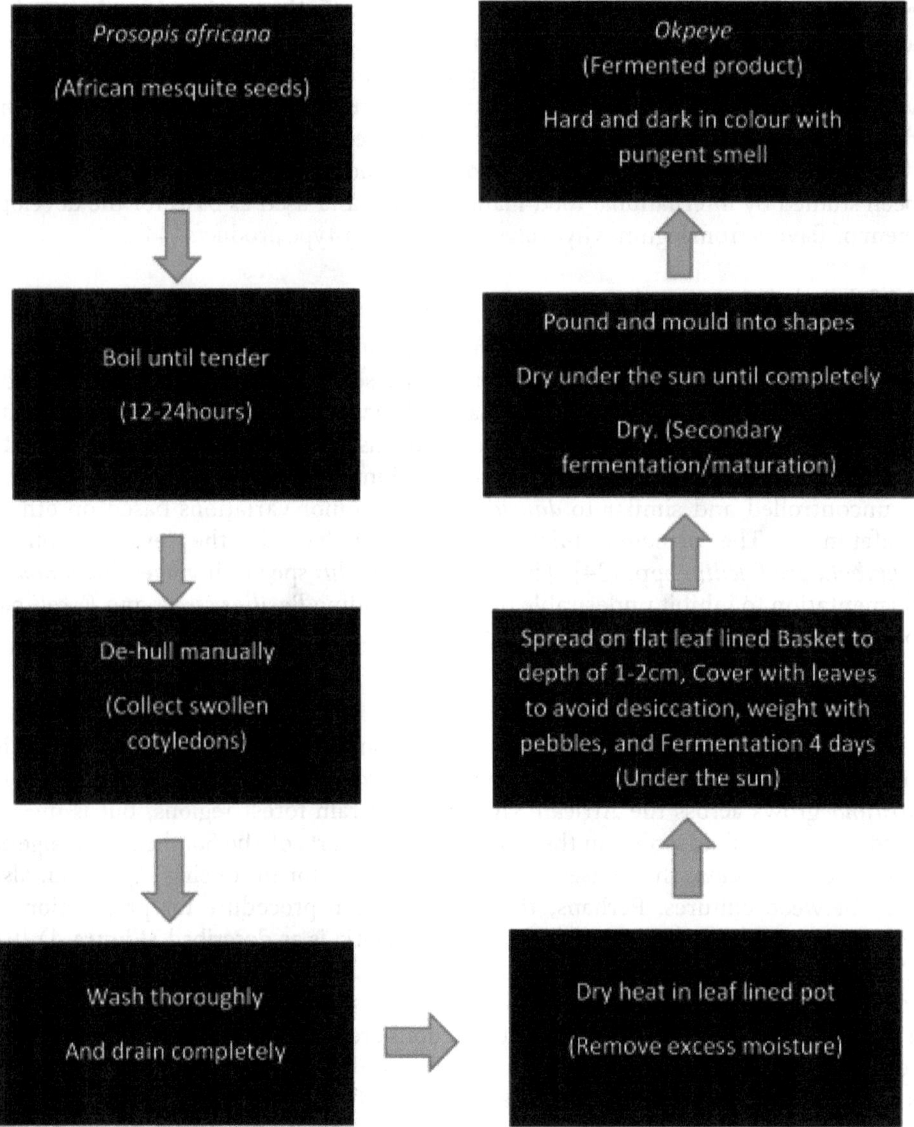

Figure 4.
Traditional process for the production of okpeye.

the primary fermentation. However, only species and strains of *Bacillus* were shown to be principal drivers of the process. Their populations increased significantly and persisted until the end. The organisms include *B. subtilis, B. velezensis*, and *B. amyloliquefaciens*. Other *Bacillus* species isolated include *B. licheniformis, B. anthra-cis, B. thuringiensis* and *B. cereus*. Apart from *Bacillus* species other bacteria that also participated in the fermentation especially at the early stages include *Enterobacter* sp., *Proteus mirabilis, Pseudomonas* sp., *Micrococcus* sp. and *Staphylococcus* sp. [49]. As in similar processes, these organisms are transient, incapable of producing condiments in pure culture and their roles in the process remain unclear.

3.1.4 Ugba

Ugba is a Nigerian-based condiment prepared by the solid state alkaline fermentation of seeds of the African oil bean (*Pentaclathra macrophylla*). It is also known

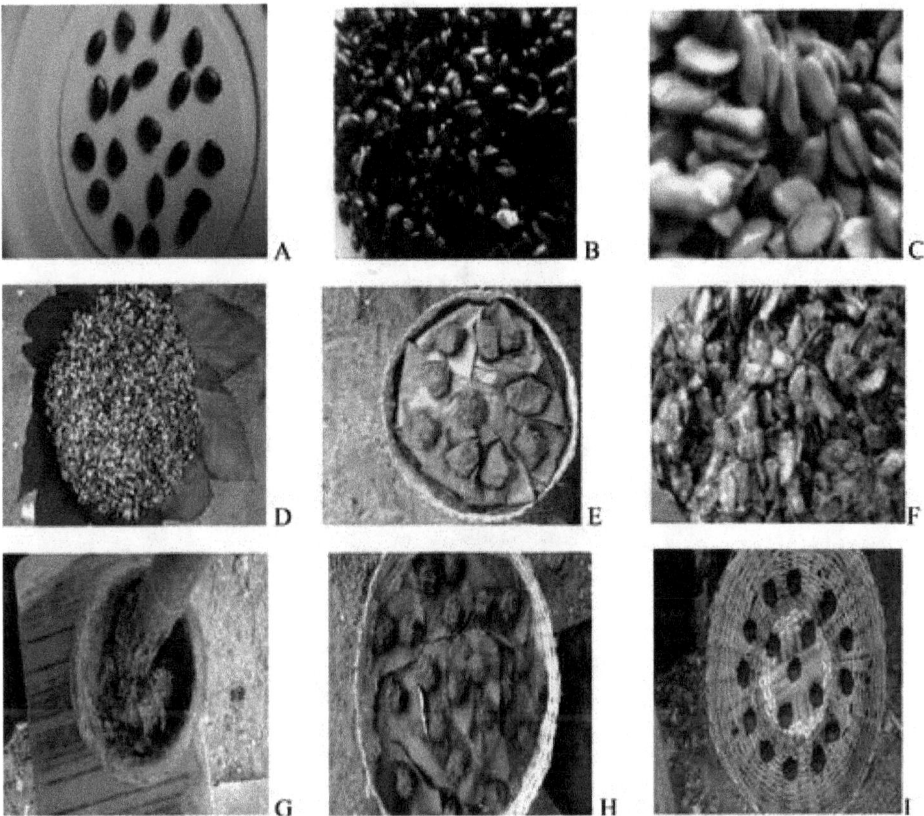

Figure 5.
Steps in the traditional fermentation of P. africana *seeds to produce* okpeye. *(A) Seeds before boiling; (B) boiled seeds; (C) de-hulled seeds before fermentation; (D) de-hulled cotyledons spread on leaf lined basket; (E) fermentation taking place outside under the sun; (F) fermented cotyledons; (G) ground paste; (H) molded seasoning undergoing drying under the sun; (I) dried* okpeye *seasoning. Source: [89].*

as *ukpaka* by the Igbos in the Southeastern part of Nigeria where it is most popular. *Ugba* is consumed as a delicacy, appetizer or used as a flavoring agent in various traditional dishes. Prepared in different ways, *ugba* is an important food product for various traditional ceremonies [9]. The production, like other traditional processes, is still carried out in various homes on small scale under uncontrolled condition resulting in products that are non-uniform in quality.

The basic procedures (**Figure 6**) involve boiling of oil bean seeds for 12 h or more, removing the seed coat and slicing the cotyledons into thin slices. The slices are then soaked in water overnight, washed thoroughly and wrapped with fresh leaves for fermentation to take place. Fermentation is usually done at ambient temperature and the duration varies depending on the intended use. Fermentation can last as short as 3 days or up to 5 days. **Figure 7** shows African oil bean seeds, fermented slices of oil bean cotyledons and fermented product (*ugba*) packaged in different ways.

Microbiological and biochemical changes that take place during the traditional process have been studied extensively [50, 51]. A diverse group of microorganisms were reported to participate in the traditional fermentation of African oil bean, with *Bacillus cereus* dominating the process [50]. Over 30 different organic compounds of varying molecular weights and volatility including alcohols, organic acids, ketone, aldehydes, hydrocarbons amines and esters have been shown to contribute to the flavor of the final product [51]. The specific contributions of these various molecules remain to be established as also their flavor threshold in the product.

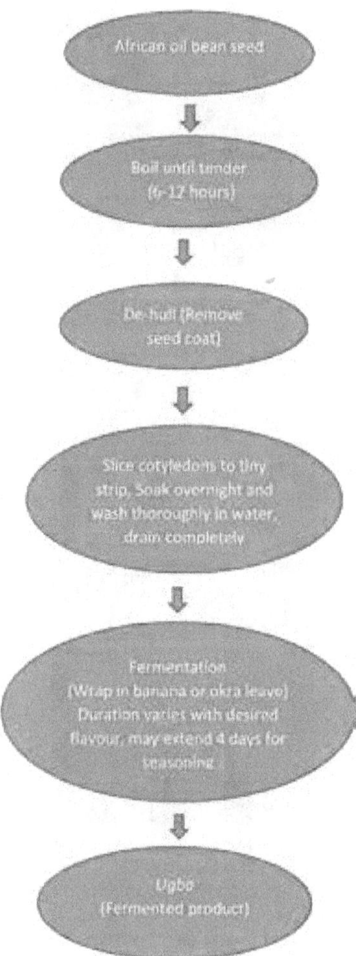

Figure 6.
Flow chart for the traditional production of ugba.

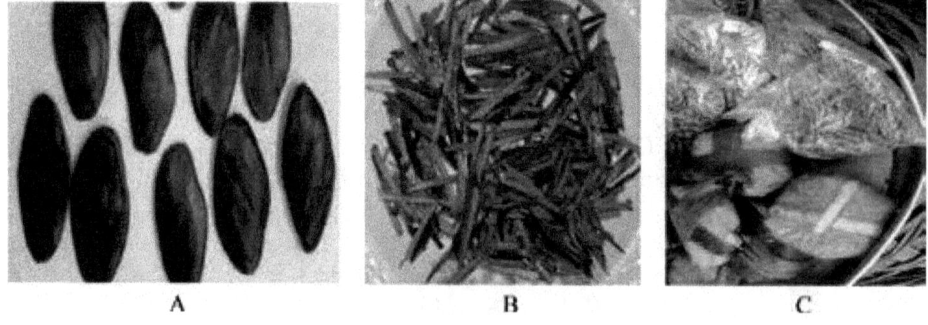

Figure 7.
African oil bean seeds (A), fermented slices of oil bean cotyledons (B) and fermented oil seeds cotyledons (ugba) packaged in polythene bags or wrapped with local leaves.

3.1.5 Ogiri

Ogiri is a popular African fermented seasoning, traditionally prepared by the solid state alkaline fermentation of castor oil seeds (*Ricinus communis*). Depending on locality, season and availability it may also obtained by fermenting melon seeds

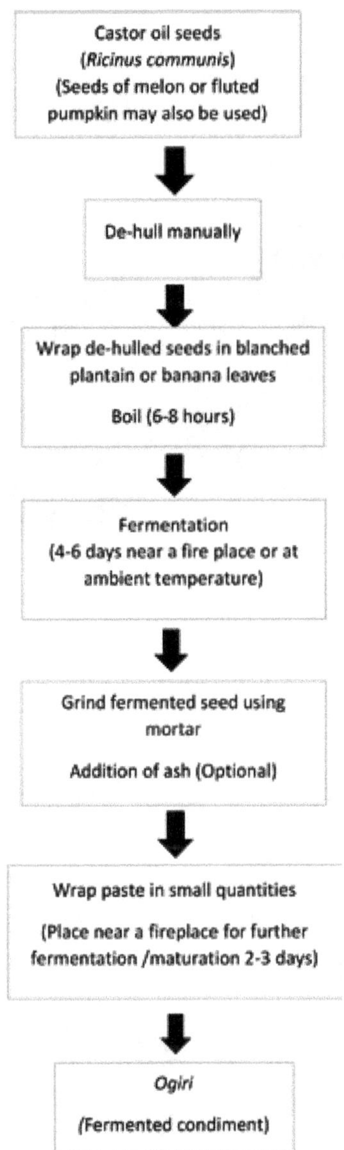

Figure 8.
Flow chart for the traditional production of Ogiri.

(*Citrullus vulgaris*) and fluted pumpkin seeds (*Telfairia occidentalis*). *Ogiri* is used in flavoring many traditional soups. In fact, it is regarded as an indispensable seasoning in the preparation of specialized soups which are highly cherished and extensively consumed by the Igbo ethnic group in the Southeastern Nigeria. Like many indigenous fermented products, production of *ogiri* is still by the traditional family-village art done on a small-scale cottage level. Details of the traditional pro-cess may vary between cultures. For production of ogiri (**Figure 8**) the shelled seeds of castor oil are wrapped in blanched banana leaves and boiled for about 8 hours until the seeds are properly cooked. The wrapped seeds are then placed near the fireplace to ferment for 4–6 days depending on the intensity of the fire. On comple-tion of this stage, the fermented seeds which are now sticky and strong smelling are ground on a grinding stone or mortar into a fine paste which is divided into small portions and packaged in blanched banana leaves (**Figure 9**). The packs are placed

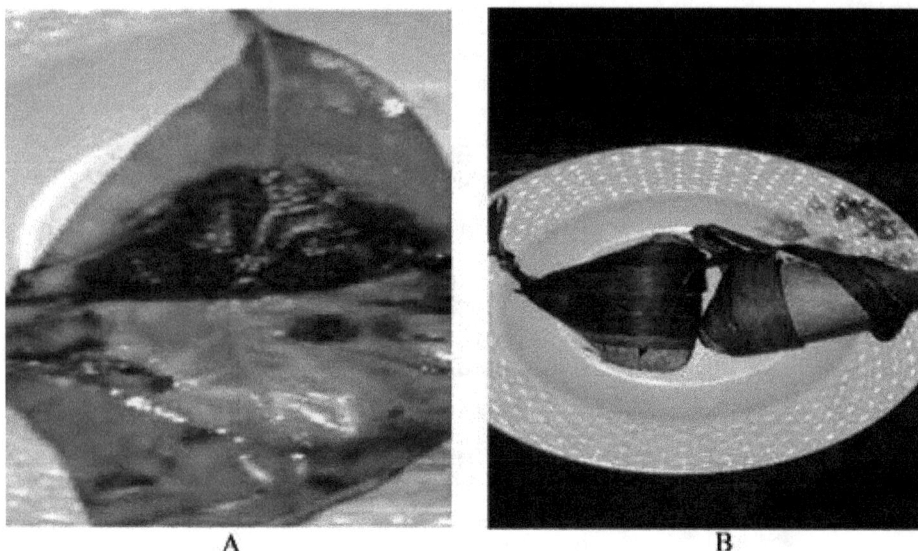

Figure 9.
Fermented castor oil seed (Ogiri) condiment (A) and fermented castor oil seeds (Ogiri) condiment wrapped with local leaves (B).

near the fireplace or in a warm place to ferment further for 1–2 days. At this stage the fermented condiment is ready to use or sale and it has a characteristic strong pungent flavor. *Bacillus* strains mainly *B. subtilis* and *B. licheniformis* have been reported as the predominant fermenting microorganisms. The pH of the fermented product is alkaline (>8.0) [4, 23, 24].

3.2 Less common legume-based alkaline fermented seasoning agents of Africa

3.2.1 *Aisa*

Aisa is a Nigerian seasoning agent processed from the solid state alkaline fermentation of *Albizia saman* (Jacq) F. Mull popularly known as monkey pod, rain tree or saman tree [2, 21]. *Albizia saman* is one of the uncommon and under-exploited legumes in the sub-Saharan regions. Like other traditional fermented seasonings, *aisa* is used to flavor various traditional dishes and soups. The production process is similar to dawadawa. The basic method involves boiling of the saman seeds until tender, followed by manual de-hulling. The cotyledons are washed and boiled again for 1–2 h, and washed in water. The cotyledons are wrapped in clean fresh leaves (banana or paw-paw) in bundles. The wrapped bundles are placed in calabashes and allowed to ferment for 1–7 days at ambient temperature. At the end of fermentation, the product is dark brown, sticky mash covered with mucilaginous coat and possessing a strong ammoniacal smell. *Bacillus* species are reported as the predominant micro flora responsible for *aisa* fermentation [21]. Other organisms that have also been reported to take part in the process include *Escherichia coli, Klebsiella, Enterobacter, Proteus* and *Staphylococcus* species.

3.2.2 *Owoh*

Owoh is another African fermented seasoning whose substrate is under-utilized. It is made by the solid state alkaline (pH 8.8 and above) fermentation of cotton seeds (*Gossypium hirsutum)* [2].*Owoh* is mainly used as a seasoning in the

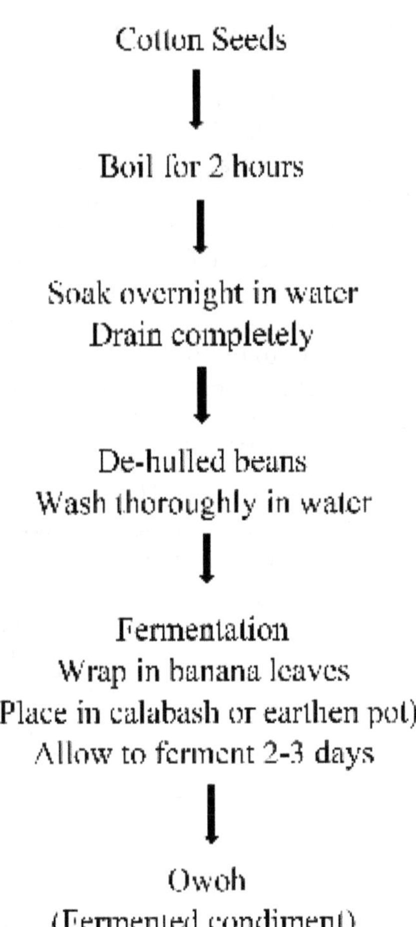

Figure 10.
Traditional process for the production of Owoh.

mid-Western Nigeria. The raw seeds are toxic and inedible. The traditional process (**Figure 10**) involves boiling of cotton seeds until they are properly cooked and become tender. The seed coats are removed manually. The cotyledons are then washed, wrapped in banana leaves and boiled again for 1–2 h. The wraps are removed from water and placed in calabashes or earthen pots, and then covered with jute sacks and placed in a warm location (often beside the fire place) to ferment. Fermentation is done at ambient temperature for 2–3 days. At the end of the fermentation, the mash is ground and molded into balls. The product may be used at this point, but preferably it is sun dried to extend the shelf life and also to develop more desirable aroma [52]. The major fermentative organisms are reported to be *Bacillus* species including *B. subtilis, B. licheniformis,* and *B. pumilus.*

3.2.3 Bikalga

Bikalga is an African alkaline fermented condiment made from *Hibiscus sabdariffa*. It is widely used to flavor various traditional dishes in Burkina Faso [37]. It is known by different ethnic names in different parts of African. In Niger it is known as *dawadawa bosto, datou* in Mali, *furundu* in Sudan and *mbuja* in Cameroon. The predominant organisms involved in the fermentation of *Hibiscus sabdariffa* are *Bacillus* spp. notably *B. subtilis* subsp. *subtilis* and *B. licheniformis*.

3.3 Fish-based fermented products used as condiment in Africa

Like oil seeds, fermentation of fish is an essential part of socio-economic life of many communities in Africa particularly in Ghana, Egypt and Nigeria. Owing to the rapid deterioration of fish, fermentation offers an easy and cost effective way of preserving it. In the tropical countries fish fermentation usually involves the use of high concentration of salt combined with drying [53, 54]. Duration of fermentation is from a few days to weeks. Fish can be fermented whole or in parts. As in many traditional processes, fermentation of fish is carried out on small scale in homes [55, 56]. As with other traditional processes, uncontrolled nature and lack of hygiene practices are the major challenges to product quality and safety. The most popular fermented fish products used as condiments in Africa include *Lanhouin, Momone* and *Feseekh* [56]. **Table 3** presents the common fermented fish-based condiments of Africa and some other regions of the World.

Substrate	Product local name	Nature of product	Region/ country	Microorganism	Reference
Cassava fish (*Pseudotolithus senegalensis*)	*Lanhouin*	Condiment	West Africa	*Bacillus, Staphylococcus, Corynebacterium, Pseudomonas, Micrococcus, Streptococcus, Achromobacterium, Alcaligenes*	[37]
Cat fish, barracuda, sea bream and African jack mackerel (*Caranx hippos*)	*Momoni*	Condiment	West Africa (Ghana)	*Micrococcus, Streptococcus, Pediococcus*	[37]
Alestes baremoze (Pebbly fish), Hydrocynus sp. (Tiger fish)	*Feseekh*	Sauce	Egypt	Not known	[37, 95]
Fish	*Azu-okpo*	Condiment	Nigeria	Not known	[31]
Shrimp, salt	*Belacan* (*Blacan*)	Condiment	Malaysia	*Bacillus, Pediococcus, Lactobacillus, Micrococcus, Sarcina, Clostridium, Brevibacterium, Flavobacterium, Corynebacterium*	[45, 96]
Fish shrimp	*Bakasang*	Condiment	Indonesia	*Micrococcus, Streptococcus, Pediococcus* sp. *Pseudomonas, Enterobacter, Moraxella, Staphylococcus Lactobacillus*	[37, 45, 97]
Marine fish salt, sugar	*Budu*	Condiment	Thailand, Malaysia	*Micrococcus luteus, Staphylococcus arlettae, Pediococcus halophilus, Staphylococcus aureus, S. epidermidis, Bacillus subtilis, B. laterosporus, Proteus* sp., *Micrococcus* sp., *Sarcina* sp., *Corynebacterium* sp.	[37, 45, 98]

Substrate	Product local name	Nature of product	Region/ country	Microorganism	Reference
Finger size fish (*Esomus danricus*)	*Hentak*	Condiment	India	*Lactococcus lactis, L. plantarum, L. fructosus, L. amylophilus, L. coryniformis, Enterobacter faecium, Bacillus subtilis, B. pumilus, Micrococcus* sp., *Candida* sp., *Saccharomycopsis* sp.	[45, 99]
Sardine, salt	*Jeotgal*	Condiment	Korea	*Staphylococcus, Bacillus* sp., *Micrococcus* sp.	[45, 100]
Shell fish	*Gulbi*	Condiment	Korea	*Bacillus licheniformis, Staphylococcus* sp., *Aspergillus* sp., *Candida* sp.	[45, 101]
Marine fish	*Nuoc mam*	Condiment	Vietnam	*Bacillus* sp., *Pseudomonas* sp., *Micrococcus* sp., *Staphylococcus* sp. *Halococcus* sp., *Halobacterium salinarum, H. cutirubrum*	[45, 102]

Table 3.
Some common fermented fish products used as condiment in Africa and other continents of the world.

3.3.1 Lanhouin

Lanhouin is a traditional fermented salted fish condiment in West Africa, prepared from whole cassava fish (*Pseudotolithus senegalensis*). It is added as a flavoring to many traditional dishes especially soups [55, 56]. The traditional preparation of *lanhouin* involve scaling of the fresh fish and removing of the gut, followed by soaking in salted water for 8–11 h to allow ripening. The quantity of salt added var-ies between 20 and 35% depending on the size of the fish. Fermentation takes place naturally for about 2–9 days. On completion of fermentation, the product is washed in water to remove excess salt and then dried under the sun for 2–4 days. The fermented fish product has characteristic strong smell. The predominant microorganisms involved in the traditional process of *lanhouin* are members of the *Bacillus* species. Other bacteria such as *Staphylococcus, Corynebacterium, Pseudomonas, Micrococcus, Streptococcus, Achromobacter, Alcaligen*es were also reported to take part in the process although their roles are not clearly established.

3.3.2 Momoni

Momoni is another type of fermented fish from West Africa. It is particularly common in Ghana. Different types of fresh fish such as catfish, barracuda, sea bream and African jack mackerel (*Caranx hippos*) are used as substrate to prepare *momoni* [57]. *Momoni* is prepared traditionally based on the experience of the processor. During preparation, the scales and gut are removed. The gill and the gut regions are heavily salted (up to 30% salt may be used). Fermentation lasts for 1–5 days. Afterwards, the fermented product is washed in brine and cut into parts, followed by sun drying for a few hours [58]. Like *lanhouin*, *momoni* is widely used as flavor intensifier to prepare traditional dishes. The main organisms associated with *momoni* production are *Bacillus* species and lactic acid bacteria (LAB). Other bacteria and fungi are also reported to be associated with the process.

3.3.3 *Feseekh*

Feseekh is a fermented fish product from Egypt. It is popularly served as an appetizer, but in some occasions such as during feasts it may be the main meal. Unlike *lanhouin* and *momoni*, *feseekh* is fermented without drying. The type of fish used for preparing *feseekh* are *Alestes baremoze* (Pebbly fish), and *Hydrocyrus* sp. (Tiger fish) [59]. The quantity of salt added during fermentation may vary from 20 to 30%. *Feseekh* is processed at a temperature of about 18–20°C for about 60 days and the product can be stored up to 3 months. Microorganisms involved in the fermentation have not yet been fully characterized.

4. Significance of food fermentation to rural communities and economies

The significance of fermented foods including seasoning agents in human nutrition, particularly among rural populations is now better appreciated. As a result research efforts are being intensified towards better understanding of the processes as well as to achieve commercialization of these foods. Fermented foods including those derived from alkaline fermentation are critical components of the human diets world-wide. Currently, it is estimated that fermentation derived foods, beverages and condiments contributes about a third of the human diets and food supply world-wide [60]. These foods are particularly important in the cultures and food ecology of the developing nations such as Africa where thy have been reported to contribute more than half the calorie, ensuring the food security of millions [61]. In the recent times, the awareness of the nutritional values and health benefits associated with eating fermented products has made them indispensable as part of food system and has also led to their being classified as functional foods.

In some African countries especially Nigeria, fermented foods are valuable in the nutrition of infants and school-age children. In the rural communities, *akamu*, a fermented cereal based product is an important weaning food as well as breakfast meal. Indigenous fermentation technologies help to reduce the problem of food insecurity in the world [27]. In this regard, fermentation increases food availability by providing different types of products in a diversity of flavors, aroma and texture. Food fermenta-tion as an enterprise is particularly useful in the economy and socio-cultural lives of many communities. The nutritional and socio-economic values, health benefits and functional attributes of fermented foods have been widely documented [62–73]. Many of the substrates used for producing fermented foods contain naturally occurring toxins and anti-nutrients and only become edible following detoxification through fer-mentation. It also increases the bioavailability of key nutrients such as essential amino acids while enriching the sensory quality and functional properties of foods [74, 75].

5. Traditional fermentation techniques

The techniques of traditional fermentation can be solid-state or submerged culture. In solid state fermentation, the microorganisms grow on solid substrate containing little or no free moisture, but enough to sustain metabolic activity of the organisms. This technique is used in the production of all the alkaline fermented seasoning agents discussed earlier. Submerged fermentation is performed on a liquid substrate or a solid substrate immersed in a solution to form a suspension or slurry. These types of fermentations are seen in most commercial processes such as those employed for the production of alcoholic beverages and several other high-volume products.

5.1 Impact of process techniques on food quality

In the past, fermented foods were important only in the regions or places of manufacture. However, due to increasing demand, urbanization and industrialization, some of the fermented products such as soy sauce, and Japanese natto are becoming globally popular [76]. The challenge though is that the majority of the indigenous fermented foods and condiments are manufactured under conditions devoid of good manufacturing practices (GMPs) and good hygiene practices (GHPs) [22]. Often, hazard analysis and critical control point principles (HACCPs) are not observed during their production and unit operations are not clearly defined. This is inconsistent with modern food practices and may hinder the adoption of such products into the international markets [77]. Obviously, the uncontrolled nature of process techniques of traditional fermentation can have fundamental impact on the quality and safety of products. In the traditional setting, fermentation associated variables such as pH, DO, temperature, inoculum and moisture are not regulated. The variation in the fermentation conditions has frequently affected product quality, resulting in products that are non-uniform in quality between successive batches. Likewise, differences in processing techniques adopted by various processors which depend so much on personal knowledge, experience and expertise of the food handler and processor can also cause variation in product quality. Besides, the equipment used in the processing of traditional alkaline fermented seasonings such as fresh leaves, jute bag, local basket and calabash are substandard and they fall short of standard food hygiene protocol. Product consistency and safety of traditional fermented foods often arise fortuitously, from the physiological pressures imposed by the microbial selection rather than by processor actions.

In any food industry, maintaining proper hygiene in the production environment should be priority [3]. Although traditional fermentations often achieved the desired products, there is need to integrate modern GMPs in the production of these valuable constituents of traditional diets. Another significant challenge in tradi-tional fermentation that can lead to poor and inconsistent product quality is the participation of undesirable microbial strains in the process [78]. Most traditional alkaline fermentations rely on chance inoculation that encourage the participation of several species of microorganisms, including desirable and undesirable strains as against modern industrial technologies that make use of a single or defined selected strains (starter culture) to effect the desired change in the substrate. Although substrate modification and environmental condition may be tailored to favor the growth of the desired organisms, total reliance on process conditions to guarantee product/consumer safety may not always result in desirable outcomes. Besides, the contribution of the transient populations to the final product flavor and quality remain unknown. It is necessary therefore to migrate these traditional processes to modern, biotechnology-based food processing to eliminate process failures, some of which can result in consumer risks. Consumer safety and product quality can be best ensured by strict compliance with GMP.

6. Modern approach to food fermentation in Africa

In recent times, scientific knowledge and modern food processing technolo-gies have found application in food fermentation particularly in Asia and South America. The result is that many traditional kitchen technologies used in the manufacturing of fermented foods in the past are now modified [79]. The same may not be said (to the same degree) of traditional fermentation in Africa, where products are still mostly produced through kitchen technologies and village art that rely on illiterate processors. However, gradually the technology of alkaline fermented foods and products derived from other fermentations is evolving to a better and more commercial status. Progressively, traditional fermentation processes are being refined and diversified through the application of molecular biology and microbial technology (MT) such as the use of improved raw materials, starter/ protective culture, process optimization/ control including the use of modern packaging. This shift from artisanal production to scientific industrial one has generated new areas of study for industrialists and food scientists and new small scale industries. Innovations and recent developments in the production of alkaline fermented foods in Africa and other regions of the world will be approached at four levels namely: Raw material development, the use of starter culture, modified fermentation processes (process optimization), and product presentation (packaging).

6.1 Raw material development

High quality raw materials should be sourced and tested in order to select the more appropriate variety for use in fermentation. Agricultural procedures that encourage increased production of the improved varieties should be adopted. The use of improved and homogenous substrate for production of fermented foods will go a long way in solving the problem of product variability. The various raw materials (legumes and oil seeds) used for producing alkaline fermented foods are seasonal crops and are not readily available round the year. In order to overcome the current challenges of periodic or non-availability of raw materials, the use of irregularly available raw materials are being replaced wholly or partially with more abundant substrates in the production of various fermented products [3, 19]. In instances, the conventional substrate for production of ogiri an African alkaline fermented seasoning agent is castor oil seed [80]. However, when the main substrate is unavailable, alternative materials such as melon seed and fluted pumpkin seeds are used for the production [81]. Similarly, a related food condiment *owoh* can be produced from African yam bean [82] and cotton seeds [52]. In many cases, such as with oil bean and African mesquite, the seeds for producing fermented season-ings are produced by wild forest trees that are not yet domesticated. The availability of these crops is being threatened by deforestation and urbanization. For sustainability and also to overcome the bottleneck associated with the non-domestication of these crops, Agricultural and forestry management should put in place policies intended to secure the availability of these raw materials in order to ensure long-term supply [3]. This is without prejudice to the biochemical prospect of producing the product using more readily available alternatives by exploiting the versatility of fermenting microorganisms.

6.2 The use of starter culture

Literature on the use of starter cultures for the production of alkaline fermented foods including seasoning agents abound (**Table 4**). Modern researches on fermented foods have begun to adopt new approaches that focus on understanding the profile and the role of associated microorganisms in alkaline fermentations. Food researchers recognize that metabolic activities of microorganisms involved in a process have considerable impact on quality attributes of the final product such as color, flavor, texture and aroma as well as nutritional quality. Equipped with this knowledge, approaches used for characterizing the microorganisms in fermented foods have evolved to a better status. Many different techniques have been adopted to study the diversity of micro flora of fermented foods and their

Microorganism	Product	References
Bacillus subtilis B7 and B15	*Soumbala*	[85, 87, 88]
B. subtilis, B. subtilis kk-2:B_{10}	*Kinema*	[134]
B. subtilis mm-4:B12	*Ugba*	[72]
B. subtilis	*Okpehe*	[22]
B. subtilis 24BP, *B. subtilis* fpdp2	*Soy-dawadawa*	[19, 20]
B. subtilis TISTRO(BIOTEC7123)	*Thuo nao*	[135]
Lactobacillus plantarum 120 *L. plantarum* 145 *Pediococcus pentosaceus*	*Shan yu*	[36]
L. plantarum *Pediococcus acidilactici* *P. pentosaceus*	*Som-fug*	[136]
Lactobacillus plantarum *L. helveticus* *Lactococcus lactis* subsp. *lactis*	Mackerel mince	[137]

Table 4.
Bacillus strains and lactic acid bacteria (LAB) suggested as potential starter cultures for production of fermented condiment.

possible roles. This may be grouped into two: cultural/physiological methods and molecular methods [83]. Molecular techniques are of great importance in studying the microbial profiles, succession and functionality in traditional fermented foods. PCR-based methods and gene sequencing are now used for proper characterization of micro-biota including pathogens in fermented foods. Functional genomics is a useful tool in improving traditional process as this enables comparisons of traits of microorganisms involve in food fermentation and enables selection of organisms with desirable traits as potential starter cultures.

"Omics" is the acronym that has arisen from the study of functional genomics and comprises transcriptomics, proteomics and metabolomics. The introduction of "Omics" technologies offer better and clearer understanding of microbial populations in food processes and provide good opportunity for process standardization. These have been applied in the study of some traditional fermented foods such as *kimchi* a Korean fermented product. Another novel approach in food fermentation technology is the application of ultrasonic waves in the production of fermented foods [6]. The use of ultrasound has been reported as an important tool for measuring changes in chemical composition during fermentation and to enhance process efficiency and rate of production by improving mass transfer, cell permeability and removal of undesirable organisms.

Recent developments allow the establishment of starters, resulting in the evolution of kitchen technology to more optimized/controlled fermentation. The microorganisms used as starter cultures in food processing are selected based on food substrate, with the objective of achieving objective and reproducible biomodification.

In the efforts towards commercialization and upgrading of African alkaline fermented foods to industrial level, different species of microorganism have been studied and screened **Table 3** [15]. Pure cultures of *B. subtilis* var. natto is used in the commercial preparation of Japanese natto [84, 85]. Species of *B. subtilis* have been studied and demonstrated as potential starters for *soumbala* [46–48]. Similarly, strains of *Bacillus* species have been screened and suggested as starters by researchers from Nigeria. These include *B. subtilis* mm: B_{12}, for *ugba* [52] and

B. subtilis for *okpehe* [54]. The use of *B. subtilis* fpdp2, *B. subtilis* 24BP2 for soy dawadawa production has been demonstrated [19, 20], while strains of *B. subtilis* KK-2:B10 and *B. subtilis* GK have been used as starter cultures for *kinema* production and a starter for thua-nao (*B. subtilis* TISTRO (BIOTECHC7123)) has been reported [86, 87].

Besides, other bacteria such as lactic acid bacteria (LAB) have also been used as starter cultures for fermented foods including fermented fish [88–91]. A combination of *Lactobacillus plantarum, Pediococcus acidilactiçi* and *P. pentosaceus* was used as starter for the production of som-fug, a Thai fermented fish product [89]. Suan yu, a fermented fish product from China has been produced using defined strains of *L. plantarum* 120, *L. plantarum* 145, and *P. pentosaceus* 220 as mixed starter cultures, and this resulted in reduction of the fermentation time and enhanced quality [92]. Likewise, combined cultures of *L. plantarum, Lactococcus lactis* subsp. *lactis* and *L. helveticus* have been used for the production of fermented mackerel mince [93]. Lactic acid bacteria have been used as starter cultures to initiate the fermentation of cassava for garri production [94]. Sanni and coworkers [95] used antimicrobial producing strains of LAB to control spoilage organisms during production of *ogi*. Mixed cultures of LAB and yeast were also used as starter culture for 'gowe' production [96, 97].

Apart from bacteria, fungal starter cultures have been applied in the production of fermented foods. A combination of *Aspergillus* and *Actinomucor* has been used to produce *surimi*, fish-based fermented product [98, 99]. Similarly fermentation of silver cap fish using fungal starter cultures has been reported [100]. The nutritional benefits and organoleptic properties of four commercially available mold starters in fermented fish paste have been documented [101]. Despite successful applications and demonstrated beneficial roles of various starter cultures in food fermentations, their use in commercial traditional food productions is still limited and a subject of controversy. However, the prospect for commercializing the production of starter cul-tures for use in production of traditional foods and seasoning agents look promising.

6.3 Modified fermentation process (process optimization and control)

The uncontrolled nature of traditional fermentations is a major hindrance to the scaling up of indigenous food fermentations [76]. Optimization may only be possible when the roles of process variables such as duration of fermentation pH, temperature, inoculum-substrate ratio, DO and mass transfer and pretreatment are understood and controlled [16, 22, 35, 42]. An improved method of producing an African fermented condiment (Dawadawa) from locust beans has been reported to reduce fermentation time and cost of energy [42]. Also, in Burkina-Faso, a novel de-hulling machine introduced in the production of *soumbala* resulted in decrease in the boiling and de-hulling time by 75%, corresponding to an appreciable saving in energy cost and time [102].

In Uganda, pasteurization and refrigeration were used to increase the shelf life and safety of *obushera* a fermented cereal-based beverage [103]. Mechanization and modernization of the various labor intense unit operations in traditional fermented foods can result in significant improvement in the processes and increase the economics of production of these important food products besides enhancing reproducibility of process. Equipment used in traditional process remain rudimentary and process modernization and improvement based on those can constitute a great challenge as these may be difficult to replicate. The development of bioreactors will enhance performance and improve productivity. Research on-going in our laboratory is working towards developing a solid-state bioreactor for use in a trial scale-up of *okpeye* process and other related seasoning agents. Although many African alkaline fermented foods are still manufactured by the traditional family art, the manufacture of some such as *dawadawa*, and *soumbala* have been improved and elevated to pilot status [43]. More recently, due to increasing demand and awareness about natural healthy diets, some traditional alkaline fermented foods have evolved from their place of manufacture onto trans-border food markets [76, 104]. Of particular interest are two Asian fermented foods kimchi (fermented vegetable product) from Korea and natto (fermented soy bean product) from Japan which have penetrated markets beyond Asia [15, 76]. The production and marketing of tempe, an Asian fermented product has crossed borders and extended to the United States where about over 16 companies are involved in production [12].

6.4 Product presentation (packaging)

Packaging is an integral part of GMPs in foods. It provides environmental condition for storage, handling and long shelf life of product. It also minimizes post-process contamination and protects against microbial spoilage including undesirable change in sensory properties as well as consumer abuse. The shelf life of processed foods can be extended by aseptic and adequate packaging. Inadequate packaging and poor presentation of product are among the challenges mitigating the global development and consumer appeal of fermented products in Africa and other developing regions [105]. Unlike modern food industries that use attractive and esthetic packaging that increases consumer appeal, all sorts of wrapping materials are used for packaging traditional fermented products in Africa [3]. On account of this, indigenous fermented foods are often considered as food for the poor [105]. The adoption of modern esthetic packaging and adequate presentation are crucial steps to overcome the challenges of kitchen technology and also for commercialization and industrialization of fermented foods and condiments. These will help to minimize the problems of post process contamination and increase consumer confidence.

The application of these basic food-control strategies in the production of fermented foods will move these products beyond the local markets. Besides the challenges of unattractive packaging, some local fermented products such as dawadawa and okpeye are not packaged at all. These are displayed at points of sale often in open non-sterile bowls and local baskets which may lead to post-production contamination [3].

7. Conclusion and the way forward

Although, the economic and food security importance of African fermented foods and seasonings remain outstanding, their continued availability in the near future in a rapidly urbanized and global setting cannot be guaranteed on the basis of the present household technologies and practices. Today, with surge in the demand for "natural" foods, there is resurgence in demand across board for traditional foods which have also somehow become synonymous with the natural foods. The surge in demand for these traditional fermented products is not matched by supply and the trend can only get worse on the basis of kitchen technologies and traditional raw materials supply. Therefore, if we are to continue to enjoy these valuable components of our cultural diets, there has to be a way to manufacture those using sustainable modern technologies and GMPs. It is a challenge to industrialists, food scientists and researchers to ensure that knowledge generated through research are used to bring new ideas and innovations in the area of food fermentation and value chain. It has been observed that much of the research findings from scientists, particularly in the developing countries, end up in the journals and never make it to the market. There is an urgent need to address and bridge this research-to-market gap. It hoped that through research innovations, many fermented foods can be developed on the basis

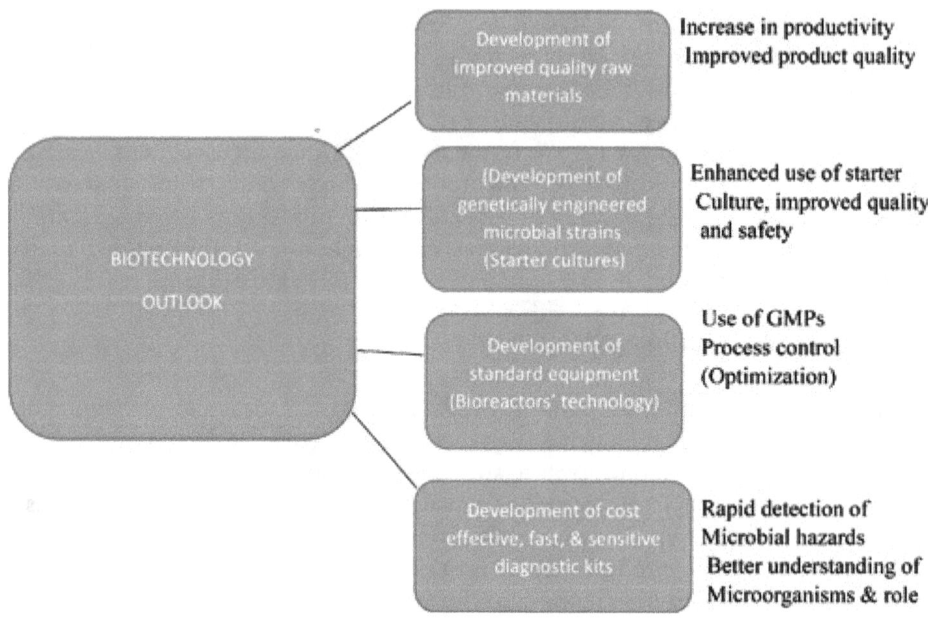

Figure 11.
Schematic presentation of prospects of biotechnology in commercialization and industrialization of fermentation process.

of good manufacturing practices (GMPs) to be able to achieve sustainable commercial marketability in the coming years. Currently, the disposition to understand traditional fermentation processes and their applications for industrialization is gaining ground in Africa and other developing regions. It is essential to recognize the critical role of microbial technology and significance of molecular biology-based applications in food processing in order to ensure quality and safety.

Recent understanding in the methods of processing fermented foods through application of scientific information has helped to improve quality of traditional foods in many ways. Microbial technology has played a key role in this aspect, being helpful for production of functional foods, bio-preservation and sensory improvement of fermented foods. With the application of sophisticated technolo-gies including genomics and proteomics, commercialization and industrialization of fermented foods look promising. **Figure 11** illustrates prospects of biotechnol-ogy in commercialization and industrialization of fermentation process. Future research will have to look at the use of improved raw materials as fermentation substrates, development and use of standard inoculums (starter cultures), and application of process control and defined unit operation, use of GMP and HACCP protocols by food processors, and use of adequate and esthetic packaging materials. These will eliminate challenges associated with food safety, achieve uniformity and reproducibility in products, enhance consumer confidence and increase product marketability across borders. A major breakthrough in the years ahead will target the evolution of viable small and medium fermentation enter-prises around traditional alkaline fermented foods and other related products in Africa and other parts of the world.

Author details

Jerry O. Ugwuanyi* and Augustina N. Okpara
Department of Microbiology, University of Nigeria, Nsukka, Nigeria

*Address all correspondence to: jerry.ugwuanyi@unn.edu.ng

References

[1] Okafor N. Modern Industrial Microbiology and Biotechnology. United States of America: Science Publishers; 2007. pp. 334-360

[2] Parkouda C, Nielsen DS, Azokpota P, Ouoba LII, Amoa-Awua WK, Thorsen L, et al. The microbiology of alkaline-fermentation of indigenous seeds used as food condiments in Africa and Asia. Critical Review in Microbiology. 2009;**35**:139-156

[3] Okpara AN, Ugwuanyi JO. Evolving status of African food seasoning agents produced by fermentation. Soft Chemistry and Food Fermentation. 2017:465-505. DOI: 10.1016/B978-0-12-81140-4.00015-1

[4] Uzogara SG, Agu LN, Uzogara EO. A review of traditional fermented foods, condiments and beverages in Nigeria: Their benefits and possible problems. Journal of Ecology, Food and Nutrition. 1990;**24**:267-288

[5] Nowak J, Kuligowski M. Functional properties of traditional food products made by mold fermentation (Chapter 3). In: Ray RC, Montet D, editors. Fermented Foods, Part II: Technological Interventions, Food Biology Series. Boca Raton, FL: Taylor and Francis Group LLC, CRC Press; 2017. pp. 46-73

[6] Mishra SS, Ray RC, Panda SK, Montet D. Technological innovations in processing of fermented foods, an overview (Chapter 1). In: Ray RC, Montet D, editors. Fermented Foods, Part II: Technological Interventions, Food Biology Series. Boca Raton, FL: Taylor and Francis Group LLC, CRC Press; 2017. pp. 21-45

[7] FAO. Traditional fermented food and beverages for improved livelihoods. In: Elaine M, Danilo M, editors. Diversification Booklet Number 21. Rural Infrastructure and Agro-industries Division, Food and Agriculture Organization of the United Nations (FAO); 2011. pp. 1-73

[8] Egwim E, Amanabo M, Yahaya A, Bello M. Nigerian indigenous fermented foods: Processes and prospects. In: Mycotoxin and Food Safety in Developing Countries. InTech Publishers; 2013. pp. 153-180. DOI: 10.57772/52877

[9] Ugwuanyi JO. Microbial technology and food security: Microbiology put safe food on our tables. In: 106th Inaugural Lecture of the University of Nigeria. 2016. 74 p

[10] Mirbach MJ, Ali EI. Industrial fermentation (Chapter 9). In: Ali MF, EI Ali BM, Speight JG, editors. Handbook of Industrial Chemistry Organic Chemicals. New York: McGraw-Hill; 2005

[11] Ray RC, Joshi VK. Fermented foods: Past, present and future scenario. In: Ray RC, Montet D, editors. Microorganisms and Fermentation of Traditional Foods. Boca Raton, Florida, USA: CRC Press; 2014. pp. 1-36

[12] Villéger R, Cachon R, Urdaci MC. Fermented foods, microbiology, biochemistry and biotechnology. In: Ray RC, Montet D, editors. Fermented Foods, Part II: Technological Interventions, Food Biology Series. Boca Raton, FL: Taylor and Francis Group LLC, CRC Press; 2017. pp. 1-20

[13] Campbell-Platt G. African locust bean (*Parkia* species) and its West African fermented products, 'dawadawa'. Journal of Ecology, Food and Nutrition. 1980;**9**:123-132

[14] Ogbadu CO, Okagbue RN. Bacterial fermentation of soybeans for 'daddawa' production. Journal of Applied Bacteriology. 1988;**65**:353-356

[15] Achi OK. Review: Traditional fermented protein condiments in Nigeria. African Journal of Biotechnology. 2005;**4**:1612-1621

[16] Omafuvbe BO, Abiose SH, Shonukan OO. Fermentation of soy bean (*Glycine max*) for soy-daddawa production by starter cultures of bacillus. Journal of Food Microbiology. 2002;**19**:561-566

[17] Sarkar PK, Hasenack B, Nout MJR. Diversity and functionality of *Bacillus* and related genera isolated from spontaneously fermented soybeans Indian 'kinema', and locust beans African soumbala. International Journal of Food Microbiology. 2002;**77**:175-186

[18] Dakwa S, Sakiyi-Dawson E, Diako C, Annan NT, Amoa-Awua WK. Effect of boiling and roasting on the fermentation of soybeans into 'soy-dawadawa'. International Journal of Food Microbiology. 2005;**104**:60-82

[19] Amoa-Awua WK, Terlabie NN, Sakyi-Dawson E. Screening of 42 *Bacillus* isolates for ability to ferment soybeans into 'dawadawa'. International Journal of Food Microbiology. 2006;**106**:343-347

[20] Terlabie NN, Sakyi-Dawson E, Amoa-Awua WK. The comparative ability of four isolates of *Bacillus subtilis* to ferment soybean into 'dawadawa'. International Journal of Food Microbiology. 2006;**106**:145-152

[21] Ogunshe AAA, Ayodele AE, Okonkwo IO. Microbial studies on 'Aisa': A potential indigenous laboratory fermented food condiment from *Albizia saman Jacq, F Mull*. Pakistan Journal of Nutrition. 2006;**5**:51-58

[22] Oguntoyinbo FA, Sanni AI, Franz CMAP, Holzapfel WH. In vitro selection and evaluation of *Bacillus* starter cultures for the production of 'okpehe', a traditional African fermented condiment. International Journal of Food Microbiology. 2007, 2007;**113**:208-218

[23] Enujiugha VN. Major fermentative organisms in some Nigerian soup condiments. Pakistan Journal of Nutrition. 2009;**8**:279-282

[24] Ibeabuchi JC, Olawuni IA, Iheagwara MC, Ojukwu M, Ofoedu CE. Microbial and sensory evaluation of 'iru' and 'ogiri-isi' as compared with commercial 'ogiri' samples. International Journal of Innovative Research and Studies. 2014;**13**:163-178

[25] Gberikon GM, Ameh JB, Ado SA, Umoh VJ. Comparative studies of the nutritional qualities of three fermented African legumes seeds using *Bacillus subtilis* and *Bacillus pumilus* as starters. Control Journal of Science and Technology. 2010;**4**:60-64

[26] Gberikon GM, Agbulu CO, Yaji ME. Nutritional composition of fermented powdered *Prosopis africana* soup condiment with and without inocula. International Journal of Current Microbiology and Applied Science. 2015;**4**:166-171

[27] Anal AK. Quality ingredients and safety concerns for traditional fermented foods and beverages from Asia: A review. Fermentation (MDPI). 2019

[28] Nwachukwu E, Achi OK, Ijeoma IO. Lactic acid bacteria in fermentation of cereals for the production of indigenous Nigerian food. African Journal of Food Science and Technology. 2010;**1**:21-26

[29] Blandino A, Al-Aseeri ME, Pandiella SS, Cantero D, Webb C. Cereal-based fermented foods and beverages. Food Research International. 2003;**36**:527-543

[30] Steinkraus KH. Classification of fermented foods: Worldwide review of

household fermentation techniques. Food Control. 1997;**8**:311-317

[31] Olusupe NA, Okorie PC. African fermented food condiments: Microbiology impacts on their nutritional values. 2019. DOI: 10.5772/intechopen.83466

[32] Zang J, Xu Y, Xia W, Regenstein JM. Quality, functionality and microbiology of fermented fish: A review. Critical Reviews in Food Science and Nutrition. 2019. DOI: 10.1080/10408398.2019.1565491

[33] Amadi EN, Barimalaa IS, Omosigho J. Influence of temperature on the fermentation of Bambara groundnut *Vigna subterranean*, to produce a 'dawadawa'-type product. Plant Foods for Human Nutrition. 1999;**54**:13-20

[34] Ndir B, Gningue RD, Keita NG, Souane M, Laurent L, Cornelius C, et al. Microbiological and organoleptic characteristics of commercial netetu. Cahiers d'etude et de recherché Francophones/Agricultures. 1997;**6**:299-304

[35] Akande FB, Adejumo OA, Adamade CA, Bodunde J. Review: Processing of locust bean fruits: Challenges and prospects. African Journal of Agricultural Research. 2010;**5**:2268-2271

[36] Rai AK, Jeyaram K. Legume-based food fermentation biochemical aspects. In: Ray RC, Montet D, editors. Fermented Foods, Part II: Technological Interventions, Food Biology Series. Boca Raton, FL: Taylor and Francis Group LLC, CRC Press; 2017. pp. 74-96

[37] Compaoré CS, Nielson DS, Ouoba LII, Berner TS, Nielson KF, Sawadogo-lingani H, et al. Co-production of surfactin and a novel bacteriocin by *Bacillus subtilis* subsp. subtilis H_4 isilated from Bikalga, an African alkaline *Hibiscuss sabdariffa* seeds fermented condiment. International Journal of Food Microbiology. 2013. DOI: 10.1016/j.ijfoodmicro.2013.01.013

[38] Odebunmi EO, Oluwaniyi OO, Bashiri MO. Comparative proximate analysis of some food condiments. Journal of Applied Science and Research. 2010;**6**:272-274

[39] Onyenekwe PC, Odeh C, Nweze CC. Volatile constituents of 'ogiri', soybean 'daddawa' and locust bean 'daddawa', three fermented Nigerian food flavor enhancers. Electronic Journal of Enviromental Agriculture and Food Chemistry. 2012;**11**:15-22

[40] Ogbadu LJ, Okagbue RN, Ahmead AA. Glutamic acid production by Bacillus isolates from Nigerian fermented vegetable proteins. World Journal of Microbiology and Biotechnology. 1990;**6**:377-382

[41] Odunfa SA. Review: African fermented foods: From art to science. MIRCEN Journal of Applied Microbiology and Biotechnology. 1988;**4**:259-273

[42] Alabi DA, Akinsulire OR, Sanyaolu MA. Qualitative determination of chemical and nutritional composition of *Parkia biglobosa Jacq, Benth*. African Journal of Biotechnology. 2005;**4**:812-815

[43] Iwuoha CI, Eke OS. Nigerian indigenous fermented foods: Their traditional process operation, inherent problems, improvements and current status. Food Research International. 1996;**29**:527-540

[44] Beaumont M. Flavoring composition prepared by fermentation with *Bacillus* spp. International Journal of Food Microbiology. 2002;**75**:187-196

[45] Ouoba LII, Diawara B, Annan NT, Poll L, Jakobson M. Volatile compounds of 'soumbala', a fermented African

locust bean, *Parkia biglobosa*, food condiment. Journal of Applied Microbiology. 2005;**99**:1413-1421

[46] Ouoba LII, Diawara B, Amoa-Awua WK, Traore AS, Moller PL. Genotyping of starter culture of *Bacillus subtilis* and *Bacillus pumilus* for fermentation of African locust bean (*Parkia biglobosa*) to produce 'soumbala'. Imternational Journal of Food Microbiology. 2004;**90**:197-205

[47] Ouoba LII, Rechinger KB, Barkholt V, Diawara B, Traore AS, Jakobsen M. Degradation of proteins during the fermentation of African locust bean *Parkia biglobosa*, by strains of *Bacillus subtilis*, and *Bacillus pumilus* for production of soumbala. Journal of Applied Microbiology. 2003;**94**:396-402

[48] Ouoba LII, Cantor MD, Diawara B, Traore AS, Jakobsen M. Degradation of African locust bean oil by *Bacillus subtilis* and *Bacillus pumilus* isolated from 'soumbala', a fermented African locust bean condiment. Journal of Applied Microbiology. 2003;**95**:862-873

[49] Okpara AN. Optimization of solid substrate fermentation of *Prosopis africana* Taub for production of *okpeye*, an African seasoning agent [PhD thesis]. University of Nigeria; 2018. pp. 1-170

[50] Ahaotu I, Anyogu A, Njoku OH, Odu NN, Sutherland JP, Ouoba LII. Molecular identification and safety of *Bacillus* species involved in the fermentation of African oil beans (*Pentaclethra macrophylla* Benth) for production of 'ugba'. International Journal of Food Microbiology. 2013;**162**:95-105

[51] Nwokeleme C, Ugwuanyi JO. Evolution of volatile flavour compounds during fermentation of African oil bean (*Pentaclethra macrophylla Benth*) seeds for 'Ugba' production. International Journal of Food Science. 2015. DOI: 10.1155/2015/706328

[52] Sanni AI, Ogbonna DN. The production of owoh—A Nigerian fermented seasoning agent from cotton seed (*Gossypium hirsitium* L). Food Microbiology. 1991;**24**:337-339

[53] El Sheikha A, Ray R, Montet D, Panda S. Worawattanamateekul, W. African fermented fish products in scope of risks. International Food Research Journal. 2014;**21**:425

[54] Oguntoyinbo FA. Safety challenges associated with traditional foods of West Africa. Food Reviews International. 2014;**30**:338-358

[55] Kindossi JM, Anihouvi VB, Vieira-Dalodé G, Akissoé NH, Jacobs A, Dlamini N, et al. Production, consumption, and quality attributes of Lanhouin, a fish-based condiment from West Africa. Food Chain. 2012;**2**:117-130. DOI: 10.3362/2046-1887.2012.009

[56] Kindossi JM, Egnonfan VB, Anihouvi O, Akpo-Djenontin EG, Vieira-Dalod H, Mathias H, et al. Microbial population and physic-chemical composition of African fish based flavouring agent and taste enhancer. African Journal of Food Science. 2016;**10**:227-237

[57] El Sheikha AF, Montet D. Fermented fish and fish products: Snapshots on culture and health. In: Microorganisms and Fermentation of Traditional Foods. Boca Raton, FL: Science Publishers Inc., CRC Press; 2014. pp. 188-222

[58] Sanni AI, Onilude A, Fadahunsi I, Ogunbanwo S, Afolabi R. Selection of starter cultures for the production of ugba, a fermented soup condiment. European Journal of Food Research and Technology. 2002;**215**:176-180

[59] Rabie M, Simin-Sarkadi L, Siliha H, El-seedy S, El Badawy A-A. Changes in free amino acids and biogenic amines of Egyptian salted-fermented fish (Feseeekh) during ripening and storage.

Food Chemistry. 2009;**115**:635-638. DOI: 10.1016/j.foodchem.2008.12.077

[60] FAO. Biotechnology applications in food processing: Can developing countries benefit? FAO Electronic Forum on Biotechnology in Food and Agriculture. 2004; Available from: http://www.fao.org/biotech/logs/C11/summary.htm

[61] Nezhad MH, Shafiabadi J, Hussain MA. Microbial resources to safeguard future food security. Advances in Food Technology and Nutritional Sciences: Open Journal. 2015;**SE1**:S8-S13. DOI: 10.17140/AFTNSOJ-SE-1-102

[62] Boudraa S, Hambaba L, Zidani S, Boudraa H. Mineral and vitamin composition of fruits of five underexploited species in Algeria: *Celtis australis* L., *Crataegus azarolus* L., *Crataegus monogyna Jacq.*, *Elaeagnus anustifolia* L., and *Zizyphus lotus* L. Fruit. 2010;**65**:75-84. DOI: 10.1051/fruits/20010003

[63] Besong EE, Balogun ME, Djobissie FA, Obu DC, Obimma JN. Medicinal and economic value of *Dialium guineense*. African Journal of Biomedical Research. 2016;**19**:63-170

[64] Ognatan K, Adi K, Lamboni C, Damorou JM, Aklikokou KA, Gbeassor M, et al. Effect of diatary intake of fermented seeds of *Parkia biglobosa* (Jacq) Benth (African locust bean) on hypertension in Bogou and Goumou-kope areas of Togo. Tropical Journal of Pharmaceutical Research. 2011;**10**:603-609

[65] Adeyemi OT, Muhammad NO, Oladiji AT. Biochemical assessment of the *Chrysophyllum* albidum seed meal. African Journal of Food Science. 2012;**6**:20-28

[66] Eme OI, Onyishi TU, Okala A, Uche IB. Challenges of food security in Nigeria: Options before government. Arabian Journal of Business and Management Review. 2014;**4**:15-25

[67] Oladejo JA, Adetunji MO. Economic analysis of maize production in Oyo state of Nigeria. Agricultural Science Research Journal. 2012;**2**:77-83

[68] Adesulu AT, Awojobi KO. Enhancing sustainable development through indigenous fermented food products in Nigeria. African Journal of Microbiology Research. 2014;**8**:1338-1343

[69] Murwan KS, Ali AA. Effects of fermentation period on the chemical composition, in-vitro protein digestibility and tannin content in two sorghum cultivars (*Dabar* and *Tabat*) in Sudan. Journal of Applied Bioscience. 2011;**39**:2602-2606

[70] Omodara TR, Olowomofe TO. Effects of fermentation on the nutritional quality of African locust bean and soy bean. International Journal of Science Research (IJSR). 2013;**4**:1069-1071

[71] Falana MB, Omemu MO, Oyewole OB. Microorganisms associated with supernatant solution of fermented maize mash (Omidun) from two varieties of maize grains. Research. 2011;**3**:1-7

[72] Mann A. Biopotency role of culinary spices and herbs and their chemical constituents in health and commonly used spices in Nigerian dishes and snacks. African Journal of Food Science. 2011;**5**:111-124

[73] Olukoya DK, Ebigwe SI, Olasupo NA, Ogunjimi AA. Production of 'Dogik': An improved 'Ogi' (Nigerian fermented weaning food) with potentials for use in diarrhoea control. Journal of Tropical Pediatrics. 2011;**40**:108-113

[74] Tamang JP. Naturally fermented ethnic soybean foods of India. Journal

of Ethnic Foods. 2015;**2**:8-17. DOI: 10.1016/j.jef2015.02.003

[75] Limón RI, Penas E, Torino MI, Martinez-Villaluenga C, Dueñas M, Frias J. Fermentation enhances the content of bioactive compounds in kidney bean extracts. Food Chemistry. 2015;**172**:342-352

[76] Byakika S, Mukisa IM, Byaruhanga YB, Male D, Muyanja C. Influence of food safety knowledge, attitudes and practices of processors on microbiological quality of commercially produced traditional fermented cereal beverages, a case of obushera in Kampala. Food Control. 2019. DOI: 10.1016/j.foodcont.2019.01.024

[77] Mukisa IM. Sensory characteristics, microbial diversity and starter culture development for "obushera", a traditional cereal fermented beverage from Uganda [PhD thesis]. Aas, Norway: Norwegian University of Life Sciences; 2012

[78] Capozzi V, Fragasso M, Romaniello R, Berbegal C, Russo P, Spono G. Spontaneous food fermentations and potential risks for human health. Fermentation. 2017;**3**:49. DOI: 10.3390/fermentation3040049

[79] Okafor N. Commercialization of fermented foods in sub-Saharan Africa. In: Application of Biotechnology to Traditional Fermented Foods. Washington, DC: National Academy Press; 1992. pp. 165-169. Chapter 4

[80] Ojinnaka MTC, Ojimelukwe PC, Ezeama CF. Effect of fermentation period on the organic and amino acid content of 'ogiri' from castor oil bean seed. Malaysian Journal of Microbiology. 2013;**9**:201-212

[81] Barber LA, Achinewhu SC. Microbiology of 'ogiri' production from melon seeds (*Citrullus vulgaris*). Nigeria Food Journal. 1992;**10**:129-135

[82] Ogbonna DN, Sokari TG, Achinewhu SC. Development of 'owoh'-type product from African yam beans (*Sphenostylis stenocarpa*) (Hoechst (ex. A.Rich) harms) by solid substrate fermentation. Journal of Plant Foods and Human Nutrition. 2001;**56**:183-194

[83] Temmerman R, Hiys G, Swings J. Identification of lactic acid bacteria: Culture-dependent and culture-independent methods. Trends in Food Science and Technology. 2004;**15**:148-359

[84] Wang J, Fung DY. Alkaline-fermented foods. A review with emphasis on 'pidan' fermentation. Critical Review in Microbiology. 1996;**22**:101-138

[85] Kiuch K. Industrialization of Japanese natto. In: Steinkraus KH, editor. Industrialization of Indigenous Fermented Foods, Second Edition, Revised and Expanded. New York: Marcel Dekker; 2004. pp. 193-246

[86] Sarkar PK, Tamang JP. Changes in the microbial profile and proximate composition during natural and controlled fermentations of soybeans to produce 'kinema'. Food Microbiology. 1995;**12**:317-325

[87] Visessanguan W, Benjagul S, Potachareon W, Panya A, Riebroy S. Accelerated proteolysis of soy proteins during fermentation of 'thua-nao' inoculated with *Bacillus subtilis*. Journal of Food Biochemistry. 2005;**29**:349-366

[88] Kose S, Hall GM. Sustainability of fermented fish-products. In: Fish Processing—Sustainability and New Opportunities. Surrey: Leatherhead Publishing. 2010. pp. 138-166

[89] Riebroy S, Benjakul S, Visessanguan W. Properties and acceptability of som-fug, a Thai fermented fish mince, inoculated with lactic acid bacteria starters. LWT-Food

Science andTechnology. 2008;**41**:569-580. DOI: 10.1016/j.Iwt.2007.04.014

[90] Semjonovs P, Auzina L, Upite D, Grube M, Shvirksts K, Linde R, et al. Application of *Bifidobacterium animalis s*ubsp lactis as starter culture for fermentation of Baltic herring (*Clupeá harengus membras*) mince. American Journal of Food Technology. 2015;**10**:184-194

[91] Speranza B, Racioppo A, Beneduce L, Bevilacqua A, Sinigaglia M, Corbo MR. Autochthonous lactic acid bacteria with probiotic aptitudes as starter cultures for fish–based products. Food Microbiology. 2017;**65**:244-253. DOI: 10.1016/j.fm.2017.03.010

[92] Zeng X, Xia W, Jiang Q , Guan L. Biochemical and sensory characteristics of whole carp inoculated with autochthonous starter cultures. Journal of Aquatic Food Product Technology. 2015;**24**:52-67. DOI: 10.1080/10498850.2012754535

[93] Yin LJ, Pan CL, Jiang ST. Effect of lactic acid bacterial fermentation on the characteristics of minced mackerel. Journal of Food Science. 2002;**67**: 786-792. DOI: 10.1111/ j.1365-2621.2002. tb10677x

[94] Kostinek M, Specht I, Edward VA, Pinto C, Egounlety M, Sossa C, et al. Characterization and biochemical properties of predominant lactic acid bacteria from fermenting cassava for selection as starter cultures. International Journal of Food Microbiology. 2007;**114**:342-351

[95] Sanni AI, Onilude AA, Ogunbanwo ST, Smith SI. Antagonistic activity of bacteriocin produced by *Lactobacillus species* from 'Ogi', an indigenous fermented food. Journal of Basic Microbiology. 1999;**39**:189-195

[96] Vieira-Dalode G, Jespersen L, Hounhouigan J, Moller PL, Nago CM, Jakobsen M. Lactic acid bacteria and yeasts associated with 'gowe' production from sorghum in Benin. Journal of Applied Microbiology. 2007;**103**:342-349

[97] Vieira-Dalode G, Madode YE, Hounhouigan J, Jespersen L, Jakobson M. Use of starter cultures of lactic acid bacteria and yeasts as inoculum enrichment for the production of 'gowe', a sour beverage from Benin. African Journal of Microbiology and Research. 2008;**2**:179-186

[98] Zhao D, Lu F, Gu S, Ding Y, Zhou X. Physicochemical characteristics, protein hydrolysis, and textual properties of surimi during fermentation with *Actinomucor elegans*. International Journal of Food Properties. 2017;**20**:538-548. DOI: 10.1080/10942912.2016.1168834

[99] Zhou X-X, Zhao D-D, Liu J-H, Lu F, Ding Y-T. Physical, chemical and microbiological characteristics of fermented surimi with *Actinomucor elegans*. LWT-Food Science and Technology. 2014;**59**:335-341. DOI: 10.1016/j.Iwt.2014.05.045

[100] Kasankala LM, Xiong YL, Chen J. Enzymatic activity and flavour compound production in fermented silver carp fish paste inoculated with douche starter culture. Journal of Agricultural and Food Chemistry. 2012;**60**:226-233. DOI: 10.1021/ jf203887x

[101] Giri A, Osako K, Ohshima T. Extractive components and taste aspects of fermented fish pastes and bean pastes prepared using different koji molds as starters. Fisheries Science. 2009;**75**: 481-489. DOI: 10.10007/ s12562-009-0069-1

[102] Sawadogo-Lingani H, Diawara B, Ganou L, Gouyahali S, Halm M, Amoa-Awua WK, et al. Effet du décorticage, mécanique sur la fermentation des grain

de néré Parkia biglobosa, en soumbala. Annales des Science Agronomique du Benin. 2003;**5**:67-84

[103] Byaruhanga Y, Ndifuna M. Effect of selected preservation methods on the shelf life and sensory quality of "obushera". Muarik-Bulletin. 2012;**5**:92-100

[104] Soni S, Dey G. Perspectives on global fermented foods. British Food Journal. 2014;**116**:1767-1787

[105] Peter-Ikechukwu AI, Kabuo NO, Alagbaoso SO, Njoku NE, Eluchie CN, Momoh WO. Effect of wrapping materials on physic-chemical and microbiological qualities of fermented melon seed (*Citrullus colocynthis* L) used as condiment. American Journal of Food Science and Technology. 2016;**4**:14-19

5

Biodegradability during Anaerobic Fermentation Process Impacted by Heavy Metals

Yonglan Tian, Huayong Zhang and Edmond Sanganyado

Abstract

In the past decades, biotechnologies for reutilizing the biomass harvested from the metal-contaminated land draw attention to many scientists. Among those technologies, anaerobic fermentation is proven as an efficient conversion process for biowaste reduction with simultaneous recovery of biogas as an energy source. During the process of anaerobic fermentation, the release of metals from the biomass will impact the growth and performance of microorganisms in reactors, which then results the variation of substrate degradation. In this chapter, the impact of metals on the degradation of substrate at different stages of fermentation process, as indicated by variations of lignocelluloses, chemical oxygen demands (COD), volatile fatty acids (VFAs), etc., will be summarized. The objective is to rationalize the relationship between metal presence and substrate degradability and give suggestions for future research on metal-contaminated biomass reutilization.

Keywords: anaerobic fermentation, heavy metal, biodegradation, lignocelluloses, chemical oxygen demands, volatile fatty acid

1. Introduction

The rapid development of industries such as electronic, mining, agrochemical, tannery, and battery industries has led to an increase in the direct and indirect discharge of metals into the environment. Some metals are potentially toxic, and unlike some organic contaminants, they are not biodegradable; thus, they may accumulate in terrestrial and aquatic organisms [1]. Hence, removal of potentially toxic metals (PTM) from contaminated environments has become an issue of urgent concern. In recent years, phytoremediation, which is defined as the use of plants to remove contaminants from contaminated environment, has drawn great attention probably because it is cost-effective and sustainable [2, 3]. However, disposing the biomass residues following phytoremediation is challenging [4–6]. Therefore, there has been a growing interest on the development of inexpensive disposal techniques and improvements in bio-resource utilization to foster sustainability in remediation systems [7].

Anaerobic fermentation is a relatively efficient conversion process for biomass waste reduction with simultaneous recovery of biogas as an energy source [8–11]. In an anaerobic reactor, there are four processes that occur simultaneously, i.e., hydrolysis, acidogenesis, acetogenesis, and methanogenesis [12]. Hydrolysis process involves the conversion of macromolecules such as proteins, polysaccharides, and fats that compose the cellular mass of the excess sludge into water-soluble molecules with a relatively small molecule (e.g., peptides, saccharides, and fatty acids) [12]. Simple molecules with a low molecular weight such as volatile fatty acids (e.g., acetic, propionic, and butyric acid), alcohols, aldehydes, and gases like CO_2, H_2, and NH_3 are produced via acidification of the hydrolyzed products (acidogenesis) [12]. The acidification products are converted into acetic acids, H_2 and CO_2, by acetogenic bacteria in a process called acetogenesis. These first three steps of anaerobic digestion are often called acid fermentation, and they help transform the waste biomass into substrates for methanogenesis [12]. In the methanogenesis process, the products of the acid fermentation (mainly acetic acid) are converted into CO_2 and CH_4.

Microorganisms responsible for anaerobic fermentation require a trace amount of metals (e.g., Ni, Co, Cu, Fe, Zn, etc.) for their optimum growth and performance [13]. Various enzymes involved in anaerobic metabolism use trace metals as their cofactors. For example, methanogenic enzymes such as CO dehydrogenase (CODH) and methyl-H4MPT:HS-CoM methyltransferase use cobalt acts as their cofactor [14].

However, waste biomass often contains varying amounts of metals depending on the source of the biomass. During the process of anaerobic fermentation, the release of metals from the biomass will influence the efficiency of fermentation by affecting the enzyme activity, microorganism community, and even degradation and metabolic pathways [15]. In this chapter, the impact of metals on the degradation of substrate, as indicated by variations of lignocelluloses, chemical oxygen demands, volatile fatty acids, etc., will be summarized. The objective is to rationalize the relationship between metal presence and substrate degradability during different fermentation stages and give suggestions for future research on metal-contaminated biomass reutilization.

2. Hydrolysis stage

Hydrolysis is oftentimes the rate-limiting step in the anaerobic digestion process probably because fermentative bacteria require an additional step of excreting extracellular enzymes, such as cellulases and lipases, to carry out the hydrolysis or solubilization process [16, 17]. It can be accelerated by enhancing the accessibility of anaerobic microorganisms to intracellular matter or cellulose using thermal, chemical, biological, and mechanical processes, as well as their combinations [18].

2.1 Lignocellulose degradation

Lignocelluloses are mainly composed of cellulose, hemicellulose, and lignin [19]. The cellulose and hemicellulose themselves are relatively easy to be broken down by microorganisms; however, their biodegradability decreases when they occur in lignocellulose complexes [20]. The impacts of metals on lignocellulose degradation vary with the metal species, concentrations, and fermentation conditions.

Previous studies showed that the presence of metals at certain concentrations may enhance the degradation of lignocelluloses [21, 22]. In one study, an average lignocellulose content of 87.49 ± 3.19%TS was obtained in a control group but decreased to 80.44 ± 3.41%TS, 77.94 ± 3.50%TS, and 79.45 ± 2.88%TS following the addition of 30, 100, and 500 mg/L Cu and 79.36 ± 3.72%TS, 79.10 ± 2.80%TS, and 76.60 ± 2.97% TS following the addition of 30, 100, and 500 mg/L Cr, respectively. Thus, Cu and Cr addition significantly enhanced the degradation of lignocellulose [21, 22].

Several studies on methanogenic bacteria found Cu, suggesting it could be a critical component for the enzymes super dismutase and hydrogenase [23]. However, at relatively high concentrations, Cu can inhibit anaerobic fermentation, which results in reduction of degradation efficiency. Cu changes the physiological steady state of the fermentation process by inhibiting the degradation of the substrate and the growth of the microbes [8, 24]. In contrast, Cu has been shown to enhance the biogas production via fermentation [9, 25]. Despite the inhibitory effects of Cu, biogas production was probably enhanced by the addition of sulfide to the digester in stoichiometrically equivalent amounts [26]. Our research suggested that the promoting effect of Cu addition on biogas yields was mainly attributable to better process stability, the enhanced degradation of lignin and hemicellulose, the transformation of intermediates into VFA, and the generation of CH_4 from VFA [22].

Cr is one of the heavy metals that have often been blamed for unsatisfactory operation or failure of anaerobic digesters [27]. Contradictory toxicity levels of Cr on anaerobic fermentation have been cited in literatures [28, 29]. This is probably because of differences in availability of Cr in fermenters (which is influenced by the precipitation and adsorption of soluble metals), differences in materials used in the studies [30], and dissimilar operational conditions (e.g., temperature, pH, hydrau-lic retention time, solid retention time, and mixed liquor volatile suspended solids) [27]. Cr in certain concentrations was found to promote the efficient generation of CH_4 by inducing better process stability, enhancing degradation of lignin and hemicellulose, transforming intermediates into VFA, and increasing coenzyme F_{420} activities [21].

According to our recent study, when compound metals were added into the fermentation reactors, the degradation of lignocelluloses performed differently (**Table 1**). It was found that the addition of Zn into the Cd- or Cu-containing reactors enhanced the degradation of lignin and cellulose significantly which resulted in a significant decrease in the total lignocellulose contents. The addition of Fe together with Cd reduced the cellulose contents and the total lignocellulose contents. In contrast, the addition of Ni into either Cd- or Cu-containing reactors did not improve the degradability of the feedstocks.

Depending on the methanogenic pathway, the general trends of metal requirements are as follows: Fe is the most abundant metal, followed by Ni and Co and smaller amounts of Mo (and/or W) and Zn [31]. Almost all metalloenzymes involved in the pathway of biogas production contain multiple Fe_2S_2, Fe_3S_4, or Fe_4S_4 clusters [17, 31, 32]. Fe is primarily present as Fe-S clusters used for electron transport and/or catalysis, as well as attenuating disturbances associated with the presence of sulfide which often results in a more stable process [31, 33–35]. Zn, like Cu, is present in relatively large concentrations in many methanogens. Zn is important in anaerobic fermentation because it is required by enzymes involved in methanogenesis such as coenzyme M methyltransferase [36]. At certain concentrations, Zn can promote biogas production [37, 38]. For example, during the swine manure anaerobic digestion, Zn concentrations in the range of 125–1250 mg/L improved significantly microbial activity [39].

Ni is an important trace element for many prokaryotic microorganisms that are in the *Bacteria* and Archaea domains [40]. It is required in the prosthetic groups of a total of eight enzymes that are found in prokaryotic microorganisms, includ-ing CODH, acetyl-CoA synthase/decarbonylase, methyl-coenzyme M reductase (MCR), [NiFe]-hydrogenases, superoxide dismutase (Ni-SOD), glyoxylase I, urease, and acireductone dioxygenase [41]. Generally, the biologically relevant oxidation states of Ni are Ni^{+}, Ni^{2+}, and Ni^{3+}, and these depend on how Ni is ligated to the protein. Ni usually functions either as a redox catalyst, for example, as in the case of hydrogenase or CODH where Ni is liganded by cysteinyl sulfurs [40]. However, Ni

Metals concentrations (mg/L)	Lignin (%TS)	Hemicellulose (%TS)	Cellulose (%TS)	Total lignocellulose (%TS)
Cd(1.0)	19.84 ± 0.94	13.14 ± 0.75	19.27 ± 1.38	52.25 ± 3.07
Cd(1.0) + Fe(10.0)	18.20 ± 0.63	11.96 ± 0.61	15.97 ± 0.90*	46.13 ± 2.14*
Cd(1.0) + Ni(2.0)	20.82 ± 1.10	15.02 ± 0.75	16.36 ± 0.60	52.20 ± 2.45
Cd(1.0) + Zn(2.0)	12.83 ± 1.07**	12.98 ± 0.64	13.55 ± 1.13**	39.36 ± 2.84**
Cu(1.0)	19.63 ± 0.85	13.23 ± 0.75	19.34 ± 1.46	52.21 ± 3.06
Cu(1.0) + Fe(10.0)	16.92 ± 0.90	11.44 ± 0.61	16.46 ± 0.83	44.83 ± 2.34**
Cu(1.0) + Ni(2.0)	19.95 ± 1.15	12.05 ± 0.69	20.21 ± 0.74	52.21 ± 2.58
Cu(1.0) + Zn(2.0)	14.56 ± 1.03**	12.34 ± 0.61	14.43 ± 1.18**	41.34 ± 2.82**

Mean ± standard error. n = 10.
$^{*}p < 0.05$.
$^{**}p < 0.01$.

Table 1.
The average contents of cellulose, hemicellulose, and lignin and total lignocellulose during the anaerobic co-digestion of corn stover and cow dung (55.0 ± 1.0°C) in the presence of different compound metals.

can act synergistically in Ni–Cu, Ni–Mo–Co, and Ni–Hg systems or antagonistically in Ni-Cd and Ni-Zn systems [42]. Ni was also found to decrease the toxicity of Cd and Cu [43]. However, the combination of Ni and Cd or Cu has been shown that they do not promote the degradation of lignocelluloses (**Table 1**). However, this line of inquiry requires further study.

2.2 Variation of chemical oxygen demands (COD)

The soluble organic components in the fermenter, shown as COD, originate from the hydrolysis process that liquefies large molecules; long-chain natural polymers of the substrate-like cellulose, hemicellulose, lignin, and polysaccharides; and proteins by extracellular enzymes [17, 44]. Previous studies on anaerobic fermentation of crops and manure showed that the COD in the reactor increased and then decreased because organic matter in the liquid was generated first and then consumed to produce the biogas. Therefore, greater COD did not cause greater biogas generation [45].

Previous studies demonstrated that COD initially increased and then decreased in the presence of Cu [22]. During the initial stage of the fermentation, the substrate was rapidly hydrolyzed into small organic molecules, bringing about an increase of COD in the first 5 days. Later, the COD of the Cu-added groups decreased. The COD of the control group decreased more slowly than those of the Cu-added groups, and the discrepancy between them increased during the fermentation. Taking the whole fermentation process into account, the COD in the Cu addition groups were relatively lower than the control group. It was suggested that Cu addition enhanced the utilization of organic molecules in the fermentation (as indicated by the decrease of COD) and the biogas production [22]. A similar promoting effect was found in the Cr-stressed anaerobic fermentation process [21]. However, Cr addition did not yield lower COD than the control group.

The COD were generally lower in Fe-added groups than in the control group [46]. Fe addition induced a stable and excellent COD conversion rate suggesting a more efficient utilization of soluble organic components in the fermenter that consequently improves biogas yields [47]. Likewise, Ni addition influenced the biogas production, and this can also be partly explained by the Ni effect on COD [48].

Treatments	Parameters	55.0 ± 1.0°C	45.0 ± 1.0°C	35.0 ± 1.0°C	25.0 ± 1.0°C
No Cd added	COD (mg/L)	10172.01 ± 1246.81	11219.30 ± 1596.13	12116.95 ± 2317.61	9517.14 ± 1199.70
	Biogas yield (mL/g TS)	67.65 ± 1.16	58.45 ± 1.01	19.42 ± 0.32	15.96 ± 0.27
Cd added	COD(mg/L)	9651.83 ± 1505.46	11052.13 ± 1612.40	13192.66 ± 2518.87	9829.71 ± 398.46
	Biogas yield (mL/g TS)	341.62 ± 5.88	68.35 ± 1.18	20.51 ± 0.34	15.84 ± 0.27

Table 2.
The average COD and cumulative biogas yields during the anaerobic co-digestion of acid-pretreated corn stover and cow dung (55.0, 45.0, 35.0, and 25.0 ± 1.0°C) with and without Cd (1.0 mg/L) addition. Mean ± standard error.

Lower COD concentrations and higher biogas yields were obtained using higher Ni concentrations. The results demonstrated the balance of different fermentation steps (from hydrolysis to methanogenic phase). Moreover, as the substrates in the Ni-added groups were better degraded in the former three stages (from 4th to 13th day), the left substrates were few, and hence the COD concentrations of Ni-added groups were not increased at the end of the experiments.

The variation of COD in the presence of metals should also be considered under different fermentation temperatures. The required amounts for Ni, Co, Zn, and Fe and in thermophilic glucose fermentation were 10 times more than those required for mesophilic acetate fermentation [49, 50]. In our anaerobic fermentation experiment with acid-pretreated corn stover mixed with fresh cow dung as feedstocks, the COD at 55, 45, 35, and 25 ± 1.0°C were analyzed with and without Cd (1.0 mg/L) addition. The cumulative biogas yields and average COD during the entire fermentation process in Cd-added and no Cd-added group are shown in **Table 2**. It was found that Cd addition resulted in higher biogas yields with the increase of temperature together with the lower COD. Overall, the biogas production should be explained by the variation and/or consumption of the COD concentrations along with VFAs during the fermentation process, rather than the values of COD concentrations.

3. Acidogenesis stage

Acidification is affected by a very diverse group of bacteria, the majority of which are strictly anaerobic. As the acidogenesis stage progresses, the acidic components, including long-chain fatty acids (LCFAs), volatile fatty acids (VFAs), etc., are generated, and they cause a change in the pH. Furthermore, during acidogenesis, organic nitrogen is converted into ammonia [51].

3.1 Variation of pH values

The optimal pH range for efficient methanogenesis ranges from 6.7 to 7.4 [52]. However, the acidogenic bacteria can metabolize organic material down to a pH of around 4. At the beginning of anaerobic fermentation, the pH values are likely to decrease due to the generation of acid components. Thus, buffer solution is suggested for preventing the dramatic pH reduction.

On the one hand, the effect of metal toxicity depends on pH [53]. In general, at high pHs metals have a tendency to form insoluble metal phosphates and carbonates [54], whereas at low pHs the initial leaching of metals from the sludge occurs and hence their solubility increases [55, 56]. Soluble levels of Ni ion were found the highest, while those of Pb ions were found the least as compared to other four heavy metals from pH 4 to 12. At extreme pH of 1, Zn, Pb, and Cd ions showed higher levels than those of Ni, Cu, and Cr. However, Cu, Ni, and Zn ion levels were found higher than those of Pb, Cd, and Cr at an extreme pH of 13. Metal ion levels showed the order of Ni > Cu > Cr > Zn > Cd > Pb between pH 8 and 12. In other pH ranges, metal ions varied with pH [57].

On the other hand, the presence of metals in the reactor during the fermentation process can modify pH values. Previous studies found that adding Cu and Cr resulted in a decrease in pH at the beginning of the experiment, but the pH later recovered [21, 22]. The average pH following addition of Cu have been shown to be generally higher than control groups as well as groups in which Cr is added. An investigation on anaerobic digestion of sewage sludge found that pH negatively related with the exchangeable (−0.838, $p < 0.01$) and residual fractions (−0.753,

$p < 0.01$) of Cu while positively related to Fe-Mn oxide-bound(0.895, $p < 0.01$) and organic-bound (0.698, $p < 0.05$) fractions of Cu [58]. In contrast, pH positively related to carbonate-bound Cr (0.768, $p < 0.01$) and organic-bound Cr (0.908, $p < 0.01$) while negatively related to Fe-Mn oxide-bound Cr (−0.899, $p < 0.01$)[58]. The results suggest the decrease of pH at the beginning of fermentation was probably beneficial for generating both the exchangeable and residual fractions of Cu. The increase of pH after the start-up of the fermentation is probably helpful for partitioning Cr to yield carbonate-bound and organic-bound fractions, thus reduc-ing the bioavailability and toxicity of Cr [21].

Addition of Fe and Ni has been shown promote an alkalescent environment for anaerobic fermentation. Previous studies found that adding 10.0 mg/L Fe into the fermenter resulted in lower pH values ($p < 0.05$) when fermentation is around its peak stage [46]. However, following fermentation peak stage, no significant change in pH has been reported even after increasing the Fe concentration from 0.5 to 5.0 mg/L or Ni concentrations from 0.2 to 2.0 mg/L [46, 48].

3.2 Variation of NH_4^+-N concentrations

Total ammonia (TAN), consisting of ammonium ions (NH_4^+) and free ammonia (FAN, NH_3), is produced during anaerobic degradation of proteins, urea, and nucleic acids [59]. At NH_4^+-N concentrations below 200 mg/L, TAN is an important nutrient for microorganism growth [60]. However, it was reported by Math-Alvarez et al. [61] and confirmed in a critical review by Chen et al. [42] that NH_4^+-N con-centrations ranging from 0.6 to 14 g/L inhibited the methanogenic activity depend-ing on different experimental conditions [62].

Previous studies showed that Cu or Cr addition induced remarkable differences in NH_4^+-N concentrations compared to a control group [21, 22]. The NH_4^+-N values in the control group fluctuated in the range 9.70–157.34 mg/L before the 21st day of fermentation; yet in Cu- and Cr-added groups, the NH_4^+-N values were relatively stable with concentrations ranging from 55.98 to 113.82 mg/L [22] and 39.85 to 105.87 mg/L [21], respectively. Thus, Cu and Cr addition contributed to the stability of the fermentation system.

Table 3 shows that further addition of other metals may increase the NH_4^+-N concentrations in the metal-stressed fermenters. According to our research, Zn addition significantly enhanced the generation of NH_4^+-N in both Cd and Cu

Metals concentrations (mg/L)	NH_4^+-N	Total VFAs
Cd(1.0)	558.39 ± 39.25	910.57 ± 273.75
Cd(1.0) + Fe(10.0)	604.50 ± 37.34	1865.18 ± 684.94
Cd(1.0) + Ni(2.0)	747.13 ± 38.21**	1003.57 ± 219.79
Cd(1.0) + Zn(2.0)	675.34 ± 36.22*	513.86 ± 195.52
Cu(1.0)	476.63 ± 37.36	369.77 ± 73.28
Cu(1.0) + Fe(10.0)	652.83 ± 61.88*	1562.24 ± 577.63*
Cu(1.0) + Ni(2.0)	569.31 ± 23.62	1029.20 ± 298.17
Cu(1.0) + Zn(2.0)	671.82 ± 43.40**	804.66 ± 286.14

Mean ± standard error. n = 10.
*$p < 0.05$.
**$p < 0.01$.

Table 3.
The NH_4^+-N and total VFA concentrations during the anaerobic co-digestion of corn stover and cow dung (55.0 ± 1.0°C) in the presence of different compound metals.

contained fermenters. Addition of Ni induced higher NH_4^+-N concentrations in Cd-stressed anaerobic fermentation process, while Fe had a similar response in Cu-stressed fermentation processes. The results suggest that metal mixtures benefited from the degradation of substrate containing nitrogen, such as proteins (**Table 3**) together with the degradation of lignocelluloses (**Table 1**).

3.3 Variation of long-chain fatty acids (LCFAs)

Long-chain fatty acids (LCFAs) are the intermediate products of lipids' hydrolysis and thus are abundant in lipid-rich substrates such as slaughterhouse wastewater and dairy industrial sludge [63, 64]. LCFAs (e.g., oleic acid) are often degraded through β-oxidation [65] to form acetate, hydrogen, and short-chain fatty acids (SCFAs). Short-chain fatty acids are further catabolized to acetate and hydrogen following cycles of β-oxidation [66]. LCFAs can inhibit the activities of the microorganisms involved in all the AD steps [67] by attaching to bacterial cell membrane, thus limiting mass transfer [68]. It was reported that LCFAs concentration of 0.2 g/L oleate had a profound inhibitory effect, while biogas production ceased when the concentration was increased to 0.5 g/ L [69]. Hwu et al. [70] reported 50% inhibition of methanogenesis in batch reactors at 0.1–0.9 g L^{-1} oleate, depending on the origin of the bacterial inoculum. It has been previously documented that LCFAs could inhibit the activity of hydrolytic, acidogenic, and acetogenic bacteria and methanogenic archaea [68, 69, 71]. However, many studies have reported an adaptation of the microbial communities during the degradation of LCFAs [70, 72]. Moreover, the archaeal community was found to be more tolerant to increased LCFA concentration levels compared to the bacterial community [73].

Metals play a major role in several metabolic pathways and thus will impact of transformation of LCFAs during fermentation process. In general, adding adequate concentrations of microelements may accelerate the degradation of short-chain fatty acids (SCFAs) and LCFAs and would be beneficial for the anaerobic mono-digestion of food waste [74]. However, there is lack of studies on the responses of LCFAs to metal stress. Further studies are necessary for revealing the underlying mechanisms.

3.4 Variation of total volatile fatty acids

VFAs are the intermediary products of the anaerobic fermentation and a precursor for methanogenesis. The concentration of VFA is an important index to evaluate the efficiency of hydrolysis, acidification, and methanogenesis [75]. Trace metal supplementation is one method to increase VFA utilization [76].

Many studies worked on the impacts of heavy metals on the degradation of VFAs [77–79]. At the beginning of the fermentation, the total VFAs often increase due to hydrolysis of substrate and the accumulation of acidic hydrolytic products [21, 22], together with the decrease of pH values. It was reported that a pH range of 5.7–6.0 was recommended as optimal to produce VFAs [80]. During this period, high concentration of Cu was found to inhibit the acidification process [22], while high concentration of Cr inhibited the methanogenesis [21], resulting in low biogas yields. Later on, the VFAs were shown to be consumed during the biogas production, and supplementing metals greatly benefited the process [21, 22].

Supplementing metals may promote the degradation of VFAs, while a metal deficiency may result in the accumulation of VFAs, which often inhibits the anaerobic processes. For example, a previous study found Fe and Ni deficiency during anaerobic digestion of wheat stillage resulted in a rapid accumulation of VFAs [81]. In another study, excluding Co, Zn, and Ni from the methanol-based feed of an

UASB reactor induced lower specific methanogenic activity (SMA) and the accumulation of VFAs [82–84].

A strong relationship has been previously reported between VFAs and differ-ent forms of metals [58]. For example, there was a strong correlation between the organic-bound Cr and VFAs ($r = -0.846$, $p < 0.01$), indicating that decrease in VFA enhanced the transformation of Cr from unstable species to organic-bound fractions, thus reducing Cr bioavailability and toxicity [58]. As a result, there was an improvement in the CH_4 yield. However, the relationships between VFA and metals have been shown to depend on the form and species of the metal. For combining metals, the addition of Fe into Cu-contained fermenters significantly increased the total VFA concentrations (**Table 3**) and resulted in higher biogas yields (data not shown).

4. Acetogenesis stage

Many factors, including substrate concentration, hydraulic retention time, temperature, pH, and process configuration, affect the performance of the acidogenesis phase [31, 85]. However, these factors are particularly susceptible to the presence and subsequent interactions with heavy metal ions [86].

The C_2-C_7 organic acids are predominant intermediates in the anaerobic digestion of organic matter. The anaerobic oxidation of the C_3-C_7 substrates is coupled to a reduction of protons (H_2 formation), and the oxidation of C_3 and C_4 organic acids is thermodynamically unfavorable (endergonic process) under standard conditions [77].

Lin studied the effects of Cr, Cd, Pb, Cu, Zn, and Ni on VFA degradation in anaerobic digestion by using serum bottle assays with acetic acid acclimated seed sludge (AASS) and mixed acid acclimated seed sludge (MASS) [87]. The rela-tive toxicity of heavy metals to degradation of acetic acid (HAc), propionic acid (HPr), and n-butyric acid (n-HBu) was Cd > Cu > Cr > Zn > Pb > Ni, Cd > Cu > ≒ Zn ≒ Cr > Pb > Ni, and Cd > Cu > Cr > Zn > Pb > Ni, respectively [87]. Cd and Cu were the most, and Pb and Ni were the least toxic heavy metals to VFA-degrading organisms. To some heavy metals, VFA-degrading acetogens were more sensitive than HAc-utilizing methanogens. The order of sensitivity of the VFA degradation to the metallic inhibition was HPr > HAc ≒ HBu for Cr, HAc > HPr ≒ HBu for Cd and Pb, HPr > HAc > HBu for Zn, HAc ≒ HPc ≒ HBu for Cu, and HAc > HPR > HBu for Ni. Mixtures of the heavy metals caused synergistic inhibi-tion on HAc degradation [87].

Lin et al. carried out a systematic study on the effect of trace metal supplementation on anaerobic degradation of butyric acid [88]. The results showed that the stimulatory effects were in the following order: $Cu^{2+} < Fe^{3+} < Zn^{2+} < Ni^{2+} < Mn^{2+}$ and the normal and isoHBu degradation activities of the methanogens increased by 14–25% and 17–43%, respectively [88]. Kim et al. reported that the supplementation of Ca, Fe, Co, and Ni to a thermophilic non-mixed reactor was required in order to achieve a high conversion of propionate at high concentrations of VFAs [89].

About 70% of CH_4 is generated from acetic acid [86]. The acetate utilization rates required per gram of VSS are used to estimate the nutrient supplementation required to prevent limitations in methanogenic activity [90]. Addition of Fe was found to have a stimulatory effect on acetate utilization by methanogens [76]. Bhattacharya et al. [91] found adding 20 mg/L Zn^{2+} resulted in a complete inhibition of acetate degradation due to Zn toxicity to methanogenesis. Ni sites in the acetyl-CoA decarboxylase/synthase enzyme complex have been identified. This enzyme seemed to have an important role in the conversion of acetate to CH_4 [92].

5. Methanogenesis stage

Methanogenesis is the microbial process, whereby CO_2, acetate, or methyl compounds are converted to CH_4 in order to generate ATP through the buildup of a sodium ion or proton gradient [31]. Methanogenesis is one of the most metal-rich enzymatic pathways in biology [93]. The contents of Cu, Fe, Ni, and Zn in methanogens (including 10 species of *Methanosarcina*, *Methanococcus*, *Methanobacterium*, *Methanobrevibacter* , etc.) were determined as <10–160 ppm, 0.07–0.28%, 65–180 ppm, and 50–630 ppm, respectively [36].The key enzyme complex in producing biogas from acetate is CODH [15, 94]. CODH cleaves the C-C and C-S bonds in the acetyl moiety of acetyl-CoA, oxidizes the carbonyl group to CO_2, and transfers the methyl group to coenzyme M. MCR catalyzes the enzymatic reduction of methyl-coenzyme M to CH_4 in methanogenesis, which includes a Ni-containing cofactor called F_{430} [15, 31, 40]. Besides, the CODH complex is also involved in the formation of acetate by acetogens from H_2/CO_2 and methanol [95]. MCR is found exclusively in methanogenic archaea [96].

5.1 Biogas yields

The stimulatory effect of trace metal supplementation in certain ranges on anaerobic digestion has been widely reported [10, 21, 46]. Depending on the methanogenic pathway, the general trends of metal requirements are as follows: Fe is the most abundant metal, followed by Ni and Co and smaller amounts of Mo (and/orW) and Zn [31].

Low concentrations of Fe have been shown to markedly increase the conversion of acetic acid to CH_4 [97]. Fe^{2+} in concentrations of up to 20 mM has been shown to increase the conversion of acetate to CH_4 [98]. It was found that Fe^{2+} marginally stimulated biogas yield and CH_4 content at 37°C and the addition of Fe^{2+} increased VFA utilization but enhanced H_2 utilization considerably [99]. The addition of Fe resulted in a stable process, and its combination with Co contributed to higher biogas production (+9%), biogas production rates (+35%), and reduced VFA concentration while simultaneously degrading the organic fraction of municipal solid waste and slaughterhouse waste [34]. The promoting effect of Fe^{2+} addition on biogas yields of mixed *Phragmites* straw and cow dung was mainly attributed to the extension of the gas production peak stage and the improvement of cellulase activities [46].

The optimum or stimulatory concentrations of Ni for batch cultures of methanogens were reported to range between 12 mg/m^3 and 5 g/m^3 [92]. Pobeheim et al. observed an increase in CH_4 production of 25% at day 25 of operation following addition of 10.6 μM Ni [100]. Ni addition of 1–200 μM enhanced the methane production from anaerobic conversion of acetate by 6.30–44.6% compared with the control, respectively [101]. Furthermore, the limitation of Ni in the fermenters led to process instability and was proven to reduce biogas generation [102]. On the other hand, the addition of Ni was found to be beneficial to the methanation process. Ni had increased the ratio of CH_4:CO_2 [103].

Besides Fe, Ni, and Co, other trace metals like Cu, Cr, and Cd were shown to promote biogas production. Lower concentrations of Cu (1.82 ± 0.01 μg/g dry wt.) and Cr (0.89 ± 0.04 μg/g dry wt.) better served as micronutrients for methanogenic bacteria and might have enhanced the process of methanogenesis and thus CH_4 content in the product biogas [9]. Cao et al. harvested five types of plant from Cu-contaminated land, including *Phytolacca americana* L., *Zea mays* L., *Brassica napus* L., *Elsholtzia splendens*, and *Oenothera biennis* L. and investigated the effects of Cu on anaerobic digestion of these plants. Compared to normal plants with low Cu content, the plants used in remediation with increased Cu levels (100 mg/kg) not only required a shorter anaerobic digestion time but also increased the CH_4 content in biogas [25]. 30 and 100 mg/L Cu^{2+} addition increased the cumulative biogas yields by up to 43.62 and 20.77%, respectively [22]. In another study, 30, 100, and 500 mg/L Cr^{6+} addition increased the cumulative biogas yields by up to 19.00, 14.85, and 7.68%, respectively, while bringing forward the daily biogas peak yield [21]. Investigations on the anaerobic fermentation of five contaminated crops showed that less than 1 mg/L of Cd in plants promoted or at least had no inhibitory effect on cumulative biogas yields [11]. Jain et al. noted that at low concentrations, Cd and Ni had a favorable effect on the rate of biogas production and its CH_4 content, but with increase in concentrations, the rate of biogas production and CH_4 content decreased [104].

5.2 Biogas compositions

Biogas is composed of CH_4, CO_2, and other trace compositions like H_2. Methanogens using H_2/CO_2 as the matrix usually contain two hydrogenases: one is a hydrogenase that uses coenzyme F_{420} as the electron acceptor called coenzyme F_{420}-reducing hydrogenase, and the other is coenzyme F_{420}-nonreducing hydrog-enase [105]. Hence, the presence of metals in the fermenters influences the biogas composition by impacting the pathways. For example, Cu^{2+} and Cr^{6+} addition stimulated biogas production and the generation of CH_4 by enhancing the activities of coenzyme F_{420} and methanogenesis [21, 22].

Previously, Fe was found in acetyl-CoA synthase, CH_4 monooxygenase, NO-reductase, and nitrite reductase [15]. Fe, together with Ni, was found in hydrogenases of *Methanosarcina barkeri*, which consumes H_2 to provide electrons for the reduction of CO_2 to CH_4 [106, 107].

As we studied, the compositions of biogas varied with temperatures in the presence of heavy metals. The impacts of Cd addition on biogas compositions of anaerobic co-digestion of acid-pretreated corn and fresh cow dung under different fermentation temperatures are shown in **Figure 1**. When temperature increased, the CH_4, CO_2, and H_2 contents also increased, but N_2 contents decreased in both Cd-added and control groups. The CH_4 contents reached plateau after the fourth day in 55°C group and seventh day in 45°C group (**Figure 1A**). The increase of CO_2 contents slowed down after the fourth day in both 55 and 45°C groups (**Figure 1B**). The tendency of CH_4 and CO_2 contents in 35 and 25°C groups was not detected fully as the biogas yields were too low to be collected by gas bag. Similar contents of H_2 were observed in 45, 35, and 25°C groups while lower in 55°C group (**Figure 1C**). The N_2 contents decreased more rapidly when fermentation temperature was increased (**Figure 1D**). The results indicated that elevated temperatures accelerated the start-up of fermentation.

Cd addition improved the CH_4 contents by approximately 6% after the fourth day in 55°C group. Taking the other biogas compositions into account, it was found that Cd addition decreased the CO_2 contents in the biogas while having little influences on H_2 and N_2 contents. However, the impact of Cd on biogas compositions in other three temperature groups was not significant. Therefore, thermophilic fermentation (55°C) promoted the CH_4 generation in the presence of Cd in the present study. Low temperature hindered the production of CH_4 which agreed with a previous study that used swine manure as substrate [108].

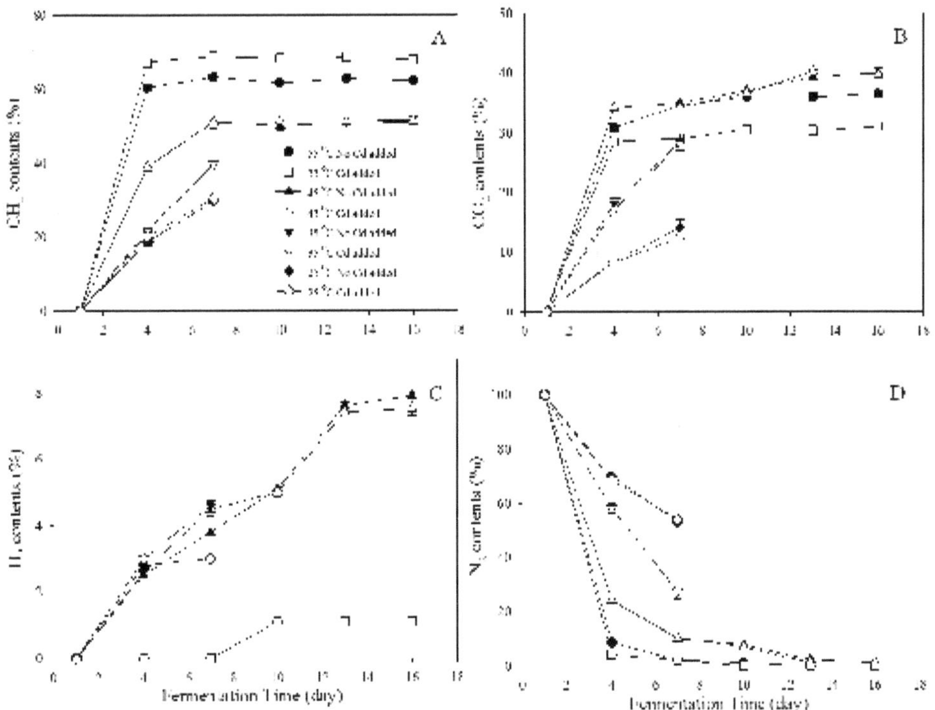

Figure 1.
Biogas composition under Cd stress with fermentation temperature of 55, 45, 35, and 25°C (A) CH_4 contents, (B) CO_2 contents, (C) H_2 contents, and (D) N_2 contents.

6. Conclusions

This book chapter reviewed the past findings in the impacts of metals on different stages of the anaerobic degradation process. The requirements of metals by the enzymes involved in the anaerobic process resulted in the different performances with varied metal species and bioavailability. In general, metals in certain concentrations were able to promote the lignocellulose degradation, the generation and consumption of organic components in the fermenters like LCFAs and VFAs, the biogas production, as well as the CH_4 contents. The mechanisms of metals were studied by many scientists, focusing on the enzyme activities, microbial communities, etc.

Although a large amount of research has been carried out on individual stages or the entire anaerobic fermentation process, there are still challenges on controlling metal-stressed anaerobic degradation process for optimal utilization of metal-contaminated biowastes. Further work on bioavailability of metals during anaerobic fermentation process and detailed compositions of intermediary products, the relationships between microbial functions and metal species, etc. are recommended for better understanding of the metal-stressed anaerobic degradation process.

Acknowledgements

This work was funded by the Major Science and Technology Program for Water Pollution Control and Treatment (No. 2017ZX07101-003, 2015ZX07204-007, 2015ZX07203-011), the Fundamental Research Funds for the Central Universities (2018MS051).

Author details

Yonglan Tian[1*], Huayong Zhang[1] and Edmond Sanganyado[2]

1 Research Center for Engineering Ecology and Nonlinear Science,
North China Electric Power University, Beijing, China

2 Marine Biology Institute, Shantou University, Shantou, Guangdong, China

*Address all correspondence to: yonglantian@ncepu.edu.cn

References

[1] Fu FL, Wang Q. Removal of heavy metal ions from wastewaters: A review. Journal of Environmental Management. 2011;**92**:407-418. DOI: 10.1016/j. jenvman.2010.11.011

[2] Bauddh K, Singh RP. Cadmium tolerance and its phytoremediation by two oil yielding plants *Ricinus communis* (L.) and *Brassica juncea* (L.) from the contaminated soil. International Journal of Phytoremediation. 2012;**14**:772-785. DOI: 10.1080/15226514.2011.619238

[3] Lakshmi PM, Jaison S, Muthukumar T, Muthukumar M. Assessment of metal accumulation capacity of *Brachiaria ramosa* collected from cement waste dumping area for the remediation of metal contaminated soil. Ecological Engineering. 2013;**60**:96-98. DOI: 10.1016/j.ecoleng.2013.07.043

[4] Sas-Nowosielska A, Kucharski R, Małkowski E, Pogrzeba M, Kuperberg JM, Kryński K. Phytoextraction crop disposal-an unsolved problem. Environmental Pollution. 2004;**128**:373-379. DOI: 10.1016/j.envpol.2003.09.012

[5] Ghosh M, Singh SP. A review on phytoremediation of heavy metals and utilization of it's by products. Applied Ecology and Environmental Research. 2005;**6**:214-231. DOI: 10.15666/aeer/0301_001018

[6] Carrier M, Loppinet-Serani A, Absalon C, Marias F, Aymonier C, Mench M. Conversion of fern (*Pteris vittata* L.) biomass from a phytoremediation trial in sub-and supercritical water conditions. Biomass and Bioenergy. 2011;**35**:872-883. DOI: 10.1016/j.biombioe.2010.11.007

[7] Kobayashi T, Wu YP, Lu ZJ, Xu KQ. Characterization of anaerobic degradability and kinetics of harvested submerged aquatic weeds used for nutrient phytoremediation. Energies. 2015;**8**:304-318. DOI: 10.3390/en8010304

[8] Yue Z-B, Yu H-Q , Wang Z-L. Anaerobic digestion of cattail with rumen culture in the presence of heavy metals. Bioresource Technology. 2007;**98**:781-786. DOI: 10.1016/j. biortech.2006.03.017

[9] Verma VK, Singh YP, Rai JPN. Biogas production from plant biomass used for phytoremediation of industrial wastes. Bioresource Technology. 2007;**98**:1664-1669. DOI: 10.1016/j. biortech.2006.05.038

[10] Zhang HY, Tian YL, Wang LJ, Zhang LY, Dai LM. Ecophysiological characteristics and biogas production of cadmium-contaminated crops. Bioresource Technology. 2013;**146**:628-636. DOI: 10.1016/j. biortech.2013.07.148

[11] Tian YL, Zhang HY. Producing biogas from agricultural residues generated during phytoremediation process: Possibility, threshold, and challenges. International Journal of Green Energy. 2016;**13**:1556-1563. DOI: 10.1080/15435075.2016.1206017

[12] van Haandel AC, der Lubbe JGM, editors. Handbook Biological Wastewater Treatment-Design and Optimization of Activated Sludge Systems. London: IWA Publishing; 2007. pp. 377-380. DOI: 10.2166/9781780400808

[13] Manyiloh CE, Mamphweli SN, Meyer EL, Okoh AI, Makaka G, Simon M. Microbial anaerobic digestion (bio-digesters) as an approach to the decontamination of animal wastes in pollution control and the generation of renewable energy. International Journal of Environmental Research and Public Health. 2013;**10**:4390-4417. DOI: 10.3390/ijerph10094390

[14] Wyman V, Serrano A, Fermoso FG, Villa Gomez DK. Trace elements effect on hydrolytic stage towards biogas production of model lignocellulosic substrates. Journal of Environmental Management. 2019;**234**:320-325. DOI: 10.1016/j.jenvman.2019.01.015

[15] Zandvoort BMH, Van HED, Fermoso FG, Lens PNL. Trace metals in anaerobic granular sludge reactors: Bioavailability and dosing strategies. Engineering in Life Sciences. 2006;**6**:293-301. DOI: 10.1002/elsc.200620129

[16] Bayer EA, Lamed R, Himmel ME. The potential of cellulases and cellulosomes for cellulosic waste management. Current Opinion in Biotechnology. 2007;**18**:237-245. DOI: 10.1016/j.copbio.2007.04.004

[17] Mudhoo A, Kumar S. Effects of heavy metals as stress factors on anaerobic digestion processes and biogas production from biomass. International Journal of Environmental Science and Technology. 2013;**10**:1383-1398. DOI: 10.1007/ s13762-012-0167-y

[18] Barua VB, Goud VV, Kalamdhad AS. Microbial pretreatment of water hyacinth for enhanced hydrolysis followed by biogas production. Renewable Energy. 2018;**126**:21-29. DOI: 10.1016/j.renene.2018.03.028

[19] Taherzadeh MJ, Karimi K. Pretreatment of lignocellulosic wastes to improve ethanol and biogas production: A review. International Journal of Molecular Sciences. 2008;**9**:1621-1651. DOI: 10.3390/ijms9091621

[20] Pokój T, Bułkowska K, Gusiatin ZM, Klimiuk E, Jankowski KJ. Semi-continuous anaerobic digestion of different silage crops: VFAs formation, methane yield from fiber and non-fiber components and digestate composition. Bioresource Technology. 2015;**190**:201-210. DOI: 10.1016/j.biortech.2015.04.060

[21] Zhang HY, Han XX, Tian YL, Li Y, Yang K, Hao H, et al. Process analysis of anaerobic fermentation of *Phragmites australis* straw and cow dung exposing to elevated chromium (VI) concentrations. Journal of Environmental Management. 2018;**224**:414-424. DOI: 10.1016/j.jenvman.2018.07.058

[22] Hao H, Tian YL, Zhang HY, Chai Y. Copper stressed anaerobic fermentation: Biogas properties, process stability, biodegradation and enzyme responses. Biodegradation. 2017;**28**:369-381. DOI: 10.1007/s10532-017-9802-0

[23] Kayhanian M, Rich D. Pilot-scale high solids thermophilic anaerobic digestion of municipal solid waste with an emphasis on nutrient requirements. Biomass and Bioenergy. 1995;**8**:433-444. DOI: 10.1016/0961-9534(95)00043-7

[24] Lenártová V, Holovská K, Javorský P. The influence of mercury on the antioxidant enzyme activity of rumen bacteria *Streptococcus bovis* and *Selenomonas ruminantium*. FEMS Microbiology Ecology. 1998;**27**:319-325. DOI: 10.1016/S0168-6496(98)00077-4

[25] Cao Z, Wang S, Wang T, Chang Z, Shen Z, Chen Y. Using contaminated plants involved in phytoremediation for anaerobic digestion. International Journal of Phytoremediation. 2015;**17**:201-207. DOI: 10.1080/15226514.2013.876967

[26] Lawrence AW, McCarty PL. The role of sulfide in preventing heavy metal toxicity in anaerobic treatment. Water Pollution Control Federation. 1965;**37**:392-406. DOI: 10.2307/25035257

[27] Alkan U, Anderson GK, Ince O. Toxicity of trivalent chromium in the anaerobic digestion process. Water

Research. 1996;**30**:731-741. DOI: 10.1016/ 0043-1354(95)00181-6

[28] Kouzeli-Katsiri A, Kartsonas N, Priftis A. Assessment of the toxicity of heavy metals to the anaerobic digestion of sewage sludge. Environmental Technology Letters. 1988;**9**:261-270. DOI: 10.1080/09593338809384566

[29] Hayes TD, Theis TL. The distribution of heavy metals in anaerobic digestion. Water Pollution Control Federation. 1978;**50**:61-72. DOI: 10.2307/25039507

[30] Karri S, Sierra-alvarez R, Field JA. Toxicity of copper to acetoclastic and hydrogenotrophic activities of methanogens and sulfate reducers in anaerobic sludge. Chemosphere. 2006;**62**:121-127. DOI: 10.1016/j. chemosphere.2005.04. 016

[31] Glass JB, Orphan VJ. Trace metal requirements for microbial enzymes involved in the production and consumption of methane and nitrous oxide. Frontiers in Microbiology. 2012;**3**:1-20. DOI: 10.3389/ fmicb.2012.00061

[32] Jackson-Moss CA, Duncan JR. The effect of iron on anaerobic digestion. Biotechnology Letters. 1990;**12**:149-154. DOI: 10.1007/bf01022433

[33] Bayr S, Pakarinen O, Korppoo A, Liuksia S, Väisänen A, Kaparaju P, et al. Effect of additives on process stability of mesophilic anaerobic monodigestion of pig slaughterhouse waste. Bioresource Technology. 2012;**120**:106-113. DOI: 10.1016/j.biortech.2012.06.009

[34] Moestedt J, Nordell E, Shakeri Yekta S, Lundgren J, Martí M, Sundberg C, et al. Effects of trace element addition on process stability during anaerobic co-digestion of OFMSW and slaughterhouse waste. Waste Management. 2016;**47**:11-20. DOI: 10.1016/j.wasman.2015.03.007

[35] Nordell E, Nilsson B, Påledal SN, Karisalmi K, Moestedt J. Co-digestion of manure and industrial waste-the effects of trace element addition. Waste Management. 2016;**47**:21-27. DOI: 10.1016/j.wasman.2015.02.032

[36] Scherer P, Lippert H, Wolff G. Composition of the major elements and trace elements of 10 methanogenic bacteria determined by inductively coupled plasma emission spectrometry. Biological Trace Element Research. 1983;**5**:149-163. DOI: 10.1007/ BF02916619

[37] Anuj K, Miglani P, Gupta RK, Bhattacharya TK. Impact of Ni(II), Zn(II) and Cd(II) on biogassification of potato waste. Journal of Environmental Biology. 2006;**27**:61-66. DOI: 10.2112/05A-0018.1

[38] Patel VB, Patel AR, Patel MC, Madamwar DB. Effect of metals on anaerobic digestion of water hyacinth-cattle dung. Applied Biochemistry and Biotechnology. 1993;**43**:45-50. DOI: 10.1007/bf02916429

[39] Zhang R, Wang X, Gu J, Zhang Y. Influence of zinc on biogas production and antibiotic resistance gene profiles during anaerobic digestion of swine manure. Bioresource Technology. 2017;**244**:63-70. DOI: 10.1016/j. biortech.2017.07.032

[40] Sawers RG. Nickel in bacteria and archaea. In: Kretsinger RH, Uversky VN, Permyakov EA, editors. Encyclopedia of Metalloproteins. New York: Springer-Verlag; 2013. pp. 1490-1496. DOI: 10.1007/978-1-4614-1533-6_86

[41] Ragsdale SW. Nickel-based enzyme systems. The Journal of Biological Chemistry. 2009;**284**:18571-18575. DOI: 10.1074/jbc.R900020200

[42] Chen Y, Cheng JJ, Creamer KS. Inhibition of anaerobic digestion process: A review. Bioresource

Technology. 2008;**99**:4044-4064. DOI: 10.1016/j.biortech. 2007.01.057

[43] Ahring BK, Westermann P. Sensitivity of thermophilic methanogenic bacteria to heavy metals. Current Microbiology. 1985;**12**:273-276. DOI: 10.1007/ bf01567977

[44] Tabatabaei M, Sulaiman A, Nikbakht AM, Yusof N, Najafpour G. Influential parameters on biomethane generation in anaerobic wastewater treatment plants. In: Manzanera M, editor. Alternative Fuel. London: IntechOpen; 2011. pp. 227-262. DOI: 10.5772/24681

[45] Chakraborty N, Chatterjee M, Sarkar GM, Lahiri SC. Inhibitory effects of the divalent metal ions on biomethanation by isolated mesophilic methanogen in AC21 medium in presence or absence of juices from water hyacinth. Bioenergy Research. 2010;**3**:314-320. DOI: 10.1007/ s12155-010-9083-5

[46] Zhang HY, Tian YL, Wang LJ, Mi XY, Chai Y. Effect of ferrous chloride on biogas production and enzymatic activities during anaerobic fermentation of cow dung and *Phragmites* straw. Biodegradation. 2016;**27**:69-82. DOI: 10.1007/ s10532-016-9756-7

[47] Vlyssides A, Barampouti EM, Mai S. Influence of ferrous iron on the granularity of a UASB reactor. Chemical Engineering Journal. 2009;**146**:49-56. DOI: 10.1016/j.cej.2008.05.011

[48] Tian YL, Zhang HY, Chai Y, Wang LJ, Mi XY, Zhang LY, et al. Biogas properties and enzymatic analysis during anaerobic fermentation of *Phragmites australis* straw and cow dung: Influence of nickel chloride supplement. Biodegradation. 2017;**28**:15-25. DOI: 10.1007/s10532-016-9774-5

[49] Takashima M, Speece RE. Mineral nutrient requirements for high-rate methane fermentation of acetate at low SRT. Research Journal of the Water Pollution Control Federation. 1989;**61**:1645-1650. DOI: 10.1016/0034-4257(89)90059-X

[50] Takashima M, Shimada K, Speece RE. Minimum requirements for trace metals (iron, nickel, cobalt, and zinc) in thermophilic and mesophilic methane fermentation from glucose. Water Environment Research. 2011;**83**:339-346. DOI: 10.2175/106143010x12780288628895

[51] Kayhanian M. Ammonia inhibition in high-solids biogasification: An overview and practical solutions. Environmental Technology. 1999;**20**:355-365. DOI: 10.1080/09593332008616828

[52] Bitton G, editor. Wastewater Microbiology. 3rd ed. Chichester: Wiley; 2005. 1132 p. DOI: 10.1146/ annurev. mi.30.100176.001403

[53] Lin CY, Chen CC. Effect of heavy metals on the methanogenic UASB granule. Water Research. 1999;**33**:409-416. DOI: 10.1016/ s0043-1354(98)00211-5

[54] Rensing C, Maier RM. Issues underlying use of biosensors to measure metal bioavailability. Ecotoxicology and Environmental Safety. 2003;**56**:140-147. DOI: 10.1016/s0147-6513(03)00057-5

[55] Gonzalez-Gil G, Lopes SI, Saikaly PE, Lens PN. Leaching and accumulation of trace elements in sulfate reducing granular sludge under concomitant thermophilic and low pH conditions. Bioresource Technology. 2012;**126**:238-246. DOI: 10.1016/j. biortech.2012.09.044

[56] Zandvoort MH, Van Hullebusch ED, Peerbolte A, Golubnic S, Lettinga G, Lens PNL. Influence of pH shocks on trace metal dynamics and performance of methanol fed granular

sludge bioreactors. Biodegradation. 2005;**16**:549-567. DOI: 10.1007/s10532-004-7789-9

[57] Lo HM, Lin KC, Liu MH, Pai TZ, Lin CY, Liu WF, et al. Solubility of heavy metals added to MSW. Journal of Hazardous Materials. 2009;**161**:294-299. DOI: 10.1016/j.jhazmat.2008.03.119

[58] Zhang M, Yang C, Jing Y, Li J. Effect of energy grass on methane production and heavy metal fractionation during anaerobic digestion of sewage sludge. Waste Management. 2016;**58**:316-323. DOI: 10.1016/j.wasman.2016.09.040

[59] Rajagopal R, Massé DI, Singh G. A critical review on inhibition of anaerobic digestion process by excess ammonia. Bioresource Technology. 2013;**143**:632-641. DOI: 10.1016/j.biortech.2013.06.030

[60] Liu T, Sung S. Ammonia inhibition on thermophilic aceticlastic methanogens. Water Science and Technology. 2002;**45**:113-120. DOI: 10.2166/wst.2002.0304

[61] Math-Alvarez J, Mtz.-Viturtia A, Llabrés-Luengo P, Cecchi F. Kinetic and performance study of a batch two-phase anaerobic digestion of fruit and vegetable wastes. Biomass and Bioenergy. 1993;**5**:481-488. DOI: 10.1016/0961-9534(93) 90043-4

[62] Tian H, Karachalios P, Angelidaki I, Fotidis IA. A proposed mechanism for the ammonia-LCFA synergetic co-inhibition effect on anaerobic digestion process. Chemical Engineering Journal. 2018;**349**:574-580. DOI: 10.1016/j.cej.2018. 05.083

[63] Wu LJ, Kobayashi T, Kuramochi H, Li YY, Xu KQ, Lv Y. High loading anaerobic co-digestion of food waste and grease trap waste: Determination of the limit and lipid/long chain fatty acid conversion. Chemical Engineering Journal. 2018;**338**:422-431. DOI: 10.1016/j.cej.2018.01.041

[64] Stoll U, Gupta H. Management strategies for oil and grease residues. Waste Management and Research. 1997;**15**:23-32. DOI: 10.1177/0734242X9701500103

[65] Jezzard P, LeBihan D, Cuenod C, Pannier L, Prinster A, Turner R. An investigation of the contributions of physiological noise in human functional MRI studies at 1.5 Tesla and 4 Tesla. In: Proceedings of Society of Magnetic Resonance in Medicine (SMRM), 12th Annual Meeting; 14-20 August 1993; New York. Berlin: Springer; 1993. p. 1392

[66] Lalman JA, Bagley DM. Anaerobic degradation and inhibitory effects of linoleic acid. Water Research. 2000;**34**:4220-4228. DOI: 10.1016/s0043-1354(00)00180-9

[67] Cirne DG, Paloumet X, Björnsson L, Alves MM, Mattiasson B. Anaerobic digestion of lipid-rich waste-effects of lipid concentration. Renewable Energy. 2007;**32**:965-975. DOI: 10.1016/j. renene.2006.04.003

[68] Pereira MA, Pires OC, Mota M, Alves MM. Anaerobic biodegradation of oleic and palmitic acids: Evidence of mass transfer limitations caused by long chain fatty acid accumulation onto the anaerobic sludge. Biotechnology and Bioengineering. 2005;**92**:15-23. DOI: 10.1002/bit.20548

[69] Angelidaki I, Ahring BK. Effects of free long-chain fatty acids on thermophilic anaerobic digestion. Applied Microbiology and Biotechnology. 1992;**37**:808-812. DOI: 10.1007/bf00174850

[70] Hwu CS, Donlon B, Lettinga G. Comparative toxicity of long-chain fatty acid to anaerobic sludges from various origins. Water Science and

Technology. 1996;**34**:351-358. DOI: 10.1016/0273-1223(96)00665-8

[71] Lalman J, Bagley DM. Effects of C18 long chain fatty acids on glucose, butyrate and hydrogen degradation. Water Research. 2002;**36**:3307-3313. DOI: 10.1016/s0043-1354(02)00014-3

[72] Alves MM, Mota Vieira JA, Alvares Pereira RM, Pereira MA, Mota M. Effects of lipids and oleic acid on biomass development in anaerobic fixed-bed reactors. Part II: Oleic acid toxicity and biodegradability. Water Research. 2001;**35**:264-270. DOI: 10.1016/S0043-1354(00)00242-6

[73] Ma J, Zhao QB, Laurens LLM, Jarvis EE, Nagle NJ, Chen S, et al. Mechanism, kinetics and microbiology of inhibition caused by long-chain fatty acids in anaerobic digestion of algal biomass. Biotechnology for Biofuels. 2015;**8**:141. DOI: 10.1186/s13068-015-0322-z

[74] Chan PC, de Toledo RA, Iu HI, Shim H. Effect of zinc supplementation on biogas production and short/long chain fatty acids accumulation during anaerobic co-digestion of food waste and domestic wastewater. Waste and Biomass Valorization. 2018:1-11. DOI: 10.1007/s12649-018-0323-9

[75] Wang CT, Yang CMJ, Chen ZS. Rumen microbial volatile fatty acids in relation to oxidation reduction potential and electricity generation from straw in microbial fuel cells. Biomass and Bioenergy. 2012;**37**:318-329. DOI: 10.1016/j.biombioe. 2011.09.016

[76] Oswald WJ, Golueke CG, Copper RC, Gee HK, Bronson JC. Water reclamation, algal production and methane fermentation in waste ponds. In: Advances in Water Pollution Research, Proceedings of the International Conference; September 1962; London. London: Pergamon Press LTD; 1964. pp. 119-157. DOI: 10.1016/ B978-1-4832-8391-3.50029-4

[77] Kong IC, Hubbard JS, Jones WJ. Metal-induced inhibition of anaerobic metabolism of volatile fatty acids and hydrogen. Applied Microbiology and Biotechnology. 1994;**42**:396-402. DOI: 10.1007/ BF00902748

[78] Yenigün O, Kizilgün F, Yilmazer G. Inhibition effects of zinc and copper on volatile fatty acid production during anaerobic digestion. Environmental Technology. 1996;**17**:1269-1274. DOI: 10.1080/09593331708616497

[79] Osuna MB, Zandvoort MH, Iza JM, Lettinga G, Lens PNL. Effects of trace element addition on volatile fatty acid conversions in anaerobic granular sludge reactors. Environmental Technology. 2003;**24**:573-587. DOI: 10.1080/09593330 309385592

[80] Lee SJ. Relationship between oxidation reduction potential (ORP) and volatile fatty acid (VFA) production in the acid-phase anaerobic digestion process [thesis]. New Zealand: The University of Canterbury; 2008

[81] Schmidt T, Nelles M, Scholwin F, Pröter J. Trace element supplementation in the biogas production from wheat stillage-optimization of metal dosing. Bioresource Technology. 2014;**168**:80-85. DOI: 10.1016/j.biortech.2014.02.124

[82] Fermoso FG, Collins G, Bartacek J, O'Flaherty V, Lens P. Role of nickel in high rate methanol degradation in anaerobic granular sludge bioreactors. Biodegradation. 2008;**19**:725-737. DOI: 10.1007/s10532-008-9177-3

[83] Fermoso FG, Collins G, Bartacek J, Lens PNL. Zinc deprivation of methanol fed anaerobic granular sludge bioreactors. Journal of Industrial Microbiology and Biotechnology.

2008;**35**:543-557. DOI: 10.1007/s10295-008-0315-z

[84] Paula LP, Bo J, Denise C, Alfons JMS, Gatze L. Effect of cobalt on the anaerobic thermophilic conversion of methanol. Biotechnology and Bioengineering. 2010;**85**:434-441. DOI: 10.1002/bit.10876

[85] Veeken A, Kalyuzhnyi S, Scharff H, Hamelers B. Effect of pH and VFA on hydrolysis of organic solid waste. Journal of Environmental Engineering. 2000;**126**:1076-1081. DOI: 10.1061/(asce)0733-9372(2000)126:12(1076

[86] Kavamura VN, Esposito E. Biotechnological strategies applied to the decontamination of soils polluted with heavy metals. Biotechnology Advances. 2010;**28**:61-69. DOI: 10.1016/j.biotechadv.2009.09.002

[87] Lin CY. Effect of heavy metals on volatile fatty acid degradation in anaerobic digestion. Water Research. 1992;**26**:177-183. DOI: 10.1016/0043-1354(92)90217-r

[88] Lin CY, Chou J, Lee YS. Heavy metal-affected degradation of butyric acid in anaerobic digestion. Bioresource Technology. 1998;**65**:159-161. DOI: 10.1016/ s0960-8524(98)00022-4

[89] Moonil K, Young-Ho A, Speece RE. Comparative process stability and efficiency of anaerobic digestion; mesophilic vs. thermophilic. Water Research. 2002;**36**:4369-4385. DOI: 10.1016/s0043-1354(02)00147-1

[90] White CJ, Stuckey DC. The influence of metal ion addition on the anaerobic treatment of high strength, soluble wastewaters. Environmental Technology. 2000;**21**:1283-1292. DOI: 10.1080/09593332108618157

[91] Bhattacharya SK, Qu M, Madura RL. Effects of nitrobenzene and zinc on acetate utilizing methanogens. Water Research. 1996;**30**:3099-3105. DOI: 10.1016/s0043-1354(96)00194-7

[92] Takashima M, Speece RE, Parkin GF. Mineral requirements for methane fermentation. Critical Reviews in Environmental Control. 1990;**19**:465-479. DOI: 10.1080/10643389009388378

[93] Zerkle AL, House CH, Brantley SL. Biogeochemical signatures through time as inferred from whole microbial genomes. American Journal of Science. 2005;**305**:467-502. DOI: 10.2475/ajs.305.6-8.467

[94] Shima S, Warkentin E, Thauer RK, Ermler U. Structure and function of enzymes involved in the methanogenic pathway utilizing carbon dioxide and molecular hydrogen. Journal of Bioscience and Bioengineering. 2002;**93**:519-530. DOI: 10.1016/s1389-1723(02)80232-8

[95] Bainotti AE, Nishio N. Growth kinetics of *Acetobacterium* sp . On methanol-formate in continuous culture. Journal of Applied Microbiology. 2000;**88**:191-201. DOI: 10.1046/j.1365-2672.2000.00854.x

[96] Thauer RK, Kaster A, Goenrich M, Schick M, Hiromoto T, Shima S. Hydrogenases from Methanogenic Archaea, nickel, a novel cofactor, and H_2 storage. Annual Review of Biochemistry. 2010;**79**:507-536. DOI: 10.1146/annurev.biochem.030508.152103

[97] Hoban DJ, BERG L. Effect of iron on conversion of acetic acid to methane during methanogenic fermentations. The Journal of Applied Bacteriology. 1979;**47**:153-159. DOI: 10.1111/j.1365-2672.1979.tb01179.x

[98] Rao PP, Seenayya G. Improvement of methanogenesis from cow dung and poultry litter waste digesters by addition of iron. World Journal of Microbiology

and Biotechnology. 1994;**10**:211-214. DOI: 10.1007/BF00360890

[99] Ram MS, Singh L, Suryanarayana MVS, Alam SI. Effect of iron, nickel and cobalt on bacterial activity and dynamics during anaerobic oxidation of organic matter. Water, Air, and Soil Pollution. 2000;**117**:305-312. DOI: 10.1023/A: 1005100924609

[100] Pobeheim H, Munk B, Johansson J, Guebitz GM. Influence of trace elements on methane formation from a synthetic model substrate for maize silage. Bioresource Technology. 2010;**101**:836-839. DOI: 10.1016/j.biortech.2009.08. 076

[101] Qing-hao H, Xiu-fen L, He L, Guo-cheng D, Jian C. Enhancement of methane fermentation in the presence of Ni^{2+} chelators. Biochemical Engineering Journal. 2008;**38**:98-104. DOI: 10.1016/j. bej.2007.07.002

[102] Pobeheim H, Munk B, Lindorfer H, Guebitz GM. Impact of nickel and cobalt on biogas production and process stability during semi-continuous anaerobic fermentation of a model substrate for maize silage. Water Research. 2011;**45**:781-787. DOI: 10.1016/j.watres.2010.09.001

[103] Aresta M, Narracci M, Tommasi I. Influence of iron, nickel and cobalt on biogas production during the anaerobic fermentation of fresh residual biomass. Chemistry and Ecology. 2003;**19**:451-459. DOI: 10.1080/0275754031000 1629134

[104] Jain SK, Gujral GS, Jha NK, Vasudevan P. Production of biogas from *Azolla pinnata* r.Br and *Lemna minor* L.: Effect of heavy metal contamination. Bioresource Technology. 1992;**41**:273-277. DOI: 10.1016/0960-8524(92)90013-N

[105] Rother M. Selenoproteins in prokaryotes. In: Kretsinger RH, Uversky VN, Permyakov EA, editors. Encyclopedia of Metalloproteins. New York: Springer-Verlag; 2013. DOI: 10.1007/978-1-4614-1533-6_466

[106] Michel R, Massanz C, Kostka S, Richter M, Fiebig K. Biochemical characterization of the 8-hydroxy- 5-deazaflavin-reactive hydrogenase from *Methanosarcina barkeri* fusaro. European Journal of Biochemistry. 1995;**233**:727-735. DOI: 10.1111/ j.1432-1033.1995.727_3.x

[107] Hausinger RP. Nickel enzymes in microbes. The Science of the Total Environment. 1994;**148**:157-166. DOI: 10.1016/0048-9697(94)90392-1

[108] Zhu Z, Cheng G, Zhu Y, Zeng H, Wei R, Wei C. The effects of different anaerobic fermentation temperature on biogas fermentation of swine manure. In: 2011 International Conference on Computer Distributed Control and Intelligent Environmental Monitoring; 19-20 February 2011; Changsha. New York: IEEE; 2011. pp. 1410-1413. DOI: 10.1109/CDCIEM.2011.117

6

Development of an Anaerobic Digestion Screening System Using 3D-Printed Mini-Bioreactors

Spyridon Achinas and Gerrit Jan Willem Euverink

Abstract

This study incorporated the concept of mini-bioreactors by employing additive manufacturing procedures. Limitations in experimental studies with large-scale equipment favor the use of mini-reactor systems and help to understand the phenomena of its large-scale counterpart better. 3D printing enables to reproduce the reaction engineering principles in a low-cost and ease of manufacture way and expedites the development of novel prototypes. Small anaerobic digesters of 40 mL were designed and fabricated to investigate the effect of downscaling on the stability and performance of the anaerobic digestion process. Baseline tests were conducted using a commercial 400-mL stirred bioreactor as reference for further comparison and validation. Miniature bioreactors showed similar stability and conversion efficiency. However, the biogas production rate and methane content of the 3D-printed bioreactors were lower than those in the baseline study bioreactors. Finally, 3D-printed systems were linked with efficient performances and are considered as an excellent opportunity for analyzing microbe-mediated bioenergy systems. This study demonstrated the high potential of miniaturized bioreactors as a process screening tool.

Keywords: miniaturization, anaerobic digester, 3D printing, biogas, screening system

1. Introduction

The transition from oil-based energy to bioenergy can strive for technological efforts in this direction. Anaerobic digesters are widely used to treat organic waste to produce high-value products (biogas and biofertilizers) with the help of micro-organisms [1, 2]. Currently, bioreactors occur in many different types. The sizes of these reactors can vary over several orders of magnitude, from mini-bioreactors (1–10 mL) to plant-scale reactors (2–500 m^3) [3–6]. The need for automated multi-parallel mini-bioreactor systems is becoming more prominent. It occurs that no device is capable yet of meeting all the challenges of miniaturizing large-scale processes while keeping the functionality of conventional bioreactors [7–9]. As the bench-scale bioreactors are expensive, 3D printing can be considered as an alternative solution for the fabrication of miniaturized bioreactor systems [10–15]. The smaller the reactor, the more efficient it can become in terms of experimental throughput [16]. This highlights the need for automated multi-parallel mini-bioreactor systems [17–19]. The pilot-scale reactors are often considered impractical since they require more feedstock, space, and energy than the mini-scale reactors. This makes the commercial bioreactors expensive, unrealistic, and inefficient as a process screening method. Additionally, the current state-of-the-art mini- bioreactors do not apply to complex microbial systems (e.g., the biogas production through anaerobic digestion) [20, 21].

A downsized approach of anaerobic digestion using mini-digesters is presented. In this study, 40 mL bioreactors were designed, fabricated, and operated to evaluate the anaerobic digestion performance and stability at a small scale. The start-up and operation of the mini-bioreactors were investigated. The results demonstrated that AD in low working volumes was feasible and efficient in terms of biogas quantity and quality. The results also established links between scale-down and process stability.

2. Materials and methods

2.1 Miniaturization concept

It may not be out of place to look into the reasons behind the ongoing technological revolution based on miniaturization. A higher degree of intelligence can be achieved by drastically increasing the amount of sensory data (by many orders of magnitude) obtained from a large number of variable fermentation experiments. High-throughput screening of fermentations demands that the bioreactors, as well as the sensors, are miniaturized so that a large number of these can be accommodated in small areas and at the same time that neither the cost nor the energy consumption exceeds acceptable limits.

When all aspects of the bioreactor scale in a similar way, the geometric integrity is maintained with the downsizing. Such type of scaling is called "isomorphic" (or "isometric") scaling [22]. On the other hand, if different elements of a system with different functionalities do not scale similarly, the scaling is called "allometric" scaling [23, 24]. Scaling laws deal with the structural and functional consequences of changes in size or scale among otherwise similar structures/organisms; thus, only through the scaling laws a designer becomes aware of physical consequences of downscaling devices and systems. Scaling effects on problems of mechanics are significant and are essential to take into account while designing systems at mini scales [24].

2.2 Miniature bioreactor design

Computer-aided design (CAD) representations were created in AutoCAD Fusion, and drawings with the external and internal dimensions and design parameters are presented in **Figure 1**. The mini-bioreactor device was designed with four ports; each one corresponds to a different function (influent, effluent, gas exit, pH electrode).

Close attention was paid to the aspect ratio of the bioreactor. In general, the aspect ratio of a vessel (the ratio between its height and its diameter) should be 1:1 at the working volume for cell culture and 2.2:1 at the working volume for microbial systems [24].

2.3 Miniature bioreactor fabrication

After the baseline design, a 40-mL bioreactor was fabricated using a stereolithography (SLA)-based 3D printer as a proof of principle (see **Figure 2**). The CAD files were converted to the STL format, which is a file type that interfaces between CAD software and additive manufacturing platforms. The PREFORM software was used to print the bioreactor. The bioreactors were printed on a Formlabs Vat Polymerization platform (Form 1) using the commercially available Formlabs Clear FLGPCL02 proprietary resin. The lowest resolution available in the machine was employed (0.1 mm) for the printing. A sturdy and rigid device was created layer by layer using a laser which initiated polymerization in the photopolymer resin. The reactor was then extensively cleaned and flushed with isopropyl alcohol (IPA) to avoid after-curing of the resin on the walls and internal channels of the bioreactor.

A post-processing step of fine polishing shortly after fabrication with this resin produced clean and semitransparent bioreactors. This offers the possibility to observe the flow patterns and enables the application of visual techniques and in-line spectroscopy for process characterization.

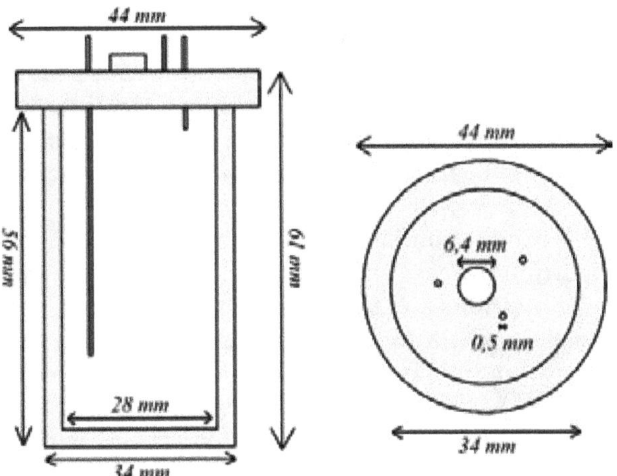

Figure 1.
Internal and external dimensions of the mini-bioreactor.

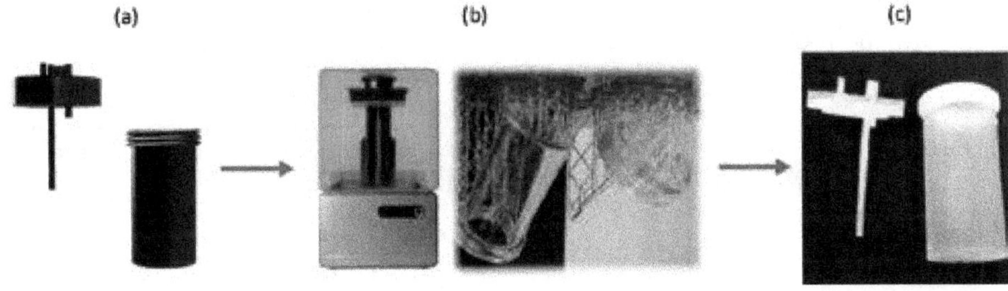

Figure2.
The fabrication steps of an ultra-scale bioreactor. (a) A CAD model was designed to meet the requirements of a suitable bioreactor. (b) The reactor device was fabricated using a Form 1 SLA printer. (c) The reactor was polished, cleaned, and extensively flushed with IPA to remove the residual resin.

In terms of fabrication, the 40-mL reactor, including support structures, required 63 mL of resin and just over 8 h to complete (**Table 1**). The large build platform of the Form 1 printer allows that both the vessel and lid are printed simultaneously reducing manufacturing time.

2.4 Inoculum and substrate

The microbial inoculum for this study was obtained from the wastewater treatment plant in Garmerwolde (Groningen, Netherlands). Anaerobic sludge was collected from an anaerobic digester degrading municipal waste and stored at 6° C. The inoculum was gently homogenized to reduce the size of big particles somewhat. The characteristics of the inoculum and substrate are shown in **Table 2**. Dried milk was used as a constant complex substrate that consists of a mixture of carbohydrates, lipids, proteins, and minerals. Dried milk powder was purchased from the local grocery market. The components of dried milk are carbohydrates (lactose) 39%, butter fat 28.2%, proteins 25.1%, moisture 3%, calcium 930 mg, phosphorus 75 mg, other minerals 3.88 g, vita-min A 636.3μg, vitamin D3 8.8μg, vitamin E 0.8 mg, vitamin B2 1.4 mg, and vitamin B12 1.8 μg.

2.5 Experimental setup

In this study, the development process of the 3D-printed mini-bioreactor consists of the vessel design and fabrication, operation test, and the baseline study. In addition to the manufacturing of the 3D-printed bioreactors, baseline studies involving commercial stirred bioreactors were carried out to examine the process in parallel with the mini-bioreactors. The two setups are schematically described in **Figure 3**.

The daily biogas production rate was determined to evaluate the behavior of the anaerobic digestion process and the stability of the miniature bioreac-tor. The experimental conditions and the content of the reactors are shown in **Table 3**. Biogas composition, pH, and COD reduction have also been employed as valuable parameters for further understanding and evaluation of the micro-r eactor performance. A single-stage semicontinuous process was performed in two 400-mL BioBLU single-use vessels (Eppendorf, USA) with a working volume of 300 mL. The vessel was placed in a temperature-controlled water bath (36°C) and fed once a day. The milk powder suspension was impelled with a syringe pump (AL-1000HP, World Precision Instruments, USA) equipped with a 30-mL syringe (Terumo, inner diameter 23.1 mm) and Teflon tubing (1.37 × 1.07 mm).

Before use, the inoculum (anaerobic sludge) was first incubated anaerobically until no methane production was observed anymore (37°C, 6–7 days). For

Parameter	Units	3D-printed reactor	Commercial reactor
Reactor volume	mL	40 mL	400 mL
Inner diameter	mm	28 mm	62 mm
Inner height	mm	56 mm	124 mm
Resin volume	mL	63 mL	—
Fabrication time	h min	(8 h 11 min)	—

Table 1.
Technical data from bioreactors used in the experiments.

Parameter	Unit	Anaerobic sludge	Milk powder
pH		7.36	—
TS	$g \cdot kg^{-1}$	39.5 ± 1.7	968.4 ± 3.5
VS	$g \cdot kg^{-1}$	27.3 ± 0.4	924.9 ± 2.9
COD	$g \cdot kg^{-1}$	40.7 ± 1.9	1147.6
TVFA	mg acetic acid $\cdot L^{-1}$	716	1400
TA	mg $CaCO_3 \cdot L^{-1}$	5884	3000

Table 2.
Physicochemical characteristics of the inoculum and substrate (influent) used in the experiments.

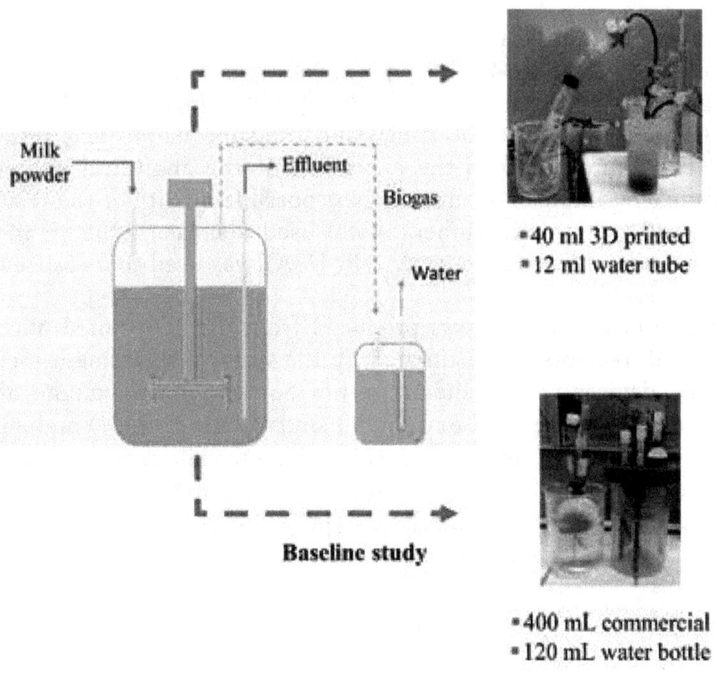

Figure 3.
The validation step of the miniature system consists of a baseline study operating commercial bioreactors and the conceptual research with the 3D-printed mini-bioreactors.

both the mini-bioreactors and the commercial reactors, no additional external nutrients/trace elements were added to the influent as it was assumed that they are sufficiently present in the inoculum and the milk powder. The reactors were mixed twice a day (2 × 5 min) by integrated magnetic stirrers (miniature bioreactors) or by propellers (commercial bioreactors) to achieve a homogenized matrix.

The experiments were carried out in a semicontinuous mode using the water displacement method to measure the biogas production for 95 days. The biogas production rate was based on the volume of biogas produced daily and is defined as mL biogas per g VS_{added} per day.

The bioreactors were filled with sieved anaerobic sludge to provide sufficient consortia of microbes to degrade organic material in the influent. The bioreactors were flushed with N_2-gas for 2 min to achieve anaerobic conditions, placed in a water bath and kept at 36 ± 1°C.

Reactor set	Time (d)	HRT (d)	Temperature (°C)	Organic load rate ($g \cdot VS \cdot (L\ reactor)^{-1} \cdot d^{-1}$)
MR1	95	20	35	0.5
MR2	95	20	35	0.5
CR1	95	20	35	0.5
CR2	95	20	35	0.5

Table 3.
Process conditions and masses of organic materials in experimental tests.

2.6 Analytical methods

Total solid (TS) and volatile solid (VS) contents were determined according to the standard method 1684 (EPA) [25]. The total volatile fatty acids (TVFA) were measured using the test kit LCK 365 (Hach Lange GmbH). The samples were centrifuged (10 min, 6000 rpm), and the supernatant was filtered. The time from the sampling up to the execution of the analytical procedure was identical for each sample to ensure the best possible quality of the results. A pH meter (HI991001, Hanna Instruments) was used to measure the pH in commercial reactors, and a mini pH meter (VWR, USA) was used to measure the pH in the miniature reactors.

The volume of biogas that was produced from the 3D-printed microreactors and the 300-mL reactors was estimated by the water displacement method, and the measuring devices were standard serum bottles with a volume of 10 and 100 mL, respectively. Chemical oxygen demand (COD; $g \bullet kg^{-1}$) and ammonium (NH_4^+-N; $g \bullet kg^{-1}$) were determined using commercial assay kits (Hach Lange GmbH, Germany) according to the manufacturer's instructions and were quantified by a spectrophotometer (DR3900, Hach, USA). Free ammonia nitrogen (FAN; $g \bullet kg^{-1}$) was calculated based on equation 1 [26]:

$$N - NH_3 = \frac{tan \times 10^{pH}}{e^{\left(\frac{6344}{273.15+T}\right)} + 10^{pH}} \quad (1)$$

The biogas volume ($mL \bullet g\ VSsubstrate^{-1} \bullet day^{-1}$) was measured with the water displacement method and was standardized according to DIN 1343 (standard conditions: temperature (T) = 0°C and pressure (P) = 1.013 bar) [27]. The biogas volume was normalized according to equation 2 [28]:

$$V_N = \frac{V \times 273 \times (760 - p_w)}{T \times 760} \quad (2)$$

where V_N is the volume of the dry biogas at standard temperature and pressure (mL_N), V is the recorded volume of the biogas (mL), p_w is the water vapor pressure as a function of ambient temperature (mmHg), and T is the ambient temperature (K).

All the experiments were carried out in duplicate (two bioreactors for the commercial reactor and two micro-bioreactors, and the experimental data from each reactor was plotted in the corresponding graphs), and the data analysis was conducted using Microsoft Excel.

3. Results

3.1 Biogas production

In this study, 3D-printed mini-bioreactors of 40 mL (MR1 and MR2) and commercial bioreactors of 400 mL (R1 and R2) were operated for 95 days (4.75 × HRT). Dried milk powder was used as a substrate, and the OLR was set to 0.5 g VS/day.

The rate of biogas production has the potential to be a valid online process condition indicator that determines the stability of a reactor. **Figure 4** clearly shows the stable production rate in the last 60 days of operation (3xHRT). In the first 20 days, the commercial reactors (R1 and R2) started with a fast production, reaching a constant rate within the range of 820–850 mL/g VS_{added}. MR1 and MR2 showed an increased biogas production for the first 60 days, reaching a similar production rate as obtained in the commercial reactor after 3xHRT.

The OLR was set at 0.5 g VS/day to avoid clogging problems during the operation system. Gou et al. [29] proposed that an OLR less than 5000 mg/L is necessary to ensure stable biogas production at mesophilic conditions. Similarly, Sun et al. [30] reported that an OLR in the range of 3000–5000 mg/L is more desirable for digester operation.

3.2 pH

The pH is a very useful indicator for the behavior of anaerobic digestion and the overall process stability. When the pH in an anaerobic reactor decreases, it is usually the first signal that the process starts to become unstable. The acidification is caused by the accumulation of short-chain fatty acids that are not efficiently converted into biogas. Typically, the pH is kept constant by the process itself. Organic substrates are hydrolyzed and converted into short-chain fatty acids and further converted into acetate, H_2, and CO_2. Specific microorganisms, archaea, convert H_2 plus CO_2 or acetate into CH_4 or CH_4 and CO_2, respectively. Different groups of microorganisms (bacteria) are responsible for the hydrolysis, acidogenesis, and acetogenesis phases. An imbalance in the ratio and activity of the bacteria and archaea may increase the concentration of acids in the reactor.

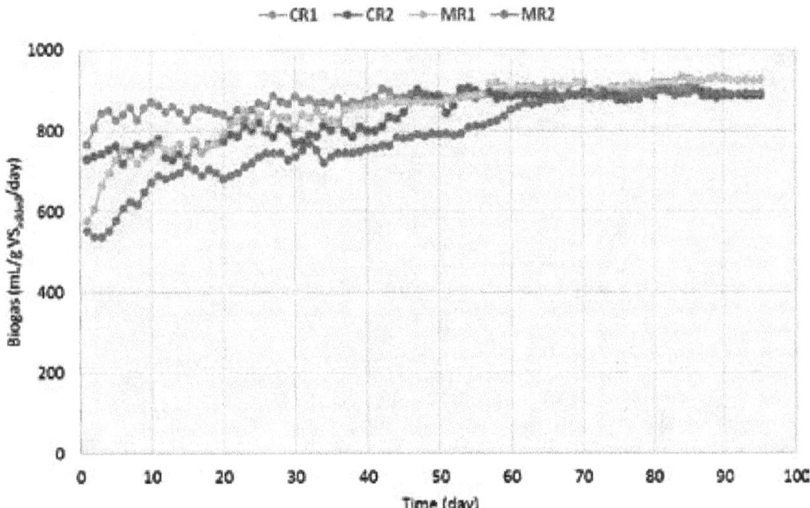

Figure 4.
Daily biogas production during the experimental period.

Methanogenic archaea consume acetate, CO_2, and H_2 but do not perform very well at a pH below 6.5, and the acidification accelerates until all microorganisms are not able to grow anymore. At this point, the anaerobic digestion comes to a halt and is not able to recover unless the pH is actively increased to pH 7. Therefore, early signs of acidification of an anaerobic reactor that produces biogas are an indication to change the process operation parameters to maintain pH neutrality (**Figure 5**).

Until day 50, the pH showed higher stability for the commercial reactors, whereas the pH of the mini-bioreactors (MR1 and MR2) was a little less stable but varied within an acceptable range. After day 50, a small but steady decrease is observed in both commercial bioreactors and the mini-bioreactors. Milk powder is mainly composed of carbohydrates, proteins, and lipids and may not result in the optimal growth conditions for especially the methanogens. An imbalance in the different processes is likely and volatile fatty acids accumulate in the reactors, and the pH decreases [31]. The similar pH profile in all reactors indicates that the mini-bioreactors behave similarly as the commercial reactors and anaerobic digestion can be downscaled and performed in 3D-printed microreactors leading to the same pH profile as in anaerobic digestion performed in commercial reactors.

3.3 FOS/TAC

The changes in VFA and TA in reactors were also monitored, and the results of the VFA/TA ratio are shown in **Figure 6**. With a ratio of less than 0.20, the microbes begin to "feel hungry," and the inoculum-to-substrate ratio must be decreased to obtain a stable process. A VFA/TA ratio greater than 0.3 indicates the beginning of "indigestion" [32, 33]. The content of the commercial bioreactors showed a significant higher buffer capacity, maintaining an optimal pH for the methanogenic bacteria. No extra alkalinity was added in the bioreactors, and the inoculum was considered as the only source of alkalinity. After 3xHRT, MR1 ranged between 0.21 and 0.28, whereas the MR2 was between 0.23 and 0.27. R1 and R2 showed lower ratios between 0.18 and 0.2, indicating better stability. It is notable that if the TVFA/TA ratio falls in the range between 0.20 and 0.3, the anaerobic digestion process is usually stable [34, 35].

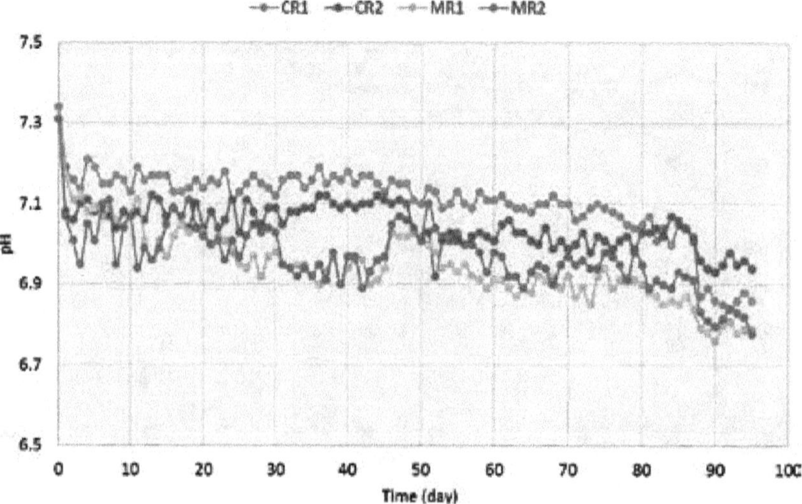

Figure 5.
pH variation during the experimental period.

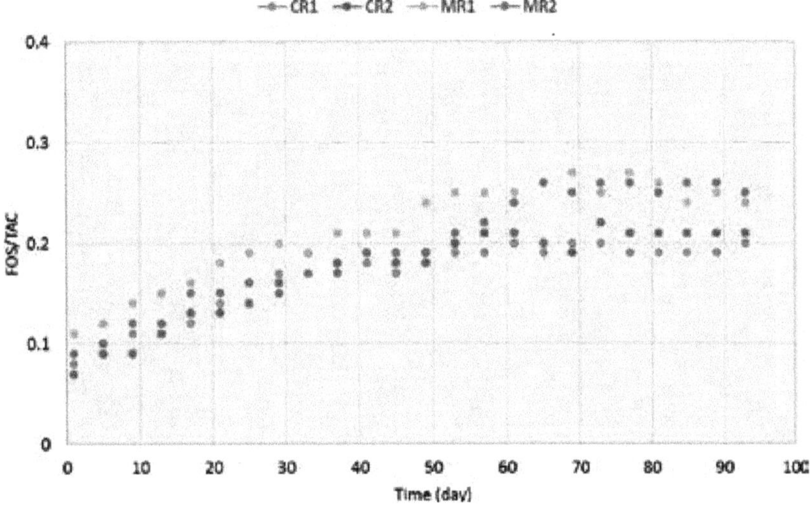

Figure 6.
FOS/TAC variation during the experimental period.

3.4 Ammonia

Ammonia is formed from the decomposition of proteins and urea in milk. It is an essential nutrient that serves as a nitrogen source for the bacteria and archaea in the reactor. Without nitrogen, the microorganisms are unable to grow and will gradually wash out of the rectors. The total ammonia nitrogen is primarily composed of ammonium ions (NH_4^+) and free ammonia (NH_3) (i.e., free ammonia nitrogen (FAN)). The predominant form of these two components mainly depends on process temperature and pH [36]. To illustrate, if the temperature or pH increases, the equilibrium between NH_3 and NH_4^+ shifts toward NH_3. Furthermore, the FAN is the most toxic species of the total ammonia nitrogen (TAN). FAN diffuses through the bacterial cell membrane and results in a proton imbalance in the cytosol. The intercellular pH increases and a rise in maintenance energy requirements inhibit the microorganisms because they will attempt to maintain their optimal intracellular pH [37].

In all reactors, the concentration of FAN increases with the same rate until day 50. After that, the rate decreases, and the level of FAN stabilized at 1.25 g/l (**Figure** 7). The degradation of the protein-rich substrate leads to the formation of FAN, and the microbial community needs to adapt to this substrate to effectively convert milk powder into biogas. The adaption seems to follow the same path in the commercial reactors and the 3D-printed mini-bioreactors.

3.5 Redox

The reduction oxidizing potential (i.e., redox potential) has been shown as a suc-cessful monitoring parameter in many AD systems due to redox-reaction-catalyzed enzymes that degrade organic materials in the anaerobic environment [38]. The strictness of the anaerobic environment is well known, which is indicated by a redox potential of ≤ -200 mV [39]. Preferably, the redox potential is between −330 and −450 mV for an optimal AD process environment. The facultative anaerobic microorganisms consume the oxygen and other oxidizing components that are dissolved in the growth medium, resulting in a sufficiently low redox potential required by the anaerobic methanogenic archaea [39].

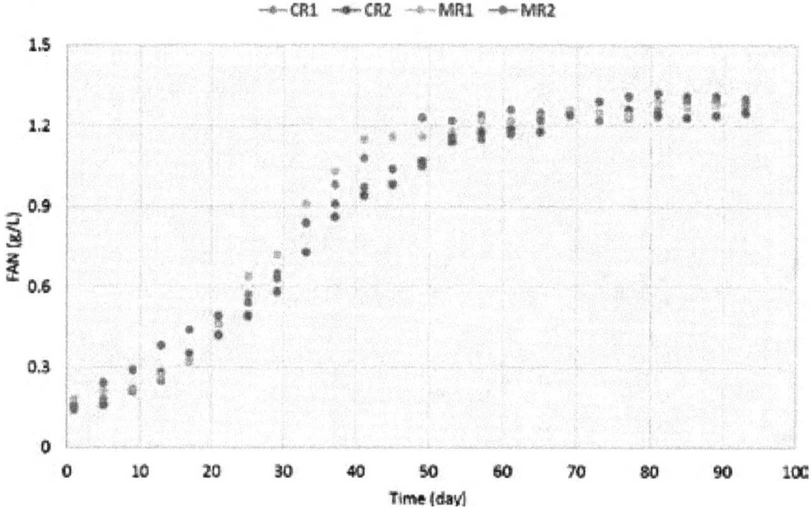

Figure 7.
FAN variation during the experimental period.

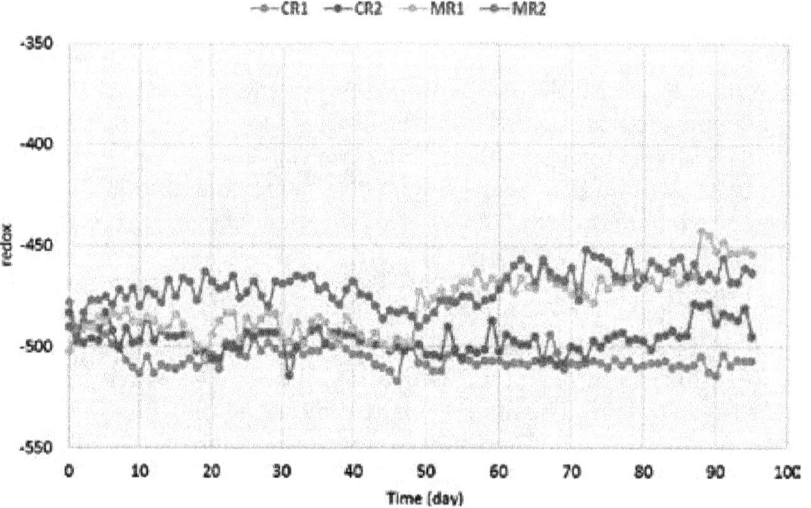

Figure 8.
Redox potential variation during the experimental period.

In **Figure 8**, a constant redox potential in the reactors CR1 and CR2 was obtained throughout the whole experimental procedure. After 50 days, the redox potential in MR1 and MR2 slowly starts to increase. The increase may be due to small oxygen leakages in, e.g., the tubing connections. Small air leakages in the commercial biore-actors can be better handled by the system because more biomass is available (the same concentration but larger volume) to consume the oxygen. In the small micro-bioreactors, a similarly sized leakage causes considerable more disturbances to the strictly anaerobic methanogenic archaea. The smaller amount of biomass is not able to metabolize all of the intruded oxygen, and the redox potential will increase.

4. Conclusion

Although there are several opportunities in the biogas sector, new challenges and barriers cannot be ignored and have to be overcome by using new process parameters and optimizing existing ones. The possibility of manufacturing bio-reactors employing 3D printing has been demonstrated in this work. In this way, miniaturized semicontinuous bioreactors have been manufactured using low-cost SLA machines for the first time. The high resolution of the printer, coupled with the satisfactory solvent compatibility of the photopolymers employed, enabled the development of reactors with the possibility to easily add advanced features, such as regularly spaced and geometry-controlled baffles, sample ports, sensor inlets, or other internal structures by simple CAD design. Furthermore, the direct printing of high-quality threads allowed working under controlled back pressure. Indeed, the micro-bioreactor manufactured here showed a similar performance as the commer-cial bioreactor in biogas production from the anaerobic digestion of milk.

The scale of the reactors demonstrated in this work adds an important step to the laboratory scale and the industrial scale, speeding up the research to obtain optimal fermentation conditions. The simplicity, low cost, and rapid uptake of 3D printing technology will enable the development of numerous applications of advanced reactor engineering in continuous-flow chemical manufacturing. The conclusions of this work justify the use of mini AD systems for high-throughput process screen-ing to improve AD systems further. The excessive amounts of biowaste and waste-water produced in our society need to be taken care of properly. Better performing AD reactors contribute considerably to the sustainable treatment of biowaste and wastewater.

Author details

Spyridon Achinas and Gerrit Jan Willem Euverink*
Engineering and Technology institute Groningen, University of Groningen, Groningen, Netherlands

*Address all correspondence to: g.j.w.euverink@rug.nl

References

[1] Deublein D, Steinhausez A. Biogas from Waste and Renewable Resources. Weinheim: Wiley VCH; 2008

[2] Zupančič GD, Grilc V. Anaerobic treatment and biogas production from organic waste. In: Kumar S, Bharti A, editors. Management of Organic Waste. Rijeka: IntechOpen; 2012. pp. 1-28. DOI: 10.5772/32756

[3] Ehrfeld W, Hessel V, et al. Potentials and realization of micro reactors. In: Ehrfeld W, editor. Microsystem Technology for Chemical and Biological Microreactors. Weinheim: Verlag Chemie; 1996. pp. 1-28

[4] Powell JB. (52c) Reactor scale-down for pilot plant, bench scale, and multi-throughput units. Section: Pilot plant design and optimization. In: AIChE Annual Meeting; San Francisco, CA; 2016

[5] Rinard IH. Mini-plant design methodology. In: Ehrfeld W, Rinard IH, Wegeng RS, editors. Process Miniaturization: 2nd International Conference on Microreaction Technology; Topical Conference Preprints. New Orleans, USA: AIChE; 1998. pp. 299-312

[6] Ponton JW. Some thoughts on the batch plant of the future. In: Proceedings of the 5th World Congress on Chemical Engineering; San Diego; 1996

[7] Benson RS, Ponton JW. Process miniaturization—A route to total environmental acceptability? Transactions of the Indian Institute of Chemical Engineers. 1993;**71**:160-168, A2

[8] Menardo S, Balsari P. An analysis of the energy potential of anaerobic digestion of agricultural by-products and organic waste. Bioenergy Research. 2012;**5**(3):759-767

[9] Burke F. Scale up and scale down of fermentation processes. In: McNeil B, Harvey LM, editors. Practical Fermentation Technology. West Sussex: John Wiley & Sons, Ltd; 2008. pp. 231-270

[10] Lu B, Li D, Tian X. Development trends in additive manufacturing and 3D printing. Engineering. 2015;**1**(1):85-89

[11] Gu D, Ma C, Xia M, Dai D, Shi Q. A multiscale understanding of the thermodynamic and kinetic mechanisms of laser additive manufacturing. Engineering. 2017;**3**:675-684

[12] Yan Q, Dong H, Su J, Han J, Song B, Wei Q, et al. A review of 3D printing technology for medical applications. Engineering. 2018;**4**:729-742

[13] Wang K, Ho CC, Zhang C, Wang B. A review on the 3D printing of functional structures for medical phantoms and regenerated tissue and organ applications. Engineering. 2017;**3**:653-662

[14] Hu G, Guan K, Lu L, Zhang J, Lu N, Guan Y. Engineered functional surfaces by laser microprocessing for biomedical applications. Engineering. 2018;**4**:822-830

[15] Wolozny D, Lake JR, Movizzo PG, Long Z, Ruder WC. An additive manufacturing approach that enables the field deployment of synthetic biosensors. Engineering. 2018;**5**:173-180

[16] Wegeng RW, Call CJ, Drost MK. Chemical system miniaturization. In: Proceedings of the AIChE Spring National Meeting; 25-29 February, 1996; New Orleans, USA; 1996. pp. 1-13

[17] Smith AL, Skerlos SJ, Raskin L. Microfabricated devices that facilitate bioenergy biosynthesis research. Environmental Science: Water Research & Technology. 2015;**1**:56-64

[18] Moffitt JR et al. The single-cell chemostat: An agarose-based, microfluidic device for high-throughput, single-cell studies of bacteria and bacterial communities. Lab on a Chip. 2012;**12**:1487-1494

[19] Kim HS et al. A high-throughput microfluidic light controlling platform for biofuel producing photosynthetic microalgae analysis. In: 14th International Conference on Miniaturized Systems for Chemistry and Life Sciences; 2010. pp. 295-297

[20] Hou H et al. Microfabricated microbial fuel cell arrays reveal electrochemically active microbes. PLoS One. 2009;**4**:e6570

[21] Szita N et al. Development of a multiplexed microbioreactor system for high-throughput bioprocessing. Lab on a Chip. 2005;**5**:819-826

[22] Ghosh A. Chapter 2: Scaling laws. In: Chakraborty S, editor. Mechanics Over Micro and Nano Scales. Vol. 61. New York, USA: Springer Science + Business Media, LLC; 2011. p. 269. DOI: 10.1007/978-1-4419-9601-5_2

[23] West GB, Brown JH. The Origin of allometric scaling laws in biology from genomes to ecosystems: Towards a quantitative unifying theory of biological structure and organization. The Journal of Experimental Biology. 2005;**208**:1575-1592

[24] Matthews G. Chapter 2: Fermentation equipment selection: Laboratory scale bioreactor design considerations. In: McNeil B, Harvey LM, editors. Practical Fermentation Technology. Chichester, UK: John Wiley & Sons, Ltd; 2008

[25] Method 1684: Total, Fixed, and Volatile Solids in Water, Solids, and Biosolids. Washington, DC, USA: U.S. Environmental Protection Agency (EPA); 2001. EPA-821-R-01-015

[26] Anthonisen AC, Loehr RC, Prakasam TBS, Srinath EG. Inhibition of nitrification by ammonia and nitrous acid. Journal - Water Pollution Control Federation. 1976;**48**:835-849

[27] VDI 4630. Fermentation of organic materials. In: Characterisation of Substrate, Sampling, Collection of Material Data, Fermentation Tests. Düsseldorf: VDI Gesellschaft Energietechnik; 2006

[28] Dinuccio E, Balsari P, Gioelli F, Menardo S. Evaluation of the biogas productivity potential of some Italian agro-industrial biomasses. Bioresource Technology. 2010;**101**:3780-3783

[29] Gou C, Yang Z, Huang J, Wang H, Xu H, Wang L. Effects of temperature and organic loading rate on the performance and microbial community of anaerobic co-digestion of waste activated sludge and food waste. Chemosphere. 2014;**105**:146-151

[30] Sun MT, Fan XL, Zhao XX, Fu SF, He S, Manasa MRK, et al. Effects of organic loading rate on biogas production from macroalgae: Performance and microbial community structure. Bioresource Technology. 2017;**235**:292-300

[31] Yu H, Fang H. Acidogenesis of dairy wastewater at various pH levels. Water Science and Technology. 2002;**45**:201-206

[32] Liotta F, Esposito G, Fabbricino M, van Hullebusch ED, Lens PNL, Pirozzi F, et al. Methane and VFA production in anaerobic digestion of rice straw under dry, semi-dry and wet conditions during start-up phase. Environmental Technology. 2016;**37**:505-512

[33] Franke-Whittle IH, Walter A, Ebner C, Insam H. Investigation into the effect of high concentrations of volatile fatty acids in anaerobic digestion on methanogenic communities. Waste Management. 2014;**34**:2080-2089

[34] Wang Y, Zhang Y, Wang J, Meng L. Effects of volatile fatty acid concentrations on methane yield and methanogenic bacteria. Biomass and Bioenergy. 2009;**33**:848-853

[35] Wang L, Zhou Q , Li F. Avoiding propionic acid accumulation in the anaerobic process for biohydrogen production. Biomass and Bioenergy. 2006;**30**:177-182

[36] Kayhanian M. Ammonia inhibition in high-solids biogasification: An overview and practical solutions. Environmental Technology. 1999;**20**:355-365

[37] Akindele A, Sartaj M. The toxicity effects of ammonia on anaerobic digestion of organic fraction of municipal solid waste. Waste Management. 2018;**71**:757-766

[38] Guwy A, Hawkes F, Wilcox S, Hawkes D. Neural network and on-off control of bicarbonate alkalinity in a fluidized-bed anaerobic digester. Water Research. 1997;**31**:2019-2025

[39] Schnurer A, Jarvis A. Microbiological handbook for biogas plants. Swedish Waste Management U2009:03. Swedish Gas Centre Report. 2010. p. 207

7

Solid-State Fermentation of Cassava Products for Degradation of Anti-Nutritional Value and Enrichment of Nutritional Value

Mohamed Hawashi, Tri Widjaja and Setiyo Gunawan

Abstract

The cassava plant is grown in tropical and subtropical countries, which represents, alongside with its by-products, an important source of food and feed. Hence, this plant has the capacity to promote the economic development of those countries and provide food security. However, cassava has some disadvantages due to the antinutrient compounds produced in its tissues. In addition, the cassava roots have a low protein content. Due to the economic and practical advantages, the solid-state fermentation (SSF) has been used as a cost-effective and efficient processing method to detoxify the cassava products and enrich them in nutrients. This chapter reviews the solid-state fermentation technique of cassava products for the production of valuable components for food and feed applications, microorganisms involved in this process, and key factors used to optimize the SSF process.

Keywords: anti-nutritional value, cassava, nutritional value, processing variables, solid-state fermentation

1. Introduction

Cassava (*Manihot esculenta Crantz*) is grown in tropical and subtropical countries. It is a vital source of food and feed and it can promote economic development and provide food security [1]. Cassava production has been promoted globally by the International Fund for Agricultural Development (IFAD) and the United Nations Food and Agriculture Organization (FAO) to develop cassava strategies [2]. Reports indicate that production rates will reach 300 million tons per year by 2020 [3]. Due to its high drought tolerance, cassava plant cultivation can take place even under critical environmental conditions, with an ideal high yield of approximately 50% for leaves and 6% for roots at plant maturity [4]. Its peel may make up 10–20% of the roots' wet weight [5]. However, cassava has some disadvantages; its tissues contain anti-nutritional compounds and very low protein content [6, 7].

Among all the antinutrients, hydrogen cyanide (HCN) is of great concern, the concentration of which is in cassava and its by-products are much higher than the World Health Organization (WHO) safe limit for human consumption (10 ppm) [8, 9]. Konzo is an irreversible neurological disease associated with intake of HCN [10]. Therefore, a detoxification process is needed to reduce anti-nutritional levels in order to consume cassava safely. Solid-state fermentation (SSF) has been used as an economical and efficient processing method for enriching and detoxifying cassava and its by-products [11, 12]. Various process parameters such as particle size, moisture content, water activity, pH, the inoculum size, incubation time, concentration of nutrient supplementation, and temperature can affect the microbial growth, enzyme production, and formation of the product during the SSF process [13].

This chapter discusses fermented cassava products through solid-state fermentation for food and feed applications, as well as microorganisms involved in solid-state fermentation and the essential processing variables used to optimize the process.

2. Fermentation processes

Fermentation has been one of the most used technologies to improve the taste and sensory properties of food and continues to be one of the most widely used methods of preserving the food for a length of time [14, 15]. The cassava fermentation process is a strategy to improve nutritional value by enriching protein and detoxifying toxic and anti-nutritional compounds, in particular by reducing toxic cyanogenic glycosides to a safe level of consumption in cassava products as well as reducing post-harvest losses [16–18].

There are two kinds of fermentation, i.e., spontaneous (natural) fermentation and controlled fermentation. For the natural fermentation, the conditions are selected so that to produce the most suitable microorganisms for the production of growth by-products characteristic of a particular type of fermentation [19]. The controlled fermentation is generally used when the natural fermentation is unstable or the bacteria are not able to grow. In this case, specific microbial strains, such as lactic acid bacteria (LAB), yeast, and fungal are isolated, characterized, and preserved for later use as starter cultures [20]. Under optimal growth conditions, these cultures can be used as single or combined starter cultures. As a result, the quality of products and their organoleptic characteristics are well controlled and predictable [20, 21].

However, the fermentation process can be broadly categorized into submerged fermentation (involving soaking in water) and solid-state fermentation (without soaking in water) [22]. The solid-state fermentation (SSF) technique has several advantages over submerged fermentation (SmF). However, the SSF has some constraints. **Table 1** illustrates the advantages and disadvantages of SSF over SmF [23].

2.1 Solid-state fermentation and its application in cassava products

In recent years, the cassava population has developed numerous processing methods (soaking, boiling, drying, and fermentation) [24–26]. SSF is one of the promising processes of enriching protein and detoxifying of cassava products [27–29].

Fermented cassava products by SSF, such as flour, gari, starch, bread, and biomass contain high protein content that can either be consumed by humans or animals, replacing expensive, conventional protein sources in different parts of Latin America, Africa, and Asia [30]. The major fermented cassava products by SSF can be derived from different parts of the cassava plant, such as roots, peels, and leaves.

2.1.1 Cassava roots

Cassava is grown in many developing countries for its roots as a primary source of carbohydrates and ranks third in the developing countries as the leading source

Parameter	Solid-state fermentation	Submerged fermentation
Substrates	Insoluble substrates (starch, cellulose, pectins, lignin)	Soluble substrates (sugars)
Aseptic techniques	Sterilization of steam and non-sterile conditions	Sterilization of heat and aseptic control
Temperature	Difficult temperature control	Easy temperature control
Water	Low water consumption	High water consumption
pH control	Difficult pH control	Easy control of pH
Industrial level	Relatively small scale, newly designed equipment is needed	The industrial level is available
Inoculation	Spore inoculation, batch process	Easy inoculation, continuous process
Contamination	Contamination risk of low-growth fungi	Contamination risk of single strain bacteria
Energy	Low consumption of energy	High consumption of energy
Equipment volumes	Low volumes and low equipment costs	High volumes and high equipment costs
Pollution (effluents)	No volumes of effluents	High volumes of effluents
Concentration/products	100/300 g/L	30–80 g/L

Table 1.
Comparative characteristics of solid-state and submerged fermentations.

of energy in human diets along with rice and wheat [31]. World production of cassava is estimated at 277 million tons of fresh root in 2017 [32]. Cassava root has several advantages compared to other crop roots, including high productivity, resistance to droughts and pests, flexible harvesting age, and it can be kept in the ground until they are needed [33]. However, cassava root also has certain disadvantages; its tissues contain toxic compounds (a cyanogenic glycoside), low protein content (1% fresh root weight), and short shelf life of 1–3 days [34].

Food processing techniques have been used to convert cassava tubers into flour as an alternative way to preserve the roots after harvesting and then further use it for industrial and traditional purposes [35, 36]. Gari and flour are the most popular fermented food products from cassava roots by SSF. In West Africa, approximately 200 million people consume gari [37, 38]. **Figure 1** shows the production of flour and gari under the solid-state fermentation [11].

The purpose of cassava root fermentation is to increase the low protein con-tent from 2% to about 7% or more than the critical crude protein content [39]. To achieve this goal, several solid-state fermentation techniques have been used. Raimbault et al. [40] reported the principle underlying the SSF procedure for the enrichment of cassava flour. This procedure led to the enrichment of crude protein from 1 to 18–20%, which improved between 1700 and 1900% after 30 h of fermentation. Oboh and Elusiyan [41] studied the effect of solid-state fermentation by *R. oryzae* and *S. cerevisiae* on the improvement of nutritional values of cassava flour produced from two different varieties of cassava root. The nutritional contents of cassava flour were assayed before and after 72 h of fermentation. This study has observed that *S. cerevisiae* was more effective than *R. oryzae* in the nutrient enrichment of cassava flour. The results of this study are presented in **Figures 2–5**. Essers et al. [42] investigated the effect of SSF on the degradation of hydrogen cyanide level in cassava root using six fungal strains, namely *Rhizopus stolonifer*, *Rhizopus*

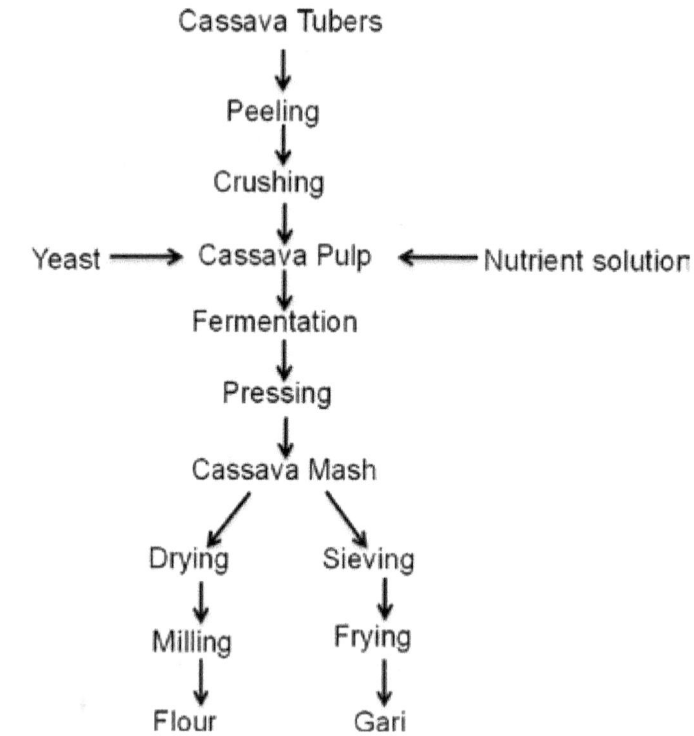

Figure 1.
The production chart of cassava products (flour and gari) under SSF.

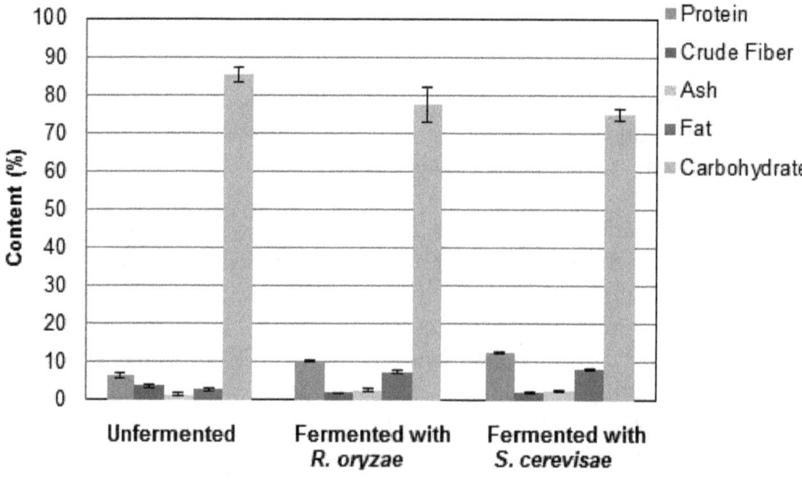

Figure 2.
Proximate composition of the cassava flour obtained from cassava varieties of low HCN subjected to SSF.

oryzae, *Mucor racemosus, Bacillus sp. Geotrichum candidum*, and *Neurospora sitophila*. The reduction in cyanide content was more than 60% after 72 h of fermentation.

In addition, Oboh and Akindahunsi [11] investigated the effect of solid-state fermentation with *S. cerevisiae* on the nutritional and antinutrient contents of cassava products (flour and gari). After 72 h of fermentation, the results revealed that the content of protein and fats in cassava flour increased by 10.9 and 4.5%, respectively. The protein and fat content of fermented gari also improved by 6.3% and 3.0%. In contrast, the content of cyanide in flour and gari decreased to 9.5 and 9.1 (mg/kg),

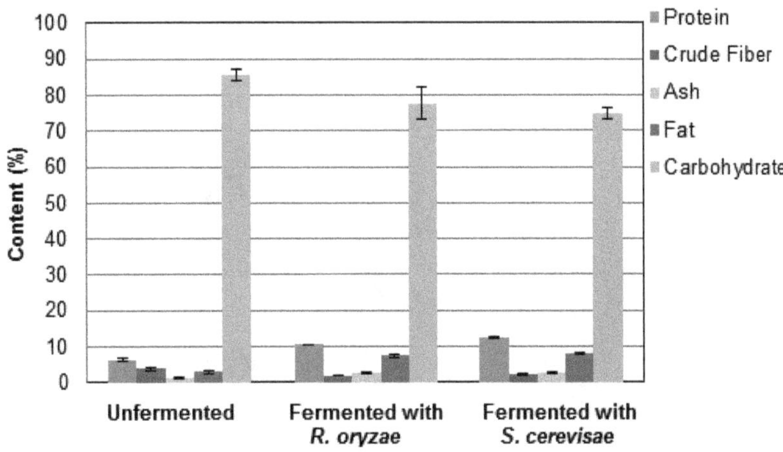

Figure 3.
Proximate composition of the cassava flour obtained from cassava varieties of medium HCN subjected to SSF.

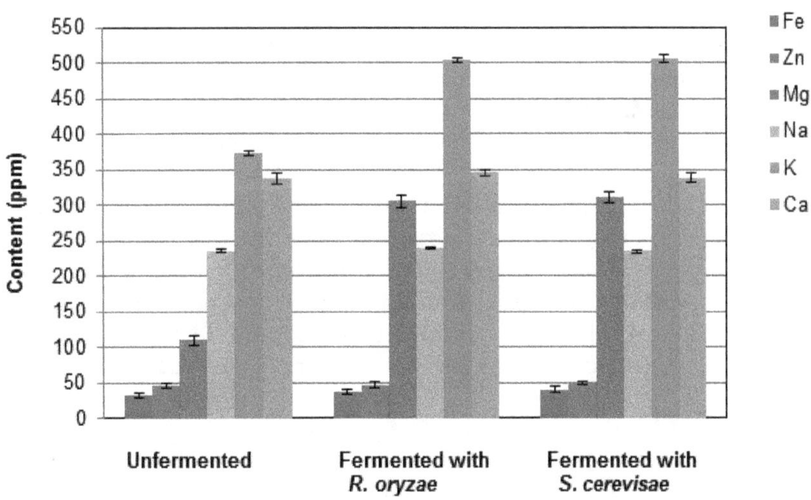

Figure 4.
Mineral contents of the cassava flour obtained from cassava varieties of low HCN subjected to SSF.

respectively. However, the tannin content, crude fiber, and ash content of the cassava products did not change significantly under SSF.

2.1.2 Cassava peels

Cassava wastes, such as peels and leaves and starch residues make up 25% of the total cassava plant [43]. Cassava peel is the leading waste from the cassava plant, but its use is limited due to the high content of cyanide and fiber as well as low protein and therefore disposed of it after cassava processing into food or other industrial products [44, 45]. Many efforts have been made using SSF techniques to enrich the protein content and degrade the cyanide level of cassava peels for animal feed.

Bayitse et al. [12] studied protein enrichment of cassava residue using *Trichoderma pseudokoningii* under solid-state fermentation for 12 days, urea, and ammonium sulfate was used as a nitrogen source, and the moisture content ranged from 60 to 70%. The result showed an improvement in crude protein content of 12.5% using urea as a nitrogen source, and a moisture content of 70%,as compared

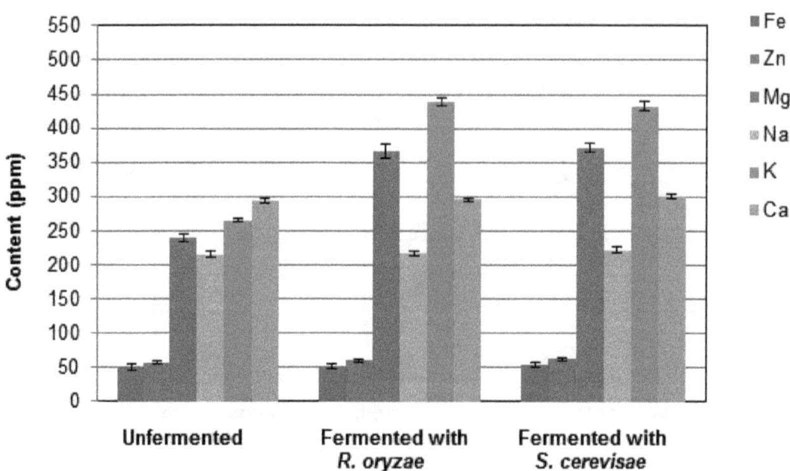

Figure 5.
Mineral contents of the cassava flour obtained from cassava varieties of medium HCN subjected to SSF.

to 8.89 and 6.37% improvement observed with ammonium sulfate as a nitrogen source, and without using nitrogen source. The study observed a decrease in cyanide content, but it did not attribute it to the fermentation effect of *Trichoderma pseudokoningii*, rather it stated that the reduction could have been as a result of the pre-processing of cassava peels.

Iyayi and Losel [43] also evaluated protein improvement of cassava peels using different types of microorganisms and fermentation time (*Saccharomyces cerevisiae, Aspergillus niger, Rhizomucor miehei,* and *Mucor strictus*). The solid-state fermentation of cassava peels by *S. cerevisiae* produced the highest protein content from 5.6 to 16.74% for 21 days. Also, they reported the maximum fermentation period for the protein enrichment of cassava peel to be from 12 to 15 days, after which no significant change was observed, which is in line with the work reported by Bayitse et al. [12].

Ezekiel and Aworh [13] evaluated the effectiveness of SSF with *Trichoderma viride* on the reduction of cyanide content and enrichment of the crude protein content of cassava peel by optimizing the fermentation conditions such as moisture content, pH, particle size, nitrogen source, and incubation temperature. The optimum SSF conditions were found at the initial moisture content of 60% (v/w), the particle size of 4.00 mm, a pH of 6.0, 30°C of temperature, and ammonium sulfate (10 g N/kg substrate) as nitrogen sources. After 8 days of fermentation, the cyanide content was reduced by 71% and improved the crude protein content from 4.2 to 10.43% at optimized conditions.

In another study by Ruqayyah et al. [45], the application of response surface methodology was used to optimize SSF conditions (moisture content, inoculum size, and pH) with *P. tigrinus* to enrich the crude protein content of cassava peel. A maximum protein content of 89.58 (mg/g) was obtained at 75% (v/w) mois-ture content, 7% (v/w) inoculum size, and pH of 5.3 with a fermentation time of 15 days. The optimum level resulted in a significant enrichment of the protein content by 55.16%.

Oboh [46] investigated the effect of solid-state fermentation of cassava peel with a mixture of *Saccharomyces cerevisiae* and two strains of lactic acid bacteria, *Lactobacillus delbrueckii* and *Lactobacillus coryniformis* to improve the nutritional value and detoxification of cassava peel. The chemical composition of cassava peel has been analyzed before and after fermentation. The results showed the effective performance of the SSF technique in removing cyanide by 86% after 7 days of fermentation. On the other hand, the mineral composition of the cassava peel did not change during the fermentation. The results of this study are presented in **Table 2**.

2.1.3 Cassava leaves

Cassava leaves are an extremely rich source of proteins, vitamins, and minerals that exceed some of the other green vegetables [47, 48]. The production of cassava leaves is estimated at 10 tons of dry leaves per hectare, which has a similar yield with the roots [49]. Cassava leaves are consumed in most Southeast Asian and African societies, such as Indonesia, Malaysia, Congo, Madagascar, and Nigeria [50, 51]. However, cassava leaves contain both nutritive (33.8–37.4% protein content) and anti-nutritional compounds [301.04–192.47 (mg/100 g) HCN content] [52]. Boiling, soaking, steaming, drying the sun, drying the oven, and cooking are the most common methods for processing cassava leaves in African and Asian countries [53].

The origin of HCN in the cassava leaves is a two-step process [54, 55]. First, the linamarin, a cyanogenic glycoside, which represent 93% of cyanogenic glycosides found in cassava (7% is lotaustralin), is hydrolyzed by linamarase (a beta-glycosidase) into glucose and cyanohydrin. Then, in the second step, the cyanohydrin is decomposed, either enzy-matically or not, to HCN and acetone. The nonenzymatic pathway depends on pH. At pH > 6, the HCN is liberated, but at an acidic pH (~5), the process is much lower, and the resulting HCN is therefore relatively lower in concentration. However, this approach did not assure full hydrolysis of cyanogens. The partial breakdown of the leaf cells only partially releases linamarase resulting in only a certain proportion of the cyanogenic compounds being converted to HCN. This implies that a proportion of the cyanogens remain present in the leaves after processing and resulting in the release of HCN directly into the human body upon consumption.

The conventional methods have been proven to be ineffective for lowering the cyanide content in cassava leaves to the safe limit, at the same time causing a significant loss of protein and essential nutrients, which is highly desired from the cassava leaves [56–60]. Hence, the establishment of a universally acceptable method that produces edible leaves with low cyanide level while maintaining

Composition	Fresh	Naturally fermented	Fermented with a mixed culture
Crude protein (%)	8.2 ± 0.1	11.1 ± 0.3	21.5 ± 1.2
Crude fiber (%)	11.7 ± 0.5	6.5 ± 0.5	11.7 ± 0.5
Fat (%)	3.1 ± 0.4	3.5 ± 0.2	2.1 ± 0.1
Ash (%)	6.4 ± 0.4	6.0 ± 0.2	7.2 ± 0.2
Carbohydrate (%)	64.6 ± 0.2	67.3 ± 0.4	51.1 ± 0.4
Moisture (%)	5.1 ± 0.3	5.7 ± 0.2	6.4 ± 0.4
Ca (ppm)	0.03 ± 0.00	0.03 ± 0.00	0.03 ± 0.00
Na (ppm)	00.04 ± 0.00	0.04 ± 0.00	0.04 ± 0.00
Zn (ppm)	0.01 ± 0.00	0.01 ± 0.00	0.01 ± 0.00
K (ppm)	0.05 ± 0.00	0.05 ± 0.00	0.05 ± 0.00
HCN (mg/kg)	45 ± 0.3	24 ± 0.2	6.1 ± 0.4

Table 2.
The effect of fermentation on the chemical composition of cassava peels.

maximum nutritional content is challenging and still far away from being established. Among the efforts made so far, Morales et al. [61] proposed a solid-state fermentation of cassava leaves, reducing the cyanide content while improving the nutritional value of the processed leaves. SSF was performed using *Rhizopus oligosporus*, and babassu mesocarp flour was the substrate used, supplemented by cassava leaf flour. The solid-state fermentation decreased the total cyanide content of the cassava leaves by 94.18%, also SSF increased the quantity and quality of crude protein content by 15%, resulting in the relative nutritional value of 98.18% for food, which is equivalent to casein (100%). Furthermore, Kobawila et al. [62] investigated the effect of alkaline fermentation on the reduction of cyanide level in cassava leaves to produce ntoba mbod. The dominant microflora in the fermen-tation of the cassava leaves was *Bacillus subtilis, Bacillus macerans,* and *Bacillus pumilus*. These bacteria can utilize cyanide acid for their nutrition [63]. Thus, they are responsible for the reduction of the cyanide content in the medium of fer-mentation (~70% removal). However, the report did not provide the effect of the fermentation process on the protein content of cassava leaves.

One of the essential criteria for the solid-state fermentation is the selection of an appropriate microorganism [64]. Several research works have explored different types of microorganisms mainly fungi, yeasts, and bacteria, as well as different sub-strates to favor the metabolism of the microorganisms in SSF of cassava products. Examples of microorganisms associated with solid-state fermentation of cassava products for food and feed applications are summarized in **Table 3**.

Microorganism	Substrate	Product	References
Rhizopus oryzae	Cassava root	Gari	[29]
Rhizopus oryzae (TISTR 3052), *Rhizopus oryzae* (TISTR 3058), *Rhizopus delemar* (TISTR 3534), and *Rhizopus delemar* (TISTR 3190)	Cassava flour	Bread	[65]
Panus tigrinus (M609RQY)	Cassava peels	Animal feed	[66]
Aspergillus niger and *Panus tigrinus*	Cassava peels	Poultry feed	[67]
Rhizopus oryzae and *Saccharomyces cerevisiae*	Cassava pulp	Animal feed	[68]
Saccharomyces cerevisiae, Aspergillus niger, Rhizomucor miehei, and *Mucor strictus*	Cassava leaves	Animal feed	[43]
Rhizopus oryzae	Cassava pulp	Lactic acid production	[69]
Lactobacillus plantarum and *Rhizopus oryzae*	Cassava root	Cellulase production	[70]
Bacillus sp., *Mucor racemosus, R. oryzae, Neurospora sitophila R. stolonifer* and *Geotrichum candidum*	Cassava root	Cassava flour	[71]
Rhizopus stolonifer LAU 07	Cassava peel	Feed supplements	[72]
Rhizopus sp.	Cassava starch and leaves	Lactic acid and ethanol productions	[73]

Table 3.
Examples of microorganisms associated with the SSF of cassava products.

3. Environmental factors

The process control of the solid-state fermentation parameters is closely related to the metabolic regulation of microorganisms [74]. Based on the metabolic needs of the fermentation microorganisms, the control of water activity, oxygen content, temperature, and pH are the main solid-state fermentation parameters [23]. In the solid-state fermentation process, the water, gas, and heat caused by the growth microbes are the dominant factors that determine the environmental changes. The environmental factors can affect the microbial growth and formation of the product during the SSF process [75, 13]. Therefore, the physical-chemical parameters must be controlled.

3.1 Water activity and moisture content

The unique feature of solid-state fermentation is that there is almost no free water in the substrate [76]. However, microorganisms can grow depending upon the water activity of the substrate [64, 75]. The growth of fungi and some yeast usually requires a water activity value between 0.6 and 0.7 [77]. In addition to meet the microbial physiological requirements, the water content level plays a decisive role in the variation of the three-phase structure relating to water retention, permeability, and thermal conductivity. The degree of swelling in the SSF system was low at a lower moisture level and hence increased water stress reduces nutrient solubility. On the contrary, the higher level of humidity results in changes in substrates that reduce porosity, thus contributing to stickiness and reduced gas exchange [78, 79]. According to Grover et al. [80], the required moisture content should range between 60 and 80% for an efficient SSF system.

3.2 Temperature

The fermentation temperature affects microbial growth, spore germination, and the formation of product [81]. Heat generation in solid-state fermentation system is more problematic than in liquid fermentation. Due to poor heat conductivity and accumulation of metabolic heat in the material combined with substrate shrink-age and decreased porosity, gas convection is severely impeded. Previous studies showed that the significant resistance to heat transfer in solid-state fermentation was low conduction efficiency [82, 83].

Therefore, moisturizing is a common measure of temperature control. In addition, routine operations (e.g., forced ventilation and jacket cooling) all can solve these problems [84]. The evaporative cooling is one of the main solid-state fermentation temperature control measures [85, 86]. In general, the aeration could reduce the temperature gradient of the medium [23]. The forced ventilation can take away more than 80% of the heat generated from the substrate [84]. From the current investigation, it is difficult to maintain the temperature at an ideal range in SSF system. To reach this aim, the main strategy used in large-scale solid-state fermentation is to combine ventilation and humidity [77].

3.3 Oxygen concentration

The gas environment is a critical factor that significantly affects the relative levels of biomass and the production of an enzyme [23]. Oxygen uptake rate (OUR) and carbon dioxide production (CDPR) can be used to assess the state of the solid-state fermentation process. However, different microorganisms cause these assessments to vary. Ghildyal et al. [87] studied the impact of the gas concentration gradient on product yield in a tray solid-state fermentation bioreactor. The results showed that the variations of O_2 and CO_2 concentration gradients were visible, which severely affected product yield. The yield decreased when gradient increases. Gowthaman et al. [88] also studied the impact of gas concentration gradient on the product in a packing bed bioreactor. The results showed that the gas concentration gradient could be eliminated and the ability of mass transfer can be enhanced by forced ventilation, which increased enzyme activity.

3.4 pH value

In general, if the initial pH value of the medium is adjusted, the variations of pH value during the solid-state fermentation process need to be considered [89]. During the fermentation process, the pH values change drastically. The reason is that organic acids including citric and lactic are secreted during the fermentation process, which decreases the pH [23]. While the increase in pH was rationalized in terms of organic acid decomposition and protein degradation in the raw materials into amino acids and peptide fractions [90]. The pH values are difficult to determine by conventional detection in SSF due to the low water content of the substrate. Nitrogen-containing inorganic salts (such as urea) are often used as sources of nitrogen to offset the pH variation in the fermentation process [91, 92].

In the study conducted by Ezekiel and Aworh [13] to evaluate the effect of pH on protein enrichment and soluble sugars of cassava peel by *Trichoderma viride*, the fungus was grown in a controlled pH medium of 4.– 6.0 with an incubation time of 8 days. The optimal growth condition was observed at pH 6.0. The protein increased in cassava peels from 230 at pH 4.0 to 270 (mg/gm) at pH 6.0. Also, the sugars yield at pH 5.0 and 6.0 was five times higher compared to pH 4.0. According to the study, the growth rate of the fungi at pH below five was affected by high acid-ity, leading to reduced bio-conversion of sugars into protein.

4. Conclusions

The results discussed in this chapter highlighted the importance of the SSF technique applied to cassava to improve its nutritional value. The solid-state fermentation using microbial protein is beneficial for the reduction of cyanide contents while the content of protein and other nutrients is increased compared to those obtained by the conventional approaches, i.e., soaking, boiling, and drying. Thus, the SSF technique for processing cassava products is better suited for developing societies and rural communities in the African and Asian countries that do not have easy access to available protein sources.

Acknowledgements

The authors are thankful to the Ministry of Research, Technology and Higher Education of the Republic of Indonesia for its financial support to this project through the grant no. 849/PKS/ITS/2018.

Author details

Mohamed Hawashi, Tri Widjaja and Setiyo Gunawan*
Department of Chemical Engineering, Institut Teknologi Sepuluh Nopember (ITS), Surabaya, Indonesia

*Address all correspondence to: gunawan@chem-eng.its.ac.id

References

[1] FAO, IFAD. The global cassava development strategy and implementation plan. In: Proceedings of the FAO and IFAD Validation Forum on the Global Cassava Development Strategy; 26-28 April 2000; Rome. Rome: FAO and IFAD; 2001. pp. 13-15

[2] Howeler RH. Endorsement of the global cassava development strategy. In: Proceedings of the FAO and IFAD Validation Forum on the Global Cassava Development Strategy; 26-28 April 2000; Rome. Rome: FAO and IFAD; 2001. p. 57

[3] Agustian A. Bioenergy development in the agricultural sector: Potential and constraints of cassava bioenergy development. Analisis Kebijakan Pertonian. 2015;**13**(1):19-38

[4] Tewe OO, Lutaladio N. Cassava for Livestock Feed in Sub-Saharan Africa. Rome: FAO; 2004. p. 64

[5] Obadina AO, Oyewole OB, Sanni LO, Abiola SS. Fungal enrichment of cassava peels proteins. African Journal of Biotechnology. 2006;**5**(3):302-304. DOI: 10.5897/AJB05.360

[6] Gunawan S, Widjaja T, Zullaikah S, Istianah N, Aparamarta HW, Prasetyoko D, et al. Effect of fermenting cassava with *Lactobacillus plantarum*, *Saccharomyces cereviseae*, and *Rhizopus oryzae* on the chemical composition of their flour. International Food Research Journal. 2015;**22**(3):1280-1287

[7] Hawashi M, Ningsih TS, Cahyani SBT, Widjaja KT, Gunawan S. Optimization of the fermentation time and bacteria cell concentration in the starter culture for cyanide acid removal from wild cassava (*Manihot glaziovii*). MATEC Web of Conferences. 2018;**156**:01004. DOI: 10.1051/matecconf/201815601004

[8] Codex Alimentarius Commission. Codex Alimentarius. Rome: Food and Agriculture Organization; 1992

[9] Hadiyat MA, Wahyudi RD. Integrating steepest ascent for the Taguchi experiment: A simulation study. International Journal of Technology. 2013;**3**:280-287. DOI: 10.14716/ijtech.v4i3.132

[10] Bradbury JH. Simple wetting method to reduce cyanogen content of cassava flour. Journal of Food Composition and Analysis. 2006;**19**(4):388-393. DOI: 10.1016/J.JFCA.2005.04.012

[11] Oboh G, Akindahunsi AA. Biochemical changes in cassava products (flour & gari) subjected to *Saccharomyces cerevisae* solid media fermentation. Food Chemistry. 2003;**82**(4):599-602. DOI: 10.1016/50308-8146(03)00016-5

[12] Bayitse R, Hou X, Laryea G, Bjerre AB. Protein enrichment of cassava residue using *Trichoderma pseudokoningii* (ATCC 26801). AMB Express. 2015;**5**(1):80. DOI: 10.1186/s13568-015-0166-8

[13] Ezekiel OO, Aworh OC. Solid state fermentation of cassava peel with *Trichoderma viride* (ATCC 36316) for protein enrichment. World Academy of Science, Engineering and Technology. 2013;**7**(3):6892-6991

[14] Motarjemi Y. Impact of small scale fermentation technology on food safety in developing countries. International Journal of Food Microbiology. 2002;**75**(3):213-229. DOI: 10.1016/S0168-1605(01)00709-7

[15] Smid EJ, Hugenholtz J. Functional genomics for food fermentation processes. Annual Review of

Food Science and Technology. 2010;**1**:497-519. DOI: 10.1146/annurev.food.102308.124143

[16] Caplice E, Fitzgerald GF. Food fermentations: Role of microorganisms in food production and preservation. International Journal of Food Microbiology. 1999;**50**(1-2):131-149. DOI: 10.1016/S0168-1605(99)00082-3

[17] Kostinek M, Specht I, Edward VA, Schillinger U, Hertel C, Holzapfel WH, et al. Diversity and technological properties of predominant lactic acid bacteria from fermented cassava used for the preparation of Gari, a traditional African food. Systematic and Applied Microbiology. 2005;**28**(6):527-540. DOI: 10.1016/j.syapm.2005.03.001

[18] Achi OK, Akomas NS. Comparative assessment of fermentation techniques in the processing of fufu, a traditional fermented cassava product. Pakistan Journal of Nutrition. 2006;**5**(3):224-229

[19] Stiles ME, Holzapfel WH. Lactic acid bacteria of foods and their current taxonomy. International Journal of Food Microbiology. 1997;**36**(1):1-29. DOI: 10.1016/S0168-1605(96)01233-0

[20] Zulu RM, Dillon VM, Owens JD. Munkoyo beverage, a traditional Zambian fermented maize gruel using *Rhynchosia* root as amylase source. International Journal of Food Microbiology. 1997;**34**(3):249-258. DOI: 10.1016/S0168-1605(96)01195-6

[21] Oguntoyinbo FA, Cho GS, Trierweiler B, Kabisch J, Rösch N, Neve H, et al. Fermentation of African kale (*Brassica carinata*) using *L. plantarum* BFE 5092 and *L. fermentum* BFE 6620 starter strains. International Journal of Food Microbiology. 2016;**238**:103-112. DOI: 10.1016/j.ijfoodmicro.2016.08.030

[22] Ray RC, Swain MR. Bio-ethanol, bioplastics and other fermented industrial products from cassava starch and flour. In: Colleen MP, editor. Cassava: Farming, Uses and Economic Impact. Hauppauge: Nova; 2011. pp. 1-32

[23] Raimbault M. General and microbiological aspects of solid substrate fermentation. Electronic Journal of Biotechnology. 1998;**1**(3):26-27. DOI: 10.4067/S0717-34581998000300007

[24] Nambisan B, Sundaresan S. Effect of processing on the cyanoglucoside content of cassava. Journal of the Science of Food and Agriculture. 1985;**36**(11):1197-1203. DOI: 10.1002/jsfa.2740361126

[25] Bradbury JH, Denton IC. Rapid wetting method to reduce cyanogen content of cassava flour. Food Chemistry. 2010;**121**(2):591-594. DOI: 10.1016/j.foodchem.2009.12.053

[26] Ezekiel OO, Aworh OC, Blaschek HP, Ezeji TC. Protein enrichment of cassava peel by submerged fermentation with *Trichoderma viride* (ATCC 36316). African Journal of Biotechnology. 2010;**9**(2):187-194. DOI: 10.5897/AJB09.620

[27] Reade AE, Gregory KF. High-temperature production of protein-enriched feed from cassava by fungi. Applied and Environmental Microbiology. 1975;**30**(6):897-904

[28] Vlavonou BM. Cassava processing technologies in Africa. In: Proceedings of the Interregional Experts' Group Meeting on the Exchange of Technologies for Cassava Processing Equipment and Food Products; 13-19 April 1988; Ibadan, Nigeria. New York: UNICEF House; 1988. pp. 19-25

[29] Akindahunsi AA, Oboh G, Oshodi AA. Effect of fermenting cassava with *Rhizopus oryzae* on the chemical

composition of its flour and Gari products. Rivista Italiana delle Sostanze Grasse. 1999;**76**:437-440

[30] Behera SS, Ray RC. Microbial linamarase in cassava fermentation. In: Ramesh RC, Christina MS, editors. Microbial Enzyme Technology in Food Applications. Boka Raton: CRC Press; 2017. pp. 337-346

[31] Food and Agricultural Organization. Food Outlook: Global Market Analysis. Rome: FAO; 2009. pp. 23-27

[32] Food and Agricultural Organization. Food Outlook: Biannual Report on Global Food Markets. Rome: FAO; 2018

[33] Gunawan S, Istighfarah Z, Aparamarta HW, Syarifah F, Dwitasari I. Utilization of modified cassava flour and its by-products. In: Klein C, editor. Handbook on Cassava. New York: Nova Science Publisher; 2017. pp. 271-295

[34] Westby A. Cassava utilization, storage and small-scale processing. In: Hillocks RJ, Thresh JM, Bellotti AC, editors. Cassava: Biology, Production and Utilization. New York: CABI Publishing; 2002. pp. 281-300

[35] Defloor I, Nys M, Delcour JA. Wheat starch, cassava starch, and cassava flour impairment of the breadmaking potential of wheat flour. Cereal Chemistry. 1993;**70**(5):526-530

[36] Dakwa S, Sakyi-Dawson E, Diako C, Annan NT, Amoa-Awua WK. Effect of boiling and roasting on the fermentation of soybeans into dawadawa (*soy-dawadawa*). International Journal of Food Microbiology. 2005;**104**(1):69-82. DOI: 10.1016/j.ijfoodmicro.2005.02.006

[37] Yao AA, Dortu C, Egounlety M, Pinto C, Edward VA, Huch M, et al. Production of freeze-dried lactic acid bacteria starter culture for cassava fermentation into gari. African Journal of Biotechnology. 2009;**8**(19):4996-5004

[38] Udoro EO, Kehinde AT, Olasunkanmi SG, Charles TA. Studies on the physicochemical, functional and sensory properties of gari processed from dried cassava chips. Journal of Food Processing & Technology. 2014;**5**(1):293. DOI: 10.4172/2157-7110.1000293

[39] Aro SO. Improvement in the nutritive quality of cassava and its by-products through microbial fermentation. African Journal of Biotechnology. 2008;**7**(25):4789-4797. DOI: 10.5897/AJB08.1005

[40] Raimbault M, Deschamps F, Meyer F, Senez JC. Direct protein enrichment of starchy products by fungal solid fermentation. In: Proceedings of the 5th International Conference on Global Impacts of Applied Microbiology; 21-26 November 1977. Bangkok; 1977

[41] Oboh G, Elusiyan CA. Changes in the nutrient and anti-nutrient content of micro-fungi fermented cassava flour produced from low-and medium-cyanide variety of cassava tubers. African Journal of Biotechnology. 2007;**6**(18):2150-2157. DOI: 10.5897/AJB2007.000-2336

[42] Essers AA, Jurgens CM, Nout MR. Contribution of selected fungi to the reduction of cyanogen levels during solid substrate fermentation of cassava. International Journal of Food Microbiology. 1995;**26**(2):251-257. DOI: 10.1016/0168-1605(94)00116-N

[43] Iyayi EA, Losel DM. Protein enrichment of cassava by-products through solid state fermentation by fungi. Journal of Food Technology in Africa. 2001;**6**(4):116-118. DOI: 10.4314/jfta.v6i4.19301

[44] Iyayi EA, Tewe OO. Effect of protein deficiency on utilization

of cassava peel by growing pigs. In: Proceedings of the IITA/ILCA/ University of Ibadan Workshop on the Potential Utilisation of Cassava as Livestock Feed in Africa; 14-18 November 1988. Ibadan: IITA; 1988. pp. 54-57

[45] Ruqayyah TI, Jamal P, Alam MZ, Mirghani ME, Jaswir I, Ramli N. Application of response surface methodology for protein enrichment of cassava peel as animal feed by the white-rot fungus *Panus tigrinus* M609RQY. Food Hydrocolloids. 2014;**42**(15):298-303. DOI: 10.1016/j. foodhyd.2014.04.027

[46] Oboh G. Nutrient enrichment of cassava peels using a mixed culture of *Saccharomyces cerevisae* and *Lactobacillus spp* solid media fermentation techniques. Electronic Journal of Biotechnology. 2006;**9**(1):46-49. DOI: 10.4067/S0717-34582006000100007

[47] Montagnac JA, Davis CR, Tanumihardjo SA. Nutritional value of cassava for use as a staple food and recent advances for improvement. Comprehensive Reviews in Food Science and Food Safety. 2009;**8**(3):181-194. DOI: 10.1111/j.1541-4337.2009.00077.x

[48] Wargiono J, Richana N, Hidajat A. Contribution of cassava leaves used as a vegetable to improved human nutrition in Indonesia. In: Proceedings of the Seventh Regional Workshop on Cassava Research and Development in Asia: Exploring New Opportunities for an Acient Crop; 28 October – 01 November 2002. Bangkok: CIAT; 2007. pp. 466-471

[49] Morgan NK, Choct M. Cassava: Nutrient composition and nutritive value in poultry diets. Animal Nutrition. 2016;**2**(4):253-261. DOI: 10.1016/j. aninu.2016.08.010

[50] Gidamis AB, O'Brien GM, Poulter NH. Cassava detoxification of traditional Tanzanian cassava foods. International Journal of Food Science and Technology. 1993;**28**(2):211-218. DOI: 10.1111/j.1365-2621.1993. tb01266.x

[51] Balagopalan C. Cassava utilization in food, feed and industry. In: Hillocks RJ, Thresh JM, Bellotti AC, editors. Cassava: Biology, Production and Utilization. New York: CABI Publishing; 2002. pp. 301-318

[52] Achidi AU, Ajayi OA, Maziya-Dixon BU, Bokanga M. The effect of processing on the nutrient content of cassava (*Manihot esculenta Crantz*) leaves. Journal of Food Processing & Preservation. 2008;**32**(3):486-502. DOI: 10.1111/j.1745-4549.2007.00165.x

[53] Fasuyi AO. Nutrient composition and processing effects on cassava leaf (*Manihot esculenta, Crantz*) antinutrients. Pakistan Journal of Nutrition. 2005;**4**(1):37-42

[54] Vetter J. Plant cyanogenic glycosides. Toxicon. 2000;**38**(1):11-36. DOI: 10.1016/S0041-0101(99)00128-2

[55] Montagnac JA, Davis CR, Tanumihardjo SA. Processing techniques to reduce toxicity and antinutrients of cassava for use as a staple food. Comprehensive Reviews in Food Science and Food Safety. 2009;**8**(1):17-27. DOI: 10.1111/ j.1541-4337.2008.00064.x

[56] Padmaja G, Steinkraus KH. Cyanide detoxification in cassava for food and feed uses. Critical Reviews in Food Science and Nutrition. 1995;**35**(4):299-339. DOI: 10.1080/10408399509527703

[57] Ngudi DD, Kuo YH, Lambein F. Amino acid profiles and protein quality of cooked cassava leaves or '*saka-saka*'. Journal of the Science of Food and Agriculture. 2003;**83**(6):529-534. DOI: 10.1002/jsfa.1373

[58] Ngudi DD, Kuo YH, Lambein F. Cassava cyanogens and free amino acids in raw and cooked leaves. Food and Chemical Toxicology. 2003;**41**(8):1193-1197. DOI: 10.1016/S0278-6915(03)00111-X

[59] Bradbury JH, Denton IC. Mild method for removal of cyanogens from cassava leaves with retention of vitamins and protein. Food Chemistry. 2014;**1**(158):417-420. DOI: 10.1016/j.foodchem.2014.02.132

[60] Latif S, Müller J. Potential of cassava leaves in human nutrition: A review. Trends in Food Science and Technology. 2015;**44**(2):147-158. DOI: 10.1016/j. tifs.2015.04.006

[61] Morales EM, Domingos RN, Angelis DF. Improvement of protein bioavailability by solid-state fermentation of babassu mesocarp flour and cassava leaves. Waste and Biomass Valorization. 2018;**9**(4):581-590. DOI: 10.1007/s12649-016-9759-y

[62] Kobawila SC, Louembe D, Keleke S, Hounhouigan J, Gamba C. Reduction of the cyanide content during fermentation of cassava roots and leaves to produce bikedi and ntoba mbodi, two food products from Congo. African Journal of Biotechnology. 2005;**4**(7):689-696. DOI: 10.5897/AJB2005.000-3128

[63] Knowles CJ. Microorganisms and cyanide. Bacteriological Reviews. 1976;**40**(3):652-680

[64] Pandey A. Recent process developments in solid-state fermentation. Process Biochemistry. 1992;**27**(2):109-117. DOI: 10.1016/0032-9592(92)80017-W

[65] Begum R, Rakshit SK, Rahman SM. Protein fortification and use of cassava flour for bread formulation. International Journal of Food Properties. 2011;**14**(1):185-198. DOI: 10.1080/10942910903160406

[66] Jamal P, Tijani RI, Alam MZ, Mirghani ME. Effect of operational parameters on solid-state fermentation of cassava peel to an enriched animal feed. Journal of Applied Sciences. 2012;**12**(11):1166-1170. DOI: 10.3923/jas.2012.1166.1170

[67] Purwadaria T. Solid substrate fermentation of cassava Peel for poultry feed ingredient. WARTAZOA. Indonesian Bulletin of Animal and Veterinary Sciences. 2013;**23**(1):15-22. DOI: 10.14334/wartazoa.v23i1.955

[68] Thongkratok R, Khempaka S, Molee W. Protein enrichment of cassava pulp using microorganisms' fermentation techniques for use as an alternative animal feedstuff. Journal of Animal and Veterinary Advances. 2010;**9**(22):2859-2862. DOI: 10.3923/javaa.2010.2859.2862

[69] Phrueksawan P, Kulpreecha S, Sooksai S, Thongchul N. Direct fermentation of L (+)-lactic acid from cassava pulp by solid-state culture of *Rhizopus oryzae*. Bioprocess and Biosystems Engineering. 2012;**35**(8):1429-1436. DOI: 10.1007/s00449-012-0731-3

[70] Roger DD, Jean-Justin EN, Francois-Xavier ET. Cassava solid-state fermentation with a starter culture of *Lactobacillus plantarum* and *Rhizopus oryzae* for cellulase production. African Journal of Microbiology Research. 2011;**5**(27):4866-4872. DOI: 10.5897/AJMR11.790

[71] Essers AJ, Bennik MH, Nout MJ. Mechanisms of increased linamarin degradation during solid-substrate fermentation of cassava. World Journal of Microbiology and Biotechnology. 1995;**11**(3):266-270. DOI: 10.1007/BF00367096

[72] Lateef A, Oloke JK, Kana EG, Oyeniyi SO, Onifade OR, Oyeleye AO, et al. Improving the quality of

agro-wastes by solid-state fermentation: Enhanced antioxidant activities and nutritional qualities. World Journal of Microbiology and Biotechnology. 2008;**24**(10):2369-2374. DOI: 10.1007/s11274-008-9749-8

[73] Azmi AS, Yusuf N, Jimat DN, Puad NI. Co-production of lactic acid and ethanol using *Rhizopus Sp*. from hydrolyzed inedible cassava starch and leaves. IIUM Engineering Journal. 2016;**17**(2):1-10. DOI: 10.31436/iiumej.v17i2.610

[74] Chen HZ, Li ZH. Bioreactor engineering. Chinese Journal of Process Biotechnology. 1998;**18**:46-49

[75] Nagel FJ. Process Control of Solid-State Fermentation: Simultaneous Control of Temperature and Moisture Content [Thesis]. Wageningen: Wageningen University; 2002

[76] Pandey A. Solid-state fermentation. Biochemical Engineering Journal. 2003;**13**(2-3):81-84. DOI: 10.1016/S1369-703X(02)00121-3

[77] Gervais P, Molin P. The role of water in solid-state fermentation. Biochemical Engineering Journal. 2003;**13**(2-3):85-101. DOI: 10.1016/S1369-703X(02)00122-5

[78] Mahanta N, Gupta A, Khare SK. Production of protease and lipase by solvent tolerant Pseudomonas aeruginosa PseA in solid-state fermentation using Jatropha curcas seed cake as substrate. Bioresource Technology. 2008;**99**(6):1729-1735. DOI: 10.1016/j.biortech.2007.03.046

[79] Mustafa SR, Husaini A, Hipolito CN, Hussain H, Suhaili N, Roslan HA. Application of response surface methodology for optimizing process parameters in the production of amylase by *Aspergillus flavus* NSH9 under solid state fermentation. Brazilian Archives of Biology and Technology. 2016;**59**. DOI: 10.1590/1678-4324-2016150632

[80] Grover A, Maninder A, Sarao LK. Production of fungal amylase and cellulase enzymes via solid state fermentation using *Aspergillus oryzae* and *Trichoderma reesei*. International Journal of Advancements in Research & Technology. 2013;**2**(8):108-124

[81] Lonsane BK, Ghildyal NP, Budiatman S, Ramakrishna SV. Engineering aspects of solid-state fermentation. Enzyme and Microbial Technology. 1985;**7**(6):258-265. DOI: 10.1016/0141-0229(85)90083-3

[82] Saucedo-Castañeda G, Gutierrez-Rojas M, Bacquet G, Raimbault M, Viniegra-González G. Heat transfer simulation in solid substrate fermentation. Biotechnology and Bioengineering. 1990;**35**(8):802-808. DOI: 10.1002/bit.260350808

[83] González-Blanco P, Saucedo-Castañeda G, Viniegra-González G. Protein enrichment of sugar cane by-products using solid-state cultures of *Aspergillus terreus*. Journal of Fermentation and Bioengineering. 1990;**70**(5):351-354. DOI: 10.1016/0922-338X(90)90150-U

[84] Manpreet S, Sawraj S, Sachin D, Pankaj S, Banerjee UC. Influence of process parameters on the production of metabolites in solid-state fermentation. Malaysian Journal of Microbiology. 2005;**1**(2):1-9

[85] Durand A, Arnous P, de Chardin OT, Chereau D, Boquien C. Protein enrichment of sugar beet pulp by solid-state fermentation. In: Ferranti MP, Fiechter A, editors. Production and Feeding of Single-Cell Protein. London: Applied Science Publisher; 1983. pp. 120-123

[86] Grajek W. Cooling aspects of solid-state cultures of mesophilic and thermophilic fungi. Journal of Fermentation Technology.

1988;**66**(6):675-679. DOI: 10.1016/0385-6380(88)90072-6

[87] Ghildyal NP, Gowthaman MK, Rao KR, Karanth NG. Interaction of transport resistances with biochemical reaction in packed-bed solid-state fermentors: Effect of temperature gradients. Enzyme and Microbial Technology. 1994;**16**(3):253-257. DOI: 10.1016/0141-0229(94)90051-5

[88] Gowthaman MK, Ghildyal NP, Rao KR, Karanth NG. Interaction of transport resistances with biochemical reaction in packed bed solid state fermenters: The effect of gaseous concentration gradients. Journal of Chemical Technology and Biotechnology. 1993;**56**(3):233-239. DOI: 10.1002/jctb.280560303

[89] Mitchell DA, Do DD, Greenfield PF, Doelle HW. A semi-mechanistic mathematical model for growth of *Rhizopus oligosporus* in a model solid-state fermentation system. Biotechnology and Bioengineering. 1991;**38**(4):353-362. DOI: 10.1002/bit.260380405

[90] Awasthi MK, Pandey AK, Bundela PS, Khan J. Co-composting of organic fraction of municipal solid waste mixed with different bulking waste: Characterization of physicochemical parameters and microbial enzymatic dynamic. Bioresource Technology. 2015;**182**:200-207. DOI: 10.1016/j.biortech.2015.01.104

[91] Raimbault M. Fermentation en milieu solid: Croissance de champignons filamentous sure substrate amylacé. Paris: ORSTOM; 1981. p. 291

[92] Correia R, Magalhaes M, Macêdo G. Protein enrichment of pineapple waste with *Saccharomyces cerevisiae* by solid state bioprocessing. Journal of Scientific and Industrial Research. 2007;**66**(3):259-262

8

A Comprehensive Overview of the Potential of Tequila Industry By-Products for Biohydrogen and Biomethane Production: Current Status and Future Perspectives

Octavio García-Depraect, Daryl Rafael Osuna-Laveaga and Elizabeth León-Becerril

Abstract

Nowadays, the use of agro-industrial by-products as alternative sustainable resources to generate bioenergy and high-value bioproducts is one of the most important research topics to tackle environmental concerns related to the excessive consumption of fossil-based fuels and rapid urbanization and industrialization. This chapter provides a broad overview of the potential of the main tequila industry by-products, agave bagasse and tequila vinasse, for biohydrogen (bioH_2) and biomethane (bioCH_4) production via dark fermentation and anaerobic digestion, respectively. First, pretreatment or conditioning steps commonly applied to tequila by-product streams before downstream biological processes are highlighted. The operational performance of bioH_2- and bioCH_4-producing reactors is subsequently reviewed, with a focus on reactor configuration and performance, microbial metabolic pathways, and the characterization of microbial communities. Additionally, the development of multi-stage anaerobic digestion processes is comprehensively discussed from a practical point of view. Finally, limitations and potential improvements in the field of bioH_2 and bioCH_4 production are presented.

Keywords: agave bagasse, tequila vinasse, dark fermentation, anaerobic digestion, biofuels

1. Tequila production process and its main by-products: agave bagasse and tequila vinasse

Tequila is a Mexican alcoholic beverage obtained from the distillation of fermented juice of the mature stems of *Agave tequilana* Weber var. azul. It possesses appellation of origin since 1974 and has received international recognition in the market. As an example, tequila-processing plants produced around 309 million liters of tequila in 2018, of which ~72% were exported, highlighting its international demand [1]. Thus, tequila production represents one of the most important activities for Mexico. In general, there are three major stages in the tequila production process, namely agave juice (must) extraction, fermentation, and distillation. In the first stage, the agave juice containing fermentable sugars is first obtained either through cooking or not-cooking processes. In the former, agave stems are cooked in ovens or autoclaves at high temperatures (95–120°C) for a long time (usually 8–12 h). Once cooked, the water-soluble carbohydrates are extracted by simultaneous shredding and pressure washing followed by pressing. In the latter, raw agave juice is obtained from previously shredded raw agave stems using hot water (80°C) through the use of equipment called diffuser. Afterward, the carbohydrates contained in the raw agave juice are hydrolyzed for 4–6 h under acidic conditions (pH 1.8–3) at high temperatures (80–85°C) [2, 3]. In the second stage, the agave juice is subjected to an alcoholic fermentation process, wherein agave sugars are transformed to ethanol, carbon dioxide, and other compounds (e.g. aldehydes, esters, furans, and ketones) by the action of different microorganisms, particularly yeasts [2, 3]. In the third stage, the fermented must is subjected to a two-step distillation process to obtain tequila [2, 4].

At this point, it must be noted that enormous quantities of solid (*Agave tequilana* bagasse, hereinafter referred to as AB) and semi-liquid (tequila vinasse, hereinafter referred to as TV) by-products are generated each year during the process of tequila manufacturing, particularly after the stages of agave juice extraction and distillation, respectively (**Figure 1**). It has been estimated that 1.4 kg of AB and 10–12 L of TV are obtained by each liter of tequila produced [4, 5]. Considering the tequila production of 264.9 ± 31.2 million liters reported in the last lustrum (2014–2018) by

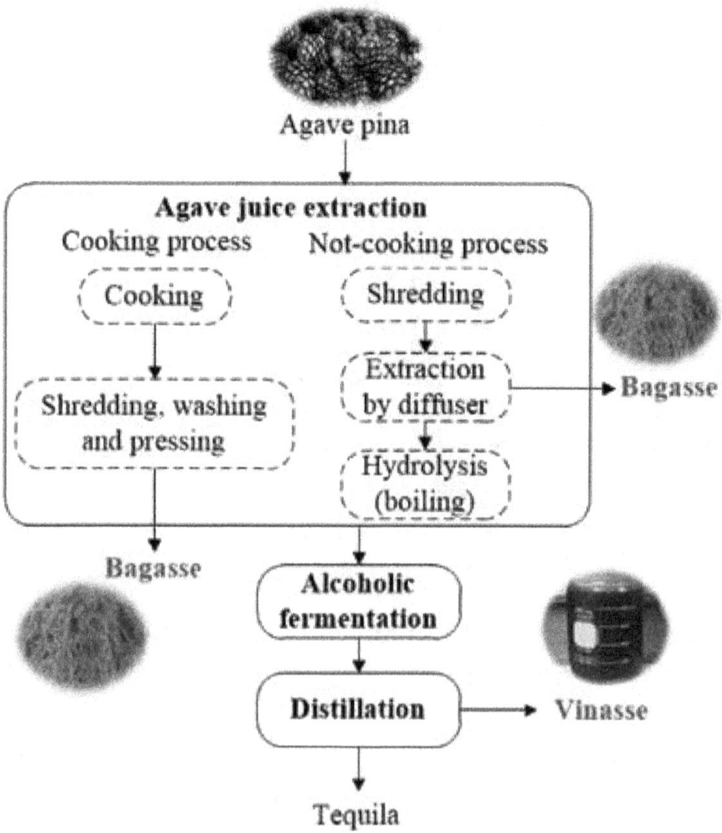

Figure 1.
Tequila manufacturing process and generation of agave bagasse and tequila vinasse.

the Tequila Regulatory Council [1], the generation of AB and TV is equivalent to 370,916 $\pm$ 43,701 tons and 2914.3 $\pm$ 343.3 million liters per year, respectively. The physicochemical composition of a given stream of AB and TV may change from batch to batch, depending mainly on the raw materials used (e.g. maturity of agave), juice extraction process (cooked and uncooked agave), and the prevailing conditions of fermentation and distillation in the case of TV [3, 6–9]. Despite such influential factors, there are some general features that can be distinguished between AB and TV. Concerning AB, it is a lignocellulosic material with a compo-sition of 11–57% hemicellulose, 31–53% cellulose, 7–15% lignin, and 19–57% extrac-tives [4, 8, 9]. Extractives are the nonstructural components of lignocellulose, including fats, phenolics, resin acids, waxes, and inorganics [10]. Regarding TV, it is a brown and acidic wastewater (pH of 3.4–4.5, total acidity of 1500–6000 mg-$CaCO_3$/L) containing high chemical oxygen demand (COD) concentration of 40–100 g/L, as well as high total solids (25–50 g/L), salts, metal ions, organic acids, phenolic compounds, and melanoidins [3, 5, 7, 11].

Regarding the management and final disposition of AB and TV, it must be highlighted that only a small part of the whole AB generated is used in the manufacturing of different products such as animal feeds, fertilizers, bricks, mattresses, furniture, and packing materials [12, 13]. Therefore, most of AB is treated as waste and returned to the fields in the form of piles that are directly exposed to outdoor conditions, where they may cause leachates, odor generation, and atmospheric pollution [12, 14]. In the case of TV, it has been reported that approximately 80% of the total volume of TV generated is discharged without receiving adequate treatment into receiving water bodies (e.g. rivers, lakes, and sewer system) or directly onto soil, which in turn can result in adverse environmental and human health impacts [5]. To valorize AB and TV and to face such disposal problems, nowadays, engineers and scientists are focusing on using them as potential substrates for the production of biofuels and value-added products in a tequila biorefinery framework. However, there are still several challenges that must be overcome before full-scale facilities could be implemented. This chapter provides an extended insight on (i) the pretreatment or conditioning steps of tequila by-product streams; (ii) the use of AB and TV to produce biogenic hydrogen (bioH_2) and methane (bioCH_4) via anaerobic fermentation processes, with a special emphasis on reactor configuration and operation, producing/competing metabolic pathways and the characterization of microbial communities; (iii) the development of multi-stage anaerobic digestion (AD) processes; and (iv) limitations and avenues for future research toward improving bioH_2 and bioCH_4 production.

2. Pretreatment/conditioning of agave bagasse and tequila vinasse

AD is the core technology for the treatment of several biodegradable organic wastes with concomitant bioenergy recovery in the form of biogas that is rich in bioCH_4, although bioH_2 may also be recovered. Besides bioCH_4 recovery, AD is advantageous due to low energy and nutrient requirements, low sludge production, and high organic loading capacity (20–35 g-COD/L-d) [15]. From a biochemical point of view, AD consists of four successive steps, namely hydrolysis, acidogenesis, acetogenesis and methanogenesis [15, 16].

It is worth mentioning that in the case of AB, the low biodegradability due to its lignocellulosic structure constitutes one of the main barriers to accelerate hydrolysis and enhance the recovery of bioH_2/bioCH_4. In the case of TV, its complex composition such as high COD, high solids content, unbalanced nutrient, presence of putative toxicants (e.g. organic acids, phenols, melanoidins) and the negligible alkalinity along with the high concentration of components with a tendency to suffer very rapid acidification constitutes the major limitations for bioH_2/bioCH_4 production. Thus, in practice, before the feedstock (AB or TV) is sent to either the hydrogenogenic or the methanogenic stage, a pretreatment/conditioning step is commonly performed as a prerequisite to improve its biodegradability as well as to prevent DF/AD processes from potential toxicants, elevated solids, and organic overloading (**Figure 2**). Unlike AB, TV is only subjected to one or more conditioning steps. Commonly, they consist of lowering temperature, rising pH (adding alkalinity), diluting, adding complementary nutrients, and removing suspended solids (**Figure 2**).

In contrast, AB is exposed to a drying step to prevent fungal and bacterial growth, mainly for long-time storage. Once AB is dried, it is subjected to a mechanical milling step devoted to reducing particle size, thereby increasing surface area, which makes carbohydrates more easily available for downstream processes. The mechanical fractionation also makes AB more homogeneous and easier to handle. After milling, the pretreatment applied to AB for either bioH_2 or bioCH_4 production may differ. For such purposes, dilute acid, alkaline hydrogen peroxide, detoxification and enzymatic hydrolysis have been evaluated in detail. Arreola-Vargas et al. [8] pretreated cooked and uncooked AB through a dilute acid hydrolysis at 5% (w/v), 56.4–123.6°C, 1.2–2.8% HCl, and 0.3–3.7 h reaction time, finding temperature as the principal factor which could increase the hydrolysis yield. Total sugars concentrations obtained were 27.9 and 18.7 g/L for cooked and uncooked AB hydrolysates, respectively. The higher yield of cooked AB was attributed to the fact that during the elaboration of tequila using cooking process, agave stems receives an in situ thermal treatment. Nevertheless, high concentrations (up to 1200 mg/L) of hydroxymethylfurfural (HMF) were detected in the cooked AB. In a further study, Arreola-Vargas et al. [17] pretreated AB through either acid or enzymatic hydrolysis for bioCH_4 and bioH_2 production. Acid hydrolysis was carried out for 1.3 h at 5% (*w/v*) of AB, 2.7% HCl and 124°C, while enzymatic hydrolysis was performed at 4% (*w/v*) of AB in 50 mM citrate buffer at pH 4.5 with Celluclast 1.5 L at 40 filter paper units (FPU) for 10 h at 45°C. As a result, 17.3 and 8.9 g-total sugars/L were obtained from acid and enzymatic hydrolysis, respectively. However, unlike enzymatic hydrolysates, acid hydrolysates promoted the generation of potential inhibitors such as formic acid (HFor), acetic acid (HAc), and phenolic and furanic compounds. In another study, Breton-Deval et al. [18] compared the type of acid catalyst (HCl vs. H_2SO_4)

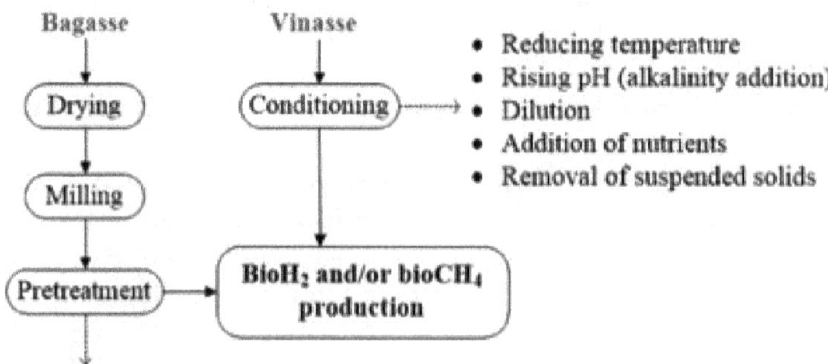

Figure 2.
Flow chart of biohydrogen and biomethane production process from agave bagasse and tequila vinasse.

on the chemical composition of hydrolysates of AB. Overall, results showed that the use of HCl induced higher sugar recoveries than the use of H_2SO_4, 0.39 versus 0.26 g-total sugars/g of AB. Furthermore, the H_2SO_4 hydrolysate contained higher concentrations of HAc and furans. To remove undesirable compounds derived from acid hydrolysis of AB (30 g AB, HCl 1.9%, 130°C, 132 min reaction time), Valdez-Guzmán et al. [19] performed detoxification of acid AB hydrolysates using 1% (*w/v*) powdered coconut shell-activated carbon. Under batch conditions (pH 0.6, 20 min reaction time, 150 rpm, room temperature), the highest removal of HAc and phenols obtained were 89 and 21%, respectively, with minimal losses of fermentable sugars (3.6%). Besides, during acid hydrolysis, a hydrolysis yield of almost 40% of total sugars, a delignification of 44%, complete hydrolysis of hemicellulose, and no detection of furfural or HMF in the hydrolysate was obtained. In another study, Contreras-Dávila et al. [20] pretreated AB for bioH_2 production using Celluclast 1.5 L during 10 h, obtaining sugar yields in the range of 0.19–0.38 g-total sugars/g of AB. Montiel and Razo-Flores [21] also pretreated AB by enzymatic hydrolysis to produce bioH_2 and bioCH_4. The conditions were 3.5% (*w/v*) of AB with Celluclast 1.5 L at 18 FPU/g of AB at 40°C during 12 h. The resulting hydrolysate had 27.2 g/L of total COD with 5.3 ± 0.8 g/L of total sugars (0.15 g-total sugars/g of AB) which contributed to 20% of the total COD, citrate buffer with 26%, enzyme with 38%, and other non-determined components with 16%. In the same year, Galindo-Hernández et al. [22] used alkaline hydrogen peroxide (AHP) as a pretreatment to remove lignin before enzymatic hydrolysis of AB. Under the experimental conditions tested (5% *w/v* of AB, 2% *w/v* of AHP, 50°C, pH 11.5 using NaOH, 120 rpm, 1.5 h reaction time), 97% of the lignin was removed and 88% of holocellulose (cellulose and hemicellulose) was recovered, promoting that the polysaccharide fractions are more available or exposed to a further enzymatic attack. The authors also demonstrated, in delignification terms, that it is better to use hydrogen peroxide and NaOH solution in a combined form than in a separate or sequential way and that using binary enzymatic hydrolysis (cellulases and hemicellulases) may improve the yield, percentage, and productivity of saccharifi-cation, which were 0.19 g-total sugars/g of AB, 26.7% and 17.1 g-total sugars/g of AB-h, respectively. The synergistic effect of using binary enzymatic hydrolysis was verified by Montoya-Rosales et al. [23], who compared the enzymatic hydrolysis of AB using a binary enzyme preparation that is composed of Celluclast 1.5 L and Viscozyme L with a single enzyme, that is, Stonezyme, which is a commercial cellulase preparation. The results showed that hydrolysis yields were higher with the binary enzymatic hydrolysis, 0.27 versus 0.22 g-carbohydrates/g of AB and 0.5 versus 0.28 g-COD/g of AB.

3. Biohydrogen production from agave bagasse and tequila vinasse

H_2 is one of the most promising alternative energy carriers to partly fulfill the growing energy demands and overcome fossil fuel dependency and has attracted global attention for its highest energy content per unit weight (142 kJ/g) and carbon-free nature since it generates only water vapor during combustion. It can be used for a variety of purposes either alone to produce energy in fuel cells and combustion engines or blended with CH_4 to produce a superior fuel known as hythane [24]. Comparing thermochemical, electrochemical, and biological ways of producing H_2, the latter is considered the most sustainable because it is eco-friendlier and less energy intensive. Among biological processes, dark fermentation (DF) is thought to be practically applicable at large commercial scales in a near time horizon owing to its capability of producing bioH_2 at higher rates and versatility of utilizing several different types of carbohydrate-rich wastes as substrate [25]. In this connection, since AB and TV are abundantly available, renewable, and have a high content of carbohydrates, they have been considered as suitable feedstocks for bioH_2 production. In the following sections, the operational performance, metabolic pathways, and microbial communities of DF systems treating either AB or TV are extensively reviewed.

3.1 Operational performance

Regarding the use of AB for bioH_2 production (**Table 1**), the first systematic study dealing with bioH_2 production from AB was conducted by Arreola-Vargas et al. (2016) [17], who assessed the use of AB hydrolysates obtained either from acid or enzymatic pretreatment for bioH_2 production. To the end, different proportions of hydrolysate (20, 40, 60, 80, and 100% *v/v*) were tested in an automatic methane potential test system (AMPTS II provided by Bioprocess control) at 37°C, 120 rpm, initial pH of 7, and using 10 g-volatile suspended solids (VSS)/L of heat-pretreated anaerobic granular sludge. Overall, the best bioH_2 production performance was achieved in the assays with enzymatic hydrolysate, obtaining the maximal bioH_2 yield (HY_2) and volumetric bioH_2 production rate (VHPR) of 3.4 mol-H_2/mol-hexose and 2.4 NL-H_2/L-d, respectively, both with the hydrolysate at 40% (*v/v*). The lower values observed with the acid hydrolysate were attributed to the feed-stock composition in terms of sugar profile, weak acids, furans, and phenolics.

In another work, Contreras-Dávila et al. [20] used an enzymatic AB hydrolysate for bioH_2 production in a continuously stirred tank reactor (CSTR) and a trickling bed reactor (TBR), which were operated up to 87 days under different organic loading rates (OLR, 17–60 g-COD/L-d) obtained by varying hydrolysate concentration and/or hydraulic retention time (HRT). The reactor configurations showed different performances. In the CSTR, the VHPR and HY_2 displayed an inverse correlation with maximum values of 2.53 L-H_2/L-d and 1.35 mol-H_2/mol-substrate, attained at OLR of 52.2 and 40.2 g-COD/L-d, respectively, both with 6 h HRT. The bioH_2 concentrations of the produced gas were between 18 and 35% (*v/v*). In contrast, in the TBR, increasing OLR up to 52.9 g COD/L-d (4 h HRT)

Pretreatment	Feeding	T (°C)	pH	YH_2 (NL/ kg AB)	VHPR (NL/L-d)	H_2 (% *v/v*)	Ref.
Acid hydrolysis	Batch	37	7[a]	1.6[b]	2.4	NR	[17]
Individual enzymatic hydrolysis	Batch	37	7[a]	140, 3.4[b]	2.4	NR	[17]
Individual enzymatic hydrolysis	Continuous	37	5.5	67	3.45	26–52	[20]
Individual enzymatic hydrolysis	Continuous	35	5.5	105	6	55	[21]
Alkaline hydrogen peroxide + binary enzymatic hydrolysis	Batch	37	7.5[a]	215	0.93	NR	[22]
Individual enzymatic hydrolysis	Semi-continuous	37	4.8	1.6[c]	0.6	49.3[d]	[26]
Acid hydrolysis + detoxification	Batch	37	8.2[a]	56.2	1.51	NR	[19]
Binary enzymatic hydrolysis	Continuous	37	5.5	117.8	13	51–60	[23]

Notes: All studies were conducted using thermally treated anaerobic granular sludge; [a]Initial pH value; [b]mol-H_2/mol hexose; [c]mol-H_2/mol of consumed sugar; [d]Value measured during the starting period; NR: not reported.

Table 1.
Comparison of the literature data on biohydrogen production efficiency using pretreated agave bagasse as feedstock.

simultaneously enhanced VHPR and HY_2, attaining values of 3.45 L-H_2/L-d and 1.53 mol-H_2/mol-substrate, respectively, with bioH_2 concentrations of the produced gas between 26 and 52% (*v/v*). The observed bioH_2 production performances were explained by differences in the liquid and gas flow rates, agitation speed, and liquid-gas interface between the CSTR and TBR configurations, which in turn may have caused distinct bioH_2 concentrations in the liquid phase.

In a further study which set up to assess the batch bioH_2 production from pretreated AB with AHP followed by binary enzymatic saccharification (hemicellulases + cellulases), Galindo-Hernández et al. [22] performed a series of experiments in the AMPTS II system at 37°C, 150 rpm, initial pH of 7.5, and using an organic load of 5 g-COD/L and 13.5 g-volatile solid (VS)/L of thermally treated anaerobic sludge. The results suggested that delignification of AB and subsequent hydrolysis with a synergistic enzymatic mixture had a beneficial effect on bioH_2 production, obtaining a YH_2 of 3 mol-H_2/mol-hexose and a VHPR of 0.93 NL-H_2/L-d.

In an investigation on the effect of OLR and agitation speed on the continuous bioH_2 production from enzymatic hydrolysates of AB, Montiel and Razo-Flores [21] operated for 84 days a mesophilic (35°C) CSTR reactor (with a working volume of 1 L) inoculated with 4.5 g-VS/L of heat-treated anaerobic granular sludge and operated at different OLRs (40–52 g-COD/L-d), which were achieved by varying hydrolysate concentration. The evaluated stirring speeds were in the range of 150–300 rpm, while the HRT was maintained at 6 h during the whole operation. The authors observed that the strategy of increasing the agitation speed from 150 to 300 rpm favored both the VHPR and bioH_2 content in the gas phase, obtaining 6 NL-H_2/L-d and 55% (*v/v*), respectively, at an OLR of 44 g-COD/L-d. Such results indicated that the increase of the agitation speed in the CSTR improved the transfer of dissolved bioH_2 from the liquid to the reactor gas phase, overcoming one of the limitations for bioH_2 production previously observed by [21].

In another study, Toledo-Cervantes et al. [26] addressed the bioH_2 production from enzymatic hydrolysates of AB using an anaerobic sequencing batch reactor (AnSBR) with a working volume of 1.25 L. The reactor was inoculated with 10 g-VS/L of thermally treated anaerobic sludge and operated at 37°C, pH 4.8, and at four OLR (10.6–21.3 g-COD/L-d), which were modified by decreasing the cycle time (from 24 to 12 h) and increasing the COD concentration (from 8 to 12 and 16 g/L). Results showed that the highest OLR promoted the highest VHPR of 0.6 NL-H_2/L-d. Conversely, the YH_2 remained constant at 1.6 mol-H_2/mol of consumed sugar.

In a similar study, Valdez-Guzmán et al. [19] showed the importance not only of optimizing pretreatment but also of removing several compounds (e.g. furfural, HMF, phenolic compounds, and organic acids) that are generated during its application. They compared the bioH_2 production potential of undetoxified and detoxified acid hydrolysates from AB. The authors reported ~39 and ~9% increases on YH_2 and VHPR, respectively, comparing detoxified AB with activated carbon and undetoxified AB, 1.71 versus 1.23 mol-H_2/mol of consumed sugar and 1.51 versus 1.38 NL-H_2/L-d. Such increments were correlated to changes in the fermentation by-products suggesting the occurrence of different pathways or changes in the microbial community, since the detoxified hydrolysate produced HAc and butyric acid (HBu), while lactic acid (HLac) was found in the undetoxified hydrolysate.

Most recently, Montoya-Rosales et al. [23] compared and evaluated the continuous bioH_2 production from individual and binary enzymatic hydrolysates of AB in two different configurations, that is, CSTR and TBR. The experiments were carried out at 37°C and pH 5.5 and at various OLRs 36–100 g-COD/L-d, which were achieved by increasing the influent concentration, while keeping the HRT constant at 6 h. The results showed that the performance was highly dependent on the type of reactor and OLR. Regarding the CSTR configuration, in general, the higher OLR resulted in higher VHPR. Nonetheless, the bioH_2 production efficiency using individual enzymatic hydrolysate (0.72–2.25 NL-H_2/L-d and 11.8–20.4 NL-H_2/kg of AB) was lower compared to that obtained with the binary enzymatic hydrolysate (3.9–13 NL-H_2/L-d and 83.3–117.9 NL-H_2/kg of AB), with the maximum VHPR and YH_2 at 100 and 60 g-COD/L-d and 90 and 52 g-COD/L-d, respectively. Regarding the TBR configuration, the binary enzymatic hydrolysate also outperformed the individual one, obtaining the maximum VHPR of 5.76 NL-H_2/L-d at an OLR of 81 g-COD/L-d and YH_2 of 72.4 NL-H_2/kg of AB at an OLR of 69 g-COD/L-d. The enhancement was attributed, on one hand, to the use of binary hydrolysis that could have contributed to produce a higher proportion of monomers of easy degradation by bioH_2-producing bacteria (HPB) and to avoid the formation/release of potential inhibitors; on the other hand, to the differences of substrate availability given by the mode of growth in each reactor.

Concerning the use of TV for bioH_2 production (**Table 2**), there are a few studies in the literature, with a particular focus on (i) optimizing pretreatments to further enhance bioH_2 production [27]; (ii) testing the effect of different operational conditions such as pH [28, 29], temperature [28, 30], substrate concentration [28, 30, 31], solid content [22, 31], nutrient formulation [22, 31], inoculum addition [22, 31], HRT [22, 30, 32], and OLR [22, 32]; (iii) producing bioH_2 in different systems, such as serum bottle [33], fixed bed reactor (FBR) [34], and CSTR [35];

Pretreatment/conditioning	Feeding	T (°C)	pH	YH_2 *	VHPR (NL/L-d)	H_2 (% *v/v*)	Ref.
Alkalinization	Batch	35	6.5–7.5	1.5^a, 2.8^{b,f}	NR	NR	[27]
None	Semi-continuous	55	5.5	13.8^{b,f}	2.8	NR	[28]
Dilution, nutrient supplementation	Semi-continuous	35	5.5	NR	2.2	29.2	[30]
Dilution	Continuous	35	4.7	1.3^a, 1.36^c	1.7	64	[34]
Dilution, nutrient supplementation	Semi-continuous	35	5.5	0.12^d	1.4	NR	[35]
Dilution	Batch	36	5.5^g	0.7^b	0.5	NR	[33]
Co-fermentation	Batch	35	5.5	1.1^b	2.6	71	[11]
Nutrient supplementation	Batch	35	6.5–5.8	4.8^c, 0.12^e	3.8	70	[37]
Solid removal (centrifugation)	Batch	35	6.5–5.8	4.3^b, 0.11^e	5.4	71	[31]
Co-fermentation	Batch	35	5.5	1.2^b	2.4	68	[36]
Co-fermentation	Batch	35	6.5–5.8	2.5^b, 2.7^c	3.7	73	[29]
Solid removal (centrifugation), nutrient supplementation	Continuous	35	5.8	3.4^c	12.3	90	[38]

*Notes: Inoculum: anaerobic digester sludge [27, 28], thermally treated anaerobic granular sludge [11, 29–31, 33–38]; *Units: ^{a}mol-H_2/mol glucose; ^{b}NL-H_2/L of reactor; ^{c}NL-H_2/L of TV; ^{d}NL-H_2/g-COD; ^{e}NL-H_2/g-VS_{fed}; ^{f}Calculated from provided information; ^{g}Initial pH value; NR: not reported.*

Table 2.
Comparison of the literature data on biohydrogen production efficiency using tequila vinasse as feedstock.

(iv) evaluating the feasibility of co-fermentation [11, 36]; and (v) exploring the microbial ecology of the process [32, 36, 37].

More particularly, Espinoza-Escalante et al. [27] evaluated the effect of three pretreatments, that is, alkalinization, cavitation, and thermal pretreatment, on the metabolic profile and the increments of COD and total reducing sugars (TRS) of TV, as well as on its bioH_2 production potential. From that study, it can be concluded that the application of such pretreatments to raw TV resulted in different degrees of solubilization of COD and TRS, depending on the applied pretreatment and combinations thereof. However, there was no apparent relation in the consumption of TRS and COD with bioH_2 production. Indeed, the optimal conditions that led to the highest solubilization of both COD and TRS did not result in a significant improvement in the YH_2, which was about 2.8 NL-H_2/L of reactor, indicating that compounds other than TRS could be involved in the mechanism of bioH_2 production.

In another report, Espinoza-Escalante et al. [28] studied the effect of pH (4.5, 5.5, and 6.5), HRT (1, 3, and 5 d), and temperature (35 and 55°C) on the semi-continuous production of bioH_2 from TV. The experiments were performed in 1-L glass vessels inoculated with 10% (*v/v*) of mesophilic anaerobic digester sludge. The results showed that all factors studied had an important effect on bioH_2 production. The highest efficiency in terms of bioH_2 production was achieved at a pH of 5.5, an HRT of 5 d and a temperature of 55°C. Based on constructed mathematical models, pH was the most influential parameter.

In a similar study, Buitrón and Carvajal [30] investigated the effect of temperature (25 and 35°C), HRT (12 and 24 h), and substrate concentration on bioH_2 production from TV using a 7-L AnSBR, with a working volume of 6 L. The exchange volume was 50% with a reaction time of 11.3 or 5.3 h depending on the applied HRT, while pH and mixing were controlled at 5.5 and 153 rpm, respectively, in all cases. It was evidenced that all parameters studied affected the efficiency of bioH_2 production. The HRT had a major influence on bioH_2 production. It was found that the shorter the HRT, the higher the bioH_2 production. Overall, the maximum VHPR of 2.2 NL-H_2/L-d and an average bioH_2 content in the biogas of $29.2 \pm 8.8\%$ (*v/v*) were obtained at 35°C, 12 h HRT, and 3 g-COD/L OLR.

Later, Buitrón et al. [34] evaluated the performance of an FBR to produce bioH_2 in a continuous mode from TV. The reactor had a working volume of 1.7 L and was packed with polyurethane rings for biomass immobilization. The temperature, pH, HRT, and OLR were kept constant at 35°C, 4.7, 4 h, and 2.15 g-COD/L-d (influent concentration of 8 g-COD/L), respectively. After an initial acclimatization period of HPB to TV, the FBR exhibited a VHPR of 1.7 NL-H_2/L-d and a YH_2 of 1.36 NL-H_2/L of TV. In a follow-up study conducted by the same research group, by using a 0.6-L AnSBR operated under mesophilic and acidophilic conditions at an HRT of 6 h, it was observed that increasing substrate concentration from 2 to 16 g-COD/L increased the VHPR up to 1.4 NL-H_2/L-d. Hence, the use of TV for bioH_2 production did not result in inhibition [35].

Another interesting advance was made by García-Depraect et al. [11], who studied the technical feasibility of using a co-fermentation approach to produce bioH_2 from TV in a well-mixed reactor operated under batch mode. Nixtamalization wastewater (NW) was chosen as the complementary substrate based on its wide availability in Mexico and high alkalinity. The TV:NW ratio of 80:20 (*w/w*) resulted in the highest VHPR of 2.6 NL-H_2/L-d with a bioH_2 content in the gas phase of 71% (*v/v*). Interestingly, the co-fermentation study allowed the identification of iron and nitrogen as essential nutrients which may be limiting in TV-fed DF reactors. This identification becomes significant to avoid nutrient-limited conditions and to prevent excessive nutrient supplementation that has been occurring in several studies at bench scale, but its practice may be prohibited on larger scales.

In this field of progressive research, the effect of pH on the bioH_2 production efficiency was subsequently studied by García-Depraect et al. [29] through macro- and micro-scale behavior analysis approaches. It was found that fixed pH of 5.8 showed a longer lag phase compared with fixed pH of 6.5, but the latter promoted bioH_2 sink through propionogenesis. Based on the above observations, a two-stage pH-shift control strategy was devised to further increase bioH_2 production. The strategy entailed the control of pH at 6.5 for first ~29 h of culture to decrease the lag time, and then the pH was maintained at 5.8 to increase the bioH_2 conversion efficiency by inhibiting the formation of propionic acid (HPr). The pH-shift strategy reduced running time and enhanced bioH_2 production by 17%, obtaining 2.5 NL-H_2/L of reactor. In a further study, the use of TV as the sole carbon source in the batch bioH_2-yielding process was evaluated through a comprehensive approach entailing the operational performance, kinetic analysis, and microbial ecology [37]. A YH_2 of 4.3 NL-H_2/L of reactor and a peak VHPR of 3.8 NL-H_2/L-d were obtained. The effects of total solids content, substrate concentration, nutrient formulation, and inoculum addition on bioH_2 production performance from TV have been also investigated in batch experiments [31]. It was observed a consistent bioH_2 production which was primarily influenced by inoculum addition followed by substrate concentration, nutrient formulation, and solids content. Maximum VHPR (5.4 NL-H_2/L-d) and YH_2 (4.3 NL-H_2/L of reactor) were achieved by removing suspended solids and enhancing nutrient content, respectively [31]. Finally, the highest VHPR (12.3 NL-H_2/L-d, corresponding to ~3.4 NL-H_2/L of TV) up to date has been achieved via a novel multi-stage process operated under continuous mode for 6 h HRT, which also resulted in high stability (VHPR fluctuations <10%) and a high bioH_2 content in the gas phase of ~90% (*v/v*) [38].

3.2 Metabolic pathways

Following the by-products formed during fermentation is of utmost importance to understand, predict, control, and optimize the behavior of DF processes. It is well known that the distribution of the fermentation by-products may change depending on culture conditions. Low bioH_2 productions matched with the presence of undesired electron sinks, such as HLac, HPr, iso-butyrate, valerate, iso-valerate, and solvents (e.g. ethanol, acetone, and butanol). For instance, the production of HPr reduces the amount of bioH_2 that may be produced, as shown in reactions 1–3 (**Table 3**). Biomass growth also represents an electron sink. Commonly bioH_2 production is growth-associated. However, higher biomass growth does not necessarily imply the achievement of the best bioH_2 production [29]. Thus, a proper balance between biomass growth and bioH_2 production is desirable. On the other hand, bioH_2 sink through the formation of bioCH_4 via the hydrogenotrophic pathway (reaction 4) seems to be less problematic in DF processes due to the application of inoculum pretreatments together with biokinetic control such as acidic pH and low HRT, even using attached-growth reactors [34]. The formation of HLac can also lead to stuck DF fermentations, as shown in reactions 5–7. Acetogenesis (reaction 8) and homoacetogenesis (reaction 9) may also occur during the process, decreasing the bioH_2 production efficiency. It has been reported that the consumption of bioH_2 and carbon dioxide due to homoacetogenesis depends on the type of reactor and OLR, being its occurrence accentuated in suspended growth systems and high OLR [20, 23].

Contrarily, bioH_2 production via DF is typically related to HBu and HAc production from carbohydrates degradation, as shown in reactions 10 and 11, respectively. Theoretically, 4 and 2 mol of H_2 derive from 1 mol of glucose when HAc and

Competing reactions	Reaction
$Glucose + 2H_2 \rightarrow 2HPr + 2H_2O$	(1)
$HLac + H_2 \rightarrow HPr + H_2O$	(2)
$3HLac \rightarrow 2HPr + H_2O$	(3)
$4H_2 + CO_2 \rightarrow CH_4 + 2H_2O$ (bioCH_4-producing reaction)	(4)
$Glucose \rightarrow 2HLac$	(5)
$Glucose \rightarrow HLac + HAc + CO_2$	(6)
$2Glucose \rightarrow 2HLac + 3HAc$	(7)
$Glucose \rightarrow 3HAc$	(8)
$4H_2 + 2CO_2 \rightarrow HAc + 2H_2O$	(9)
BioH_2-producing reactions	
$Glucose + 2H_2O \rightarrow 2HAc + 2CO_2 + 4H_2$	(10)
$Glucose \rightarrow HBu + 2CO_2 + 2H_2$	(11)
$HLac + 0.5HAc \rightarrow 0.75HBu + CO_2 + 0.5H_2 + 0.5H_2O$	(12)
$HLac + H_2O \rightarrow HAc + CO_2 + 2H_2$	(13)
$2HLac \rightarrow HBu + 2CO_2 + 2H_2$	(14)
$HFor \rightarrow H_2 + CO_2$	(15)
$Glucose + H_2O \rightarrow C_2H_5OH + HAc + 2CO_2 + 2H_2$	(16)

Table 3.
Metabolic reactions occurring in dark fermentation systems treating tequila processing by-products.

HBu are the end-products, respectively. However, from published studies in the field of DF, it seems reasonable to conclude that, in mixed cultures, a high bioH_2 production efficiency is rather related with the formation of HBu than HAc because the latter may come from acetogenesis/homoacetogenesis.

At this point, it must be noted that bioH_2 can also come from the degradation of HLac, as shown in reactions 12–14 [37]. The HLac-type fermentation could provide the basis for the design of stable bioH_2-producing reactors whose feedstocks are rich in HLac and HAc such as distillery wastewater (including TV), food waste, dairy wastewater, ensiled crops, lignocellulosic residues, and their hydrolysates (including AB), among others [36]. The amount of bioH_2 obtained from the HLac-type fermentation may vary significantly depending on several factors such as pH, temperature, HRT, OLR, operation mode, substrate type, mixing, and prevailing microorganisms [31]. Also, it has been observed that the HLac-type fermentation in vinasse-fed DF reactors could be induced by low carbohydrate-available conditions [31, 36, 37]. On the other hand, the formation of HFor also can yield bioH_2(reaction 15) via the action of HFor hydrogenase complexes [37]. In addition, ethanol-type fermentation (reaction 16) generates ethanol, HAc, bioH_2, and carbon dioxide. According to Ren et al. [39], the ethanol-type fermentation is favored by a pH of 4.0–5.0 and oxidation-reduction potential (ORP) of < -200 mV. In comparison to the HAc-HBu-mixed type fermentation, which has been ascertained as the most common bioH_2-producing pathway, the latter two reactions have been less frequently found in DF reactors fed with AB/TV.

3.3 Microbial communities

Another pertinent point is that the performance of bioH_2-producing reactors strongly depends on the selection and maintenance of HPB. However, this is a difficult task because DF processes treating unsterilized feedstocks under continuous conditions are open systems, meaning that several microbial interactions may take place. In the literature, it has been used defined mixed cultures to inoculate DF reactors treating complex feedstocks such as AB and TV. In most cases, heat-shock pretreatment has been used as the selective method for the enrichment of HPB (based on their ability in forming spores), while killing bioH_2 consumers. However, other aspects such as biological/physiological (e.g. growth rate, microbial interactions, auto/allochthonous bacteria, adaptation to environmental stress conditions, and nutrients requirements), the composition of broth culture (e.g. availability of substrate/nutrients, organic acids, and toxicants), process parameters (e.g. pH, temperature, HRT, OLR, and ORP) and reactor configurations (e.g. suspended and attached biomass, mixing, and liquid-gas interface mass transfer capacity) are also selective pressure factors to determine prevailing microbial community structure during operation. At this point, it must be noted that the application of the heat-shock pretreatment decreases the diversity eliminating not only microorganisms with a negative effect on the overall bioH_2 production, but also with a potentially positive role. Besides having a high capacity to produce bioH_2, the biocatalyst must be able to thrive on the presence of putative toxic by-products such as HFor, HAc, phenols, and furans which are commonly detected in pretreated AB and raw TV.

Interestingly, molecular biology tools reveal that HPB (e.g. *Clostridium*, *Klebsiella*, and *Enterobacter*) are, in almost all DF systems, accompanied by lactic acid bacteria (LAB) (e.g. *Lactobacillus* and *Sporolactobacillus*) [40]. This co-occurrence could be attributed to the fact that LAB are ubiquitous in the environment, the physicochemical characteristics of feedstocks could sustain the proliferation of LAB, and LAB possess complex adaptation mechanisms that confer their ecological advantages over other bacteria [31]. *Streptococcus* and *Lactobacillus* have actually been detected in TV [31]. Bearing in mind such explanations, it is reasonable to assume that DF reactors fed with TV will naturally undergo the proliferation of LAB. Indeed, this assumption was verified by [11, 29, 31, 36, 37].

Except for capnophilic HLac pathway, it is well known that HLac is produced through zero-bioH_2-producing pathways. Moreover, the proliferation of LAB is commonly associated with the deterioration of bioH_2 production, mainly due to substrate competition, acidification of cultivation broth, and excretion of antimicrobial peptides known as bacteriocins [41]. At this point, another important constraint to be mentioned is that methods devoted to preventing the growth of LAB such as pretreatment of inoculum and sterilization of feedstock may be expensive, thus imposing a high economic burden on the process. Besides, the application of pretreatments does not always hinder the proliferation of LAB [42]. Therefore, there is an urgent need for novel technical solutions to ensure a maximum VHPR and YH_2.

Fortunately, the activity of LAB may also have positive effects on the overall DF process, mainly through the aforementioned HLac-type fermentation (HLac-driven bioH_2 production). Indeed, it is noteworthy mentioning that, under certain conditions, a DF process mediated by beneficial trophic links between HPB and LAB may be highly stable and consequently of high relevance for practical applications. In this case, LAB may help in the production of bioH_2 by pH regulation, substrate hydrolysis, biomass retention, oxygen depletion, and substrate detoxification [36]. Nevertheless, to exploit these advantages, a thorough understanding of the mechanisms underlying the HLac-type fermentation is essential. In this context, molecular analyses have depicted a possible syntrophy between LAB, acetic acid bacteria (AAB) and HPB [11, 29, 31, 36, 37]. For instance, Illumina MiSeq sequencing has revealed that *Clostridium beijerinckii*, *Streptococcus* sp., and *Acetobacter* lovaniensis were the most abundant species at the highest bioH_2 production activity [37]. The possible changes of metabolites and microbial communities through time have also been investigated to understand the potential mechanism of $bioH_2$ production from HLac and HAc [36]. In this regard, the microbial structure showed coordinated dynamic behavior over time, identifying three stages throughout the process: (i) a first stage (corresponding to the lag phase in relation to $bioH_2$ production) in which the major part of TRS were consumed by dominant LAB and AAB, (ii) a second stage (corresponding to the exponential $bioH_2$ production phase) during which the HLac-type fermentation was catalyzed by emerging HPB, and (iii) a third stage (corresponding to the stationary $bioH_2$ production phase) in which non-HPB regrown while HPB became subdominant [36]. Interestingly, it has been also shown that an operating strategy based on pH-control may stimulate the syntrophy between *Clostridium* and *Lactobacillus*, and reduced the proliferation of *Blautia* and *Propionibacterium* (which are undesirable microorganisms due to their homoacetogenic and propionogenic activity, respectively), trending $bioH_2$ production to enhanced efficiency [29].

4. Biomethane production from agave bagasse and tequila vinasse

The operational performance, metabolic pathways, and microbial communities of the AD of AB and TV are extensively reviewed in the following sections.

4.1 Operational performance

In recent years, there have been several efforts to improve the AD performance of AB and TV (**Table 4**). Regarding the use of AB, the first study reported in this

Pretreatment	Feeding	Stage	T (°C)	pH	YCH_4 *	VMPR (NL/L-d)	CH_4 (% *v/v*)	Ref.
Acid hydrolysis	Semi-continuous	Single	32	7.5	0.26^b	0.3	70–74	[8]
Acid hydrolysis	Batch	Single	37	8^a	0.16^b	0.78^d	NR	[17]
Individual enzymatic hydrolysis	Batch	Single	37	8^a	0.09^b	0.6^d	NR	[17]
Acid hydrolysis	Batch	Two	37	8^a	0.24^b	0.75^d	NR	[17]
Individual enzymatic hydrolysis	Batch	Two	37	8^a	0.24^b	0.96	NR	[17]
Individual enzymatic hydrolysis	Semi-continuous	Two	37	7	NR	0.41	NR	[7]
Acid hydrolysis	Semi-continuous	Single	35	7	0.28^b, 130^c	NR	NR	[18]
Alkaline hydrogen peroxide + binary enzymatic hydrolysis	Batch	Single	37	7.5^a	0.2^b, 393^c	0.67	NR	[22]
Individual enzymatic hydrolysis	Continuous	Two	22–25	7.5	0.32^b, 225^c	6.4	70–76	[21]

*Notes: All studies were conducted using anaerobic granular sludge; [a]Initial pH value; *Units: [b]NL-CH_4/g-$COD_{removed}$, [c]NL-CH_4/kg of AB; [d]Calculated from provided information; NR: not reported.*

Table 4.
Comparison of the literature data on biomethane production efficiency using pretreated agave bagasse as feedstock.

field was conducted by Arreola-Vargas et al. [8], who evaluated the feasibility of producing bioCH_4 from acid uncooked AB hydrolysates under two conditions, that is, with and without nutrient addition. The experiments were conducted in a mesophilic (32°C) AnSBR (with recirculation) at an OLR of 1.3 g-COD/L-d (influent concentration of 5 g-COD/L). The reactor had a working volume of 3.6 L and was inoculated with 5.8 g-VSS/L of anaerobic granular sludge collected from a full-scale UASB reactor treating brewery wastewater. The total cycle time was 72 h with a reaction time of 71 h and an exchange ratio of 80% (*v/v*). Unexpectedly, the best performance was obtained without additional supplementation of nutrients, achieving a volumetric bioCH_4 production rate (VMPR) of 0.3 NL-CH_4/L-d and a bioCH_4 yield (YCH_4) of 0.26 NL-CH_4/g-$COD_{removed}$ with a CH_4 content in the biogas of 70–74% (*v/v*).

In a later study, Arreola-Vargas et al. [17], assessed the use of AB hydrolysates (20, 40, 60, 80, and 100% *v/v*) obtained either from acid or enzymatic pretreatment for bioCH_4 production in single- and two-stage AD processes. The experiments were conducted in the AMPTS II system at 37°C, 120 rpm, initial pH of 8, and using 10 g-VSS/L of anaerobic granular sludge collected from a full-scale UASB reactor treating TV as inoculum. The highest VMPR for single- (0.84 NL-CH_4/L-d) and two-stage (0.96 NL-CH_4/L-d) processes were achieved in the assays with enzymatic hydrolysates at 100% and 20%, respectively. Regarding YCH_4 results, the highest value with the single-stage process of 0.16 NL-CH_4/g-$COD_{removed}$ was obtained in the assays with 20% hydrolysate from enzymatic pretreatment, while the two-stage process attained up to 0.24 NL-CH_4/g-$COD_{removed}$, also at 20% hydrolysate regardless of the type of pretreatment used. Although both hydrolysates harbor potential fermentation inhibitors (i.e. organic acids, furan derivatives, and polyphenols) in different concentrations, results showed no negative effects in the AD performance. Toledo-Cervantes et al. [7] also evaluated the bioCH_4 production from the spent medium of DF of enzymatic hydrolysate of AB. The authors found that bioCH_4 production in an AnSBR was severely inhibited likely because the remaining catalytic activity of the enzyme used may have contributed to the degradation of CH_4 biocatalyst. In the same year, Breton-Deval et al. [18] contrasted the bioCH_4 production from acid AB hydrolysates previously obtained using two different acid catalysts, that is, HCl and H_2SO_4. The experiments were carried out in the AMPTS II at 35°C, 120 rpm, initial pH of 7.5, an organic load of 8 g-COD/L, and using 10 g-VSS/L of anaerobic granular sludge collected from a full-scale UASB reactor treating TV as inoculum. The results showed that HCl hydrolysate outperformed the H_2SO_4 one by obtaining a four-fold increase on YCH_4, that is, 0.17 versus 0.04 NL-CH_4/g-$COD_{removed}$, respectively. The impairment of the methanogenic activity was attributed to the fact that the addition of sulfate ions favored the activity of sulfate-reducing bacteria (SRB). However, when using optimized HCl hydrolysates based on bioCH_4 production (1.8% HCl, 119°C, and 103 min) rather than sugar recovery (1.9% HCl, 130°C, and 133 min), the highest YCH_4 of 0.19 NL-CH_4/g-$COD_{removed}$ (0.09 NL-CH_4/g-VS of AB) was obtained indicating that other components of the hydrolysates besides sugars may influence bioCH_4 production, for example, extractives, potential microbial inhibitors.

In another study, Galindo-Hernández et al. [22] evaluated the bioCH_4 production potential from AB previously pretreated with AHP followed by enzymatic saccharification with hemicellulases and cellulases. The experiments were performed in the AMPTS II system at 37°C, 150 rpm, initial pH of 7.0, and using an organic load of 5 g-COD/L, 10 g-VS/L of inoculum (anaerobic granular sludge from a mesophilic full-scale TV treatment plant) and a defined mineral solution. Under such conditions, the YCH_4 and VMPR were found as 0.2 NL-CH_4/g-$COD_{removed}$

(0.39 NL-CH_4/g of AB) and 0.67 NL-CH_4/L-d, respectively, indicating the potential advantage of integrating a delignification pretreatment and the use of synergistic enzymatic mixtures before the AD process.

Regarding continuous processes, Montiel and Razo-Flores [21] studied the effect of OLR on the VMPR using a mesophilic (23–25°C) 1.5-L UASB reactor (with a working volume of 1.25 L) feeding with diluted (and supplemented with nutrients) acidogenic effluent generated during the DF of enzymatic hydrolysates of AB. The reactor was inoculated with 20 g-VS/L of anaerobic granular sludge from a full-scale UASB reactor treating TV and operated for 80 d to achieve OLRs between 1.35 and 24 g-COD/L-d by increasing the COD concentration of the influent and then by decreasing the HRT from 21 to 10 h. The highest VMPR and YCH_4 of 6.4 NL-CH_4/L-d and 0.32 NL-CH_4/g-COD_{fed} (225 NL-CH_4/kg of AB) were achieved at an OLR of 20 g-COD/L-d (14 h HRT). Under such conditions, the COD removal efficiency was above 90% and the CH_4 content in the gas phase was of 73% (*v/v*).

Regarding the use of TV for bioCH_4 production (**Table 5**), Méndez-Acosta et al. [43] assessed the mesophilic AD of TV in a lab-scale CSTR reactor for 250 d at HRTs of 14–5 d corresponding to increments in the OLR from 0.7 to 6 g-COD/L-d (influent COD concentrations of 10–33 g/L). The highest YCH_4 of 0.32 L-CH_4/g-$COD_{removed}$ and VMPR of 2.8 L-CH_4/L-d with bioCH_4 concentrations in the biogas greater than 65% (*v/v*) and COD removal efficiencies over 90% were obtained, even with an unbalanced COD/N/P ratio, at 6 g-COD/L-d OLR. However, a relatively long start-up of 50 d and continuous supplementation of external alkalinity were needed in order to provide stability to the process.

With the aim of enhancing the stability of the AD of TV, López-López et al. [44] investigated the influence of alkalinity and volatile fatty acids (VFAs) on the performance of a 2-L UASB reactor. The UASB reactor was inoculated with anaerobic granular sludge and operated under mesophilic conditions during 235 d at OLRs from 2.5 to 20 g-COD/L-d with recirculation of the treated effluent at recycling flow rate to influent flow rate ratios of 1:1 to 10:1 in one-unit increments. In that study, it was found that, by maintaining a VFAs to alkalinity ratio ≤ 0.5 with recirculation 1:10, the recirculation of the effluent could induce stable performances by reducing the impact of VFAs and organic matter concentration present in the effluent, attaining a COD removal efficiency higher than 75% with a YCH_4 of 0.33

Pretreatment/ conditioning	Feeding	Stage	T (°C)	pH	YCH_4 (NL/g-$COD_{removed}$)	VMPR (NL/L-d)	CH_4 (% *v/v*)	Ref.
Dilution	Continuous	Single	35	7.4	0.32[a]	1.7[a]	65	[43]
Dilution	Continuous	Single	35	7.4	0.32	1.9[a]	75	[45]
Dilution, nutrient supplementation	Semi-continuous	Two	35	6.8–7.5	0.26	0.29	68	[35]
Dilution, solid removal (centrifugation)	Continuous	Single	35	~7	0.33	NR	60–65	[44]
Dilution	Semi-continuous	Single	32	8	0.28	2.3[a]	90	[46]
Dilution	Continuous	Single	35	7	0.24	3.03	65	[47]
Dilution	Continuous	Two	35	7.7	0.29	2.3[a]	80	[7]

Notes: All studies were conducted using anaerobic granular sludge; [a]Calculated from provided information; NR: not reported;

Table 5.
Comparison of the literature data on biomethane production efficiency using tequila vinasse as feedstock.

NL-CH_4/g-$COD_{removed}$. However, even though the high recirculation ratio led to the recovery of alkalinity without any addition of external alkalinity, the granular sludge tended to become flocculent with a reduction in the average size from 2.5 to 1.5 mm.

In another study conducted by Jáuregui-Jáuregui et al. [45], after a start-up period of 28 d, a mesophilic up-flow FBR inoculated with anaerobic granular sludge withdrawn from a full-scale UASB reactor treating brewery wastewater exhibited a YCH_4 of 0.27 NL-CH_4/g-$COD_{removed}$ with a CH_4 content of 75% (*v/v*) and COD removal efficiencies of up to 90% under an OLR of 8 g-COD/L-d and an HRT of 4 d. However, the authors also reported the inhibition of biogas production due to digester clogging, which led to an excessive VFAs accumulation. In the same year, Buitrón et al. [35] reported the performance of a UASB reactor treating the resulting effluent of a DF stage at three different COD concentrations, that is, 0.4, 1.08, and 1.6 g/L, and two HRTs, that is, 24 and 18 h. The maximal content of CH_4 in the gas phase (68% *v/v*) and COD removal (67%) were achieved at the concentration of 1.6 g-COD/L with an HRT of 24 h. A further decrease in HRT resulted in lower efficiencies, that is, 40% CH_4 content and 52% removal efficiency.

In a further study, Arreola-Vargas et al. [46] achieved YCH_4 ranging from 0.25 to 0.29 NL-CH_4/g-$COD_{removed}$ with 75–90% (*v/v*) CH_4 content and 85% COD removal using a bench scale AnSBR inoculated with anaerobic granular sludge and fed with diluted TV (8 g-COD/L), the reaction time varied within 3–9 d. Interestingly, later, the same research group performed a pilot scale study for the mesophilic AD treatment of TV using a 445-L packed bed reactor (PBR) which was operated for 231 d under increasing OLRs, from 4 to 12.5 g-COD/L-d [47]. The PBR showed a stable performance exhibiting COD removals and YCH_4 in the range of 86–89% and 0.24–0.28 NL-CH_4/g-$COD_{removed}$, respectively. Meanwhile, the highest VMPR of 3.03 NL-CH_4/L-d was reached at the highest OLR of 12.5 g-COD/L-d [47].

More recently, in two-stage PBRs operated over 335 d, Toledo-Cervantes et al. [7] achieved the highest YCH_4 of 0.29 NL-CH_4/g-$COD_{removed}$ at OLRs in the range of 2.7–6.8 g-COD/L-d (6–2.4 d HRT) with COD removal efficiencies between 81 and 95%, and with average CH_4 contents around 80% (*v/v*). However, further increasing the OLR to 12 g-COD/L-d (2.2-d HRT) decreased the removal efficiency of COD (from 81 to 74%) accompanied with HAc and HPr accumulation.

4.2 Metabolic pathways

As shown in **Table 6**, the majority of bioCH_4 produced in AD systems occurs from the use of HAc and bioH_2 via acetoclastic (reaction 17) and hydrogenotrophic (reaction 4) pathways, respectively. However, bioCH_4 can also be evolved from HFor (reaction 18), compounds with the methyl group like methanol (reaction 19),

$4H_2 + CO_2 \rightarrow CH_4 + 2H_2O$	(4)
$HAc \rightarrow CH_4 + CO_2$	(17)
$4HFor \rightarrow CH_4 + 3CO_2 + 2H_2O$	(18)
$3CH_3OH + H_2 \rightarrow CH_4 + H_2O$	(19)
$4HPr + 2H_2O \rightarrow 4HAc + CO_2 + 3CH_4$ (syntrophic conversion)	(20)
$HBu + 2H_2O \rightarrow 4HAc + CO_2 + CH_4$ (syntrophic conversion)	(21)

Table 6.
Biomethane-producing reactions.

and from the syntrophic degradation of HBu (reaction 20) and HPr (reaction 21) [48]. Thus, an even production and consumption rate of organic acids is a sign of healthy single-stage AD processes. Contrarily, excessive accumulation of organic acids in the effluent has been related to reactor upset and failure, causing a drop in biogas production and COD removal efficiency. For instance, the presence of HPr in a HPr/HAc ratio ≥ 1 is usually matched with operational instability [43]. The alkalinity ratio, α = intermediate alkalinity (pH = 5.75)/partial alkalinity (pH = 4.3), roughly relates the amounts of VFAs and bicarbonate alkalinity in anaerobic reactors, measuring the buffer potential of the systems [49]. Values ≤ 0.3 are reported as adequate for achieving stable operation; however, in the case of TV-fed anaerobic reactors, stable processes have been achieved at slightly higher range of α between 0.2 and 0.5 [44, 47]. Moreover, bioCH_4 production can be disrupted by the formation of certain by-products such as long chain fatty acids or solvents, which may jeopardize the suitable availability of bioCH_4 precursors. In this regard, in the case of integrated DF-AD schemes, special attention must be also paid to the concentra-tion and composition of organic acids coming from the DF stage. At this point, it should be mentioned that the redirection of carbon through HLac has been reported as a strategy to enhanced AD processes due to its thermodynamic advantages [50–52].

4.3 Microbial communities

AD reactors contain mixed microbial populations [15]. BioCH_4 formation from AB and TV has been related with the coexistence of syntrophic bacteria (*Anaerolineaceae*, *Candidatus*, *Cloacamonas, Syntrophobacter*, *Syntrophomonas*, and *Syntrophus*), hydrogenotrophic (*Methanobacterium* and *Methanocorpusculum*) and acetoclastic (*Methanosaeta* and *Methanosarcina*) methanogens [7, 18, 47]. It has been previously observed that the two-stage AD of TV at low concentrations of VFAs (low OLRs) favored the acetoclastic pathway, in contrast, hydrogenotrophic methanogens enriched at high concentrations (high OLRs) [7]. This change in diversity has been also observed in an AnSBR digester fed with acid AB hydrolysates [53]. However, the opposite trend was observed during the single stage AD of TV using a pilot-scale PBR [47]. Regardless of the tequila by-product used, loss of syntrophic relationships for interspecies H_2/HFor transfer and interspecies HAc transfer has been associated with microbial imbalance, which subsequently affects negatively bioCH_4 production [8, 53]. However, in the case of multi-stage AD processes, unsuitable concentrations of hydrolytic/acidogenic bacteria in DF effluent may be quite detrimental for the granular methanogenic sludge [15]. In addition, other bacteria which can compete with the methanogens for bioCH_4 precursors may also be present in AD reactors, for example, SRB [15, 18].

5. Multi-stage anaerobic digestion

Since TV has negligible levels of alkalinity and high concentrations of components with a tendency to suffer very rapid acidification [43, 44], two-stage AD processes have emerged as important operational strategies to provide enhanced stability of the CH_4-producing stage [7, 24]. However, the multi-stage AD approach seems to be also applicable for pretreated AB [17, 21]. In fact, a two-stage AD process fed with AB hydrolysates showed up to 3.3-fold higher energy recovery than a single-stage process [17]. Indeed, according to Lindner et al. [16], two-stage systems seem to be only recommendable for digesting sugar-rich feed stocks, which undergo a quick hydrolysis/acidogenesis. This approach allows to provide optimal environmental conditions for the different groups of microorganisms which have differences in terms of physiology, nutrient intake, nutritional requirements, growth rate, optimum growth conditions such as pH, and adaptation to environmental stress conditions [16]. The acidogenesis and methanogenesis separated in space may also produce bioH_2 via DF process [17, 24, 35]. However, it is not necessarily desirable to produce bioH_2 in all cases [7]. In the latter case, a stream rich in HLac can be obtained through the HLac-type fermentation which can be further fed to the methanogenic stage [36, 37], where hydrogenotrophic may be benefited for the conversion of HLac to HAc by consuming the intermediate bioH_2 gas immediately [52]. The possibility of operating at higher organic loading capacity (in the methanogenic stage), reducing alkali addition, and increasing COD removal efficiency are additional advantages of the two-stage AD as compared to single-stage AD [7, 21, 24]. A small number of reactor configurations devoted to bioH_2/bioCH_4 production from AB/TV can be found in the literature (**Figure 3**). Among them, for both AB and TV, the CSTR and UASB configurations have shown the highest performance to date for producing bioH_2 and bioCH_4, respectively, that is,

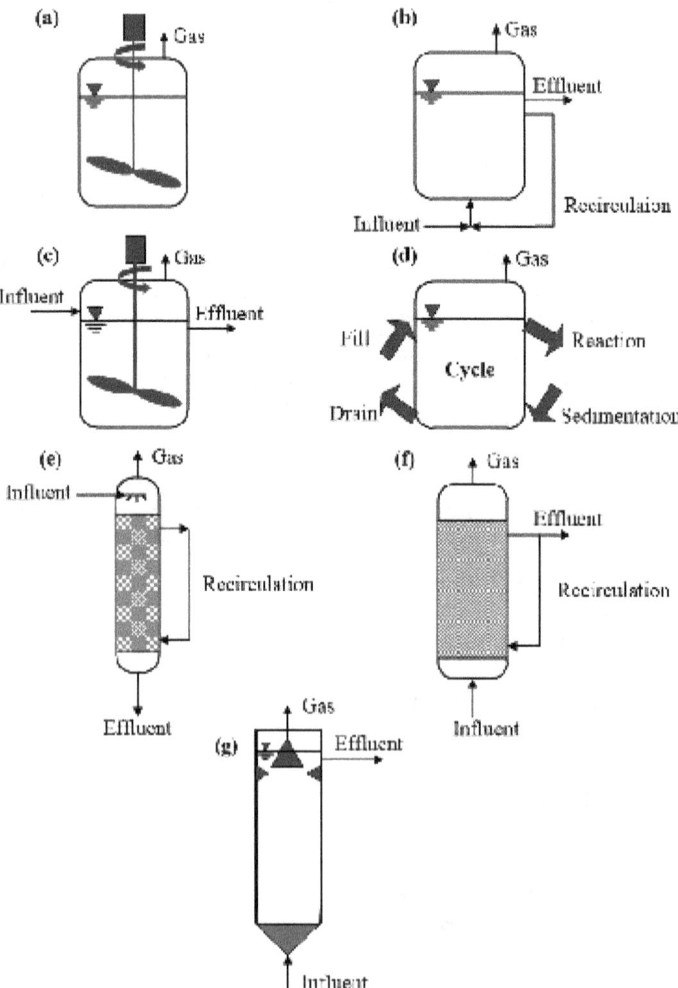

Figure 3. *Types of reactor configurations used for biohydrogen and biomethane production from tequila processing by-products. (a) Batch reactor, (b) continuously stirred tank reactor (CSTR) with recirculation, (c) CSTR, (d) anaerobic sequencing batch reactor (AnSBR), (e) trickling bed reactor with recirculation, (f) packed bed reactor, (g) up-flow anaerobic sludge blanket (UASB) reactor. AnSBR can integrate mechanical or hydraulic mixing. UASB can operate with effluent recycle.*

13 NL-H_2/L-d from AB [23] and 12.3 NL-H_2/L-d from TV [38] and 6.4 NL-CH_4/L-d from AB [21] and 3.5 NL-CH_4/L-d from TV [54].

6. Current limitations and potential improvements

Notwithstanding the enormous efforts made to achieve a better understanding of the DF/AD process of AB/TV, it is still necessary to improve not only bioH_2 or bioCH_4 productivities and yields but also the (long-term) stability of processes for commercialization purposes. TV is a highly complex wastewater that besides high COD and negligible alkalinity, harbors recalcitrant compounds such as phenols, which may act as inhibitors in DF/AD. While the main limitation to use AB as the feedstock is its recalcitrant structure. As mentioned earlier, some of the pretreatment/conditioning steps used in AB have been optimized not only in terms of hydrolysis yield, reaction time, the generation/release and effect of putative fermentation inhibitory compounds, cost-effectiveness but also in terms of bioH_2/bioCH_4 production efficiency. However, there is still a need to explore other pretreatments that have not been yet embraced in the field of DF/AD of AB but they have been ascertained as potentially useful in releasing sugars for other applications like the production of bioethanol, such as ammonia fiber explosion (AFEX), autohydrolysis, organosolv, high-energy radiation, ozonolysis, alkaline, ionic liquids, or any combination of those pretreatments. It could be also interesting to explore consolidated processes (direct fermentation) which combine into a single operation the enzymatic hydrolysis of (pretreated) biomass and biological conversion to the desired by-product (in this case bioH_2/bioCH_4) by mixed consortia.

Besides the features described before, from practical purposes, the highly variable composition of AB/TV constitutes another constraint to produce bioH_2 since DF systems are commonly unable to overcome perturbations in feedstock composition. One of the most significant challenges is to assure consistency in the prevailing metabolic pathways during the DF process and favor bioH_2-producing pathways over other unwanted routes, for example, homoacetogenesis and methanogenesis. Very little is known about the microbial community structure of DF/AD processes treating AB/TV. In this regard, it is not clear the role of microorganisms and their association with operational parameters (e.g. pH, HRT, and OLR) and process indicators (e.g. VHPR, VMPR, and metabolic composition). Also, much less is known about how microbial assemblage may change through time, and what factors (operating parameters) govern its dynamics. It is worth noticing that HLac monitoring has been disregarded limiting the understanding of integrated DF-AD processes since it, as an intermediate, has a vital role in the carbon flux.

Another concern worth to mention is that most of the previous studies were carried out in batch or semi-continuous reactors. Thus, it is vital to transfer the kinetic knowledge gained from such studies to the expansion of continuous systems. In this context, the development of integrated DF-AD schemes for the continuous production of bioH_2 and bioCH_4 using AB/TV as feed stocks requires intensive research on interlinking side streams for producing high added-value bioproducts in a biorefinery framework (e.g. HLac-bioH_2-bioCH_4) for better sustainability of the existing tequila industries.

7. Conclusions

Tequila industry generates huge amounts of AB and TV, which could be subjected to integrated DF-AD processes to produce bioH_2 and bioCH_4 while reducing their pollution potential. This chapter focused on the state-of-the-art of configurations and process parameters, metabolic pathways, and microbial ecology of bioH_2- and bioCH_4-producing reactors. The pretreatment/conditioning steps applied to enhance the valorization of AB/TV were also reviewed. It has been suggested that the HLac-type fermentation coupled to DF and AD can boost the development of cascading design in multi-stage AD processes. This multiproduct approach using AB/TV as resources in the biorefinery scheme may facilitate sustainability to the tequila industry.

Acknowledgements

This work was financially supported by Consejo Nacional de Ciencia y Tecnología (CONACYT) through the Project-PN-2015-2101-1024. Osuna-Laveaga D.R. acknowledges CONACYT for the Ph.D. scholarship: 267499.

Acronyms and abbreviations

HAc	acetic acid
AAB	acetic acid bacteria
AB	agave bagasse
AHP	alkaline hydrogen peroxide
AD	anaerobic digestion
AnSBR	anaerobic sequencing batch reactor
AMPTS II	automatic methane potential test system
bioH_2	biohydrogen
YH_2	biohydrogen yield
HPB	biohydrogen-producing bacteria
bioCH_4	biomethane
YCH_4	biomethane yield
HBu	butyric acid
COD	chemical oxygen demand
CSTR	continuously stirred tank reactor
DF	dark fermentation
FPU	filter paper units
FBR	fixed bed reactor
HFor	formic acid
HRT	hydraulic retention time
HPB	hydrogen-producing bacteria
HMF	hydroxymethylfurfural
HLac	lactic acid
LAB	lactic acid bacteria
NW	nixtamalization wastewater
ORP	oxidation-reduction potential
OLR	organic loading rate
PBR	packed bed reactor
HPr	propionic acid
SRB	sulfate-reducing bacteria
VFAs	volatile fatty acids
VS	volatile solid
VSS	volatile suspended solids
VHPR	volumetric biohydrogen production rate

VMPR	volumetric biomethane production rate
TV	tequila vinasse
TRS	total-reducing solids
TBR	trickling bed reactor
UASB	up-flow anaerobic sludge blanket reactor

Author details

Octavio García-Depraect, Daryl Rafael Osuna-Laveaga and Elizabeth León-Becerril*
Department of Environmental Technology, Centro de Investigación y Asistencia en Tecnología y Diseño del Estado de Jalisco, A.C., Guadalajara, Jalisco, México

*Address all correspondence to: eleon@ciatej.mx

References

[1] CRT. Producción total Tequila y Tequila 100% [Internet]. 2019. Available from: https://www.crt.org.mx/EstadisticasCRTweb [Accessed: 14-05-2019]

[2] Villanueva-Rodríguez SJ, Rodríguez-Garay B, Prado-Ramírez R, Gschaedler A. Tequila: Raw material, classification, process, and quality parameters. Encyclopedia of Food and Health. Academic Press; 2016:283-289. DOI: 10.1016/B978-0-12-384947-2.00688-7

[3] Rodríguez-Félix E, Contreras-Ramos SM, Davila-Vazquez G, Rodríguez-Campos J, Marino-Marmolejo EN. Identification and quantification of volatile compounds found in vinasses from two different processes of tequila production. Energies. 2018;**11**:1-18. DOI: 10.3390/en11030490

[4] Cedeño-Cruz M. Tequila production from agave: Historical influences and contemporary processes. In: Jacques KA, Lyons TP, Kelsall DR, editors. The Alcohol Textbook. 4th ed. Oxford UK: Nottingham University Press; 2003. pp. 223-245

[5] López-López A, Davila-Vazquez G, León-Becerril E, Villegas-García E, Gallardo-Valdez J. Tequila vinasses: Generation and full scale treatment processes. Reviews in Environmental Science and Biotechnology. 2010;**9**: 109-116. DOI: 10.1007/s11157-010-9204-9

[6] del Real-Olvera J, López-López A. Biogas production from anaerobic treatment of agro-industrial wastewater. In: Kumar S, editor. Biogas. Rijeka: InTech; 2012. pp. 91-112

[7] Toledo-Cervantes A, Guevara-Santos N, Arreola-Vargas J, Snell-Castro R, Méndez-Acosta HO. Performance and microbial dynamics in packed-bed reactors during the long-term two-stage anaerobic treatment of tequila vinasses. Biochemical Engineering Journal. 2018; **138**:12-20. DOI: 10.1016/j.bej.2018.06.020

[8] Arreola-Vargas J, Ojeda-Castillo V, Snell-Castro R, Corona-González RI, Alatriste-Mondragón F, Méndez-Acosta HO. Methane production from acid hydrolysates of Agave tequilana bagasse: Evaluation of hydrolysis conditions and methane yield. Bioresource Technology. 2015;**181**:191-199. DOI: 10.1016/j.biortech.2015.01.036

[9] Hernández C, Escamilla-Alvarado C, Sánchez A, Alarcón E, Ziarelli F, Musule R, et al. Wheat straw, corn Stover, sugarcane, and agave biomasses: Chemical properties, availability, and cellulosic-bioethanol production potential in Mexico. Biofuels, Bioproducts & Biorefinering. 2019:1-17. DOI: 10.1002/bbb.2017

[10] Pecha B, Garcia-Perez M. Pyrolysis of lignocellulosic biomass: Oil, char, and gas. In: Dahiya A, editor. Bioenergy Biomass to Biofuels. Oxford UK: Academic Press; 2015. pp. 413-442. DOI: 10.1016/B978-0-12-407909-0.00026-2

[11] García-Depraect O, Gómez-Romero J, León-Becerril E, López-López A. A novel biohydrogen production process: Co-digestion of vinasse and *Nejayote* as complex raw substrates using a robust inoculum. International Journal of Hydrogen Energy. 2017;**42**:5820-5831. DOI: 10.1016/j.ijhydene.2016.11.204

[12] Iñiguez-Covarrubias G, Lange SE, Rowell RM. Utilization of byproducts from the tequila industry: Part 1: Agave bagasse as a raw material for animal feeding and fiberboard production. Bioresource Technology. 2001;**77**:25-32. DOI: 10.1016/S0960-8524(00)00137-1

[13] Moran-Salazar RG, Marino-Marmolejo EN, Rodríguez-Campos J,

Dávila-Vázquez G, Contreras-Ramos SM. Use of agave bagasse for production of an organic fertilizer by pretreatment with *Bjerkandera adusta* and vermicomposting with *Eisenia fetida*. Environmental Technology. 2015;**37**: 1-12. DOI: 10.1080/09593330.2015.1108368

[14] Palomo-Briones R, López-Gutiérrez I, Islas-Lugo F, Galindo-Hernández KL, Munguía-Aguilar D, Rincón-Pérez JA, et al. Agave bagasse biorefinery: Processing and perspectives. Clean Technologies and Environmental Policy. 2018;**20**:1423-1441. DOI: 10.1007/s10098-017-1421-2

[15] van Lier JB, Mahmoud N, Zeeman G. Anaerobic wastewater treatment. In: Henze M, Loosdrecht v, Ekama GA, Brdjanovic, editors. Biological Wastewater Treatment: Principles Modeling and Design. London UK: IWA Publishing; 2008. pp. 401-442

[16] Lindner J, Zielonka S, Oechsner H, Lemmer A. Is the continuous two-stage anaerobic digestion process well suited for all substrates? Bioresource Technology. 2016;**200**:470-476. DOI: 10.1016/j.biortech.2015.10.052

[17] Arreola-Vargas J, Flores-Larios A, González-Álvarez V, Corona-González RI, Méndez-Acosta HO. Single and two-stage anaerobic digestion for hydrogen and methane production from acid and enzymatic hydrolysates of Agave tequilana bagasse. International Journal of Hydrogen Energy. 2016;**41**:897-904. DOI: 10.1016/j.ijhydene.2015.11.016

[18] Breton-Deval L, Méndez-Acosta HO, González-Álvarez V, Snell-Castro R, Gutiérrez-Sánchez D, Arreola-Vargas J. *Agave tequilana* bagasse for methane production in batch and sequencing batch reactors: Acid catalyst effect, batch optimization and stability of the semi-continuous process. Journal of Environmental Management. 2018;**224**: 156-163. DOI: 10.1016/j.jenvman.2018.07.053

[19] Valdez-Guzmán BE, Rios-Del Toro EE, Cardenas-López RL, Méndez-Acosta HO, González-Álvarez V, Arreola-Vargas J. Enhancing biohydrogen production from *Agave tequilana* bagasse: Detoxified vs. Undetoxified acid hydrolysates. Bioresource Technology. 2019;**276**: 74-80. DOI: 10.1016/j.biortech.2018.12.101

[20] Contreras-Dávila CA, Méndez-Acosta HO, Arellano-García LA, Alatriste-Mondragón F, Razo-Flores E. Continuous hydrogen production from enzymatic hydrolysate of *Agave tequilana* bagasse: Effect of the organic loading rate and reactor configuration. Chemical Engineering Journal. 2017; **313**:671-679. DOI: 10.1016/j.cej.2016.12.084

[21] Montiel CV, Razo-Flores E. Continuous hydrogen and methane production from *Agave tequilana* bagasse hydrolysate by sequential process to maximize energy recovery efficiency. Bioresource Technology. 2018;**249**:334-341. DOI: 10.1016/j.biortech.2017.10.032

[22] Galindo-Hernández KL, Tapia-Rodríguez A, Alatriste-Mondragón F, Celis LB, Arreola-Vargas J, Razo-Flores E. Enhancing saccharification of *Agave tequilana* bagasse by oxidative delignification and enzymatic synergism for the production of hydrogen and methane. International Journal of Hydrogen Energy. 2018;**43**:22116-22125. DOI: 10.1016/j.ijhydene.2018.10.071

[23] Montoya-Rosales JJ, Olmos-Hernández DK, Palomo-Briones R, Montiel-Corona V, Mari AG, Razo-Flores E. Improvement of continuous hydrogen production using individual and binary enzymatic hydrolysates of agave bagasse in suspended-culture and biofilm reactors.

Bioresource Technology. 2019;**283**: 251-260. DOI: 10.1016/j.biortech.2019.03.072

[24] Ruggeri B, Tommasi T, Sanfilippo S. Two-step anaerobic digestion process. In: Ruggeri B, Tommasi T, Sanfilippo S, editors. BioH_2 & BioCH_4 through Anaerobic Digestion. From Research to Full-Scale Applications. Green Energy and Technology. Springer London Heidelberg New York Dordrecht; Springer-Verlag London: 2015. pp. 161-191. DOI: 10.1007/978-1-4471-6431-9

[25] Ghimire A, Frunzo L, Pirozzi F, Trably E, Escudie R, Lens PNL, et al. A review on dark fermentative biohydrogen production from organic biomass: Process parameters and use of by-products. Applied Energy. 2015;**144**: 73-95. DOI: 10.1016/j.apenergy.2015.01.045

[26] Toledo-Cervantes A, Arreola-Vargas J, Elias-Palacios SE, Marino-Marmolejo EN, Davila-Vazquez G, González-Álvarez V, et al. Evaluation of semi-continuous hydrogen production from enzymatic hydrolysates of *Agave tequilana* bagasse: Insight into the enzymatic cocktail effect over the co-production of methane. International Journal of Hydrogen Energy. 2018;**43**: 14193-14201. DOI: 10.1016/j.ijhydene.2018.05.134

[27] Espinoza-Escalante FM, Pelayo-Ortiz C, Gutiérrez-Pulido H, González-Álvarez V, Alcaraz-González V, Bories A. Multiple response optimization analysis for pretreatments of Tequila's stillages for VFAs and hydrogen production. Bioresource Technology. 2008;**99**:5822-5829. DOI: 10.1016/j.biortech.2007.10.008

[28] Espinoza-Escalante FM, Pelayo-Ortíz C, Navarro-Corona J, González-García Y, Bories A, Gutiérrez-Pulido H. Anaerobic digestion of the vinasses from the fermentation of Agave tequilana weber to tequila: The effect of pH, temperature and hydraulic retention time on the production of hydrogen and methane. Biomass and Bioenergy. 2009;**33**: 14-20. DOI: 10.1016/j.biombioe.2008.04.006

[29] García-Depraect O, Rene ER, Gómez-Romero J, López-López A, León-Becerril E. Enhanced biohydrogen production from the dark co-fermentation of tequila vinasse and nixtamalization wastewater: Novel insights into ecological regulation by pH. Fuel. 2019;**253**:159-166. DOI: 10.1016/j.fuel.2019.04.147

[30] Buitrón G, Carvajal C. Biohydrogen production from tequila vinasses in an anaerobic sequencing batch reactor: Effect of initial substrate concentration, temperature and hydraulic retention time. Bioresource Technology. 2010; **101**:9071-9077. DOI: 10.1016/j.biortech.2010.06.127

[31] García-Depraect O, Rene ER, Diaz-Cruces VF, León-Becerril E. Effect of process parameters on enhanced biohydrogen production from tequila vinasse via the lactate-acetate pathway. Bioresource Technology. 2019;**273**: 618-626. DOI: 10.1016/j.biortech.2018.11.056

[32] Marino-Marmolejo EN, Corbalá-Robles L, Cortez-Aguilar RC, Contreras-Ramos SM, Bolaños-Rosales RE, Davila-Vazquez G. Tequila vinasses acidogenesis in a UASB reactor with *Clostridium* predominance. Springerplus. 2015;**4**:1-8. DOI: 10.1186/s40064-015-1193-2

[33] Moreno-Andrade I, Moreno G, Kumar G, Buitrón G. Biohydrogen production from industrial wastewaters. Water Science and Technology. 2014;**71**: 105-110. DOI: 10.2166/wst.2014.471

[34] Buitrón G, Prato-Garcia D, Zhang A. Biohydrogen production from tequila

vinasses using a fixed bed reactor. Water Science and Technology. 2014; **70**:1919-1925. DOI: 10.2166/wst.2014.433

[35] Buitrón G, Kumar G, Martinez-Arce A, Moreno G. Hydrogen and methane production via a two-stage processes (H_2-SBR + CH_4-UASB) using tequila vinasses. International Journal of Hydrogen Energy. 2014;**39**:19249-19255. DOI: 10.1016/j.ijhydene.2014.04.139

[36] García-Depraect O, Valdez-Vázquez I, Rene ER, Gómez-Romero J, López-López A, León-Becerril E. Lactate- and acetate-based biohydrogen production through dark co-fermentation of tequila vinasse and nixtamalization wastewater: Metabolic and microbial community dynamics. Bioresource Technology. 2019;**282**:236-244. DOI: 10.1016/j.biortech.2019.02.100

[37] García-Depraect O, León-Becerril E. Fermentative biohydrogen production from tequila vinasse via the lactate-acetate pathway: Operational performance, kinetic analysis and microbial ecology. Fuel. 2018;**234**: 151-160. DOI: 10.1016/j.fuel.2018.06.126

[38] García-Depraect O, van Lier JB, Muñoz R, Rene ER, Diaz-Cruces VF, León-Becerril E. Interlinking lactate-type fermentation in a side stream anaerobic digestion process of tequila vinasse: An alternative for highly stable and efficient biohydrogen production. (Unpublished)

[39] Ren N, Zhao D, Chrn X, LI J. Mechanism and controlling strategy of the production and accumulation of propionic acid for anaerobic wastewater treatment. Science In China. 2002;**45**: 319-327

[40] Sikora A, Błaszczyk M, Jurkowski M, Zielenkiewicz U. Lactic acid bacteria in hydrogen-producing consortia: On purpose or by coincidence? In: Kongo J, editor. Lactic Acid Bacteria. R & D for Food, Health and Livestock Purposes. Rijeka: InThech; 2013. pp. 487-514. DOI: 10.5772/50364

[41] Elbeshbishy E, Dhar BR, Nakhla G, Lee HS. A critical review on inhibition of dark biohydrogen fermentation. Renewable and Sustainable Energy Reviews. 2017;**79**:656-668. DOI: 10.1016/j.rser.2017.05.075

[42] Cabrol L, Marone A, Tapia-Venegas E, Steyer JP, Ruiz-Filippi G, Trably E. Microbial ecology of fermentative hydrogen producing bioprocesses: Useful insights for driving the ecosystem function. FEMS Microbiology Reviews. 2017;**41**:158-181. DOI: 10.1093/femsre/fuw043

[43] Méndez-Acosta HO, Snell-Castro R, Alcaraz-González V, González-Álvarez V, Pelayo-Ortiz C. Anaerobic treatment of tequila vinasses in a CSTR-type digester. Biodegradation. 2010;**21**: 357-363. DOI: 10.1007/s10532-009-9306-7

[44] López-López A, León-Becerril E, Rosales-Contreras ME, Villegas-García E. Influence of alkalinity and VFAs on the performance of an UASB reactor with recirculation for the treatment of tequila vinasses. Environmental Technology. 2015;**36**:2468-2476. DOI: 10.1080/09593330.2015.1034790

[45] Jáuregui-Jáuregui JA, Méndez-Acosta HO, González-Álvarez V, Snell-Castro R, Alcaraz-González V, Godonc JJ. Anaerobic treatment of tequila vinasses under seasonal operating conditions: Start-up, normal operation and restart-up after a long stop and starvation period. Bioresource Technology. 2014;**168**:33-40. DOI: 10.1016/j.biortech.2014.04.006

[46] Arreola-Vargas J, Jaramillo-Gante NE, Celis LB, Corona-González RI, González-Álvarez V, Méndez-Acosta HO. Biogas production in an anaerobic

sequencing batch reactor by using tequila vinasses: Effect of pH and temperature. Water Science and Technology. 2016;**73**:550-556. DOI: 10.2166/wst.2015.520

[47] Arreola-Vargas J, Snell-Castro R, Rojo-Liera NM, González-Álvarez V, Méndez-Acosta HO. Effect of the organic loading rate on the performance and microbial populations during the anaerobic treatment of tequila vinasses in a pilot-scale packed bed reactor. Journal of Chemical Technology & Biotechnology. 2018;**93**:591-599. DOI: 10.1002/jctb.5413

[48] Gerardi MH. Anaerobic food chain. In: Gerardi MH, editor. The Microbiology of Anaerobic Digesters. Canada: John Wiley & Sons; 2003. p. 39-41

[49] Vuitik GA, Fuess LT, Del Nery V, Bañares-Alcántara R, Pires EC. Effects of recirculation in anaerobic baffled reactors. Journal of Water Process Engineering. 2019;**28**:36-44. DOI: 10.1016/j.jwpe.2018.12.013

[50] Detman A, Mielecki D, Pleśniak Ł, Bucha M, Janiga M, Matyasik I, et al. Methane-yielding microbial communities processing lactate-rich substrates: A piece of the anaerobic digestion puzzle. Biotechnology for Biofuels. 2018;**11**:1-18. DOI: 10.1186/s13068-018-1106-z

[51] Wu Y, Wang C, Liu X, Ma H, Wu J, Zuo J, et al. A new method of two-phase anaerobic digestion for fruit and vegetable waste treatment. Bioresource Technology. 2016;**211**:16-23. DOI: 10.1016/j.biortech.2016.03.050

[52] Pipyn P, Verstraete W. Lactate and ethanol as intermediates in two-phase anaerobic digestion. Biotechnology and Bioengineering. 1981;**23**:1145-1154

[53] Snell-Castro R, Méndez-Acosta HO, Arreola-Vargas J, González-Álvarez V, Pintado-González M, González-Morales MT, et al. Active prokaryotic population dynamics exhibit high correlation to reactor performance during methane production from acid hydrolysates of *Agave tequilana* var. Azul bagasse. Journal of Applied Microbiology. 2019; **126**:1618-1630. DOI: 10.1111/jam.14234

[54] Diaz-Cruces VF, García-Depraect O, León-Becerril E. Performance of two-stage anaerobic digestion of tequila vinasse with acidogenic lactate-type fermentation. (Unpublished)

9

Continuous Beer Production

Mark Strobl

Abstract

Although the barley and hop harvest is a batch process, the ingredients are storable to a certain extent, so malting and brewing can be performed continuously. The more expensive machinery and energy are becoming, the more continuous production is becoming efficient. The advantages are smaller capacities, less energy consumption and more recuperation. Most filling processes run semi continuously, and energy-consuming processes like malting can also run continuously around the clock. Disadvantages are necessary buffers, microbiological contamination and less flexibility in producing different types of beer or adjusting the production to seasonal fluctuations. Mistakes and errors increase, even if accuracy and in-line sensors help to keep quality stable for some time. This article discusses benefits and limits of continuous technology.

Keywords: continuous process, malting, brewing, fermentation, filtration, bottling, error progress, multisensor behaviour

1. History of continuous beer production

The wish to feel good is the wish to feel good continuously. Transferring this to the consumption of beer may lead to the wish to produce beer continuously (**Table 1**).

During the last 100 years, more and more continuous steps have been integrated into the brewing process. Lots of energy can be saved during malting and wort boiling. Filling of the beer needs lots of expensive and labour-intensive machines, which work efficiently, if pursued in a continuous manner.

Brewing processes are mainly mixing and clarification processes. During malting, mashing, wort boiling and fermentation, mixing can easily be performed in a continuous manner. Clarification and separation processes like lautering, wort clarification, yeast settling and maturation are traditionally performed by gravity and settling in a batch process. For the continuous process, these have to be altered into separation processes by decanters, centrifuges and crossflow filtrations. The intersection between a batch process and a continuous process needs buffer vats, tanks before and after the continuous process. Buffers are capacities that cost money, space and energy, have to be cleaned and maintained and limit the advantages of a pure continuous process. In bottling lines conveyer buffers need space before and after each machine, which cost a lot of money and need to be controlled by computers that link the aggregates.

Steps in beer production			
Process	**Aims**	**Batch**	**Continuous**
Steeping	Mixing	Steep tank	Screw conveyers
Germination	Mixing	Malting box, tower malting	Worm conveyers
Kilning	Mixing	Floor kiln	Air conveyers
Milling	Mixing		All mills
Mashing	Mixing	Mash tun	Tube heat exchanger
Lautering	Separation	Lauter tun, mash filter	Centrifuge, decanter, Nessie
Wort boiling	Mixing	Copper	Heat exchanger
Wort clarification	Separation	Whirlpool, flotation	Decanter, centrifuge
Wort aeration	Mixing	Flotation	Inline nozzle
Fermentation	Mixing	Fermentation tub or tank	Immobilised yeast
Maturation	Separation	Storage tank	Centrifuge
Filtration	Separation	Dead-end cave filtration	Crossflow filter
Filling	Avoiding contaminations and oxygen	Tank lorries, kegs, stationary fillers	Rotary-type machines for kegs and bottles

Table 1.
Main processes in beer making, their aims and the batchwise and continuous machines.

2. Pros and cons of continuous production versus batch process

Continuous process advantages	Batch process: advantages
• Less volume, limited space requirements [1]	• Are easier to control
• Less energy and water consumption	• Are easier to change, if other beer varieties are needed
• Reduced peak consumption of utilities [1]	• Quantity demands can be respected
• Reduced extract losses [1]	• Containers are better and more often to be cleaned
• Reduced waste disposal [1]	• Processes are less automised and complicated
• Higher yields	• Machinery has less moving parts and less ware parts
• Higher degree of automatisation	• If things go wrong, it means less stress for the brewers (and yeasts)

Table 2.
Continuous brewing processes versus batch processes—the advantages.

3. Continuous malting

As malting needs lots of water and energy and only a few malts are produced in big amounts, continuous or semi continuous production is used more often than in brewing (**Table 2**). This means steeping, germination and kilning during the continuous transportation of the grains [3].

In 1960 the Domalt system was built in Toronto (Toronto malting). Barley was transported by water with a pump into a slope malt conditioning screw. The barley is slowly transported upward against rinsing water. Then the barley fell onto an endless filter belt [4]. Water was sprayed on the barley and aerated, and the degree of steeping was adjusted. It fell on a conveyer situated below for the germination. The belt is moved at 0.7 m/h, the green malt being 0.9 m high. Stationary turning machines homogenised the germinating grains. Withering and kilning were also conducted on conveyers by tempered air blowing through the debris. Finally the malt was cooled down by air. Only one person controlled the process. Yields were 1–1.5% above per batch systems. Energy and water were saved [3]. Production size was 11,600 t/a. The steeping and germination time was 100–110 h for two rowed and 70–80 h with multi-rowed barley [4]. Advantages were the quick pregermination at less moisture, the better cytolysis, the shorter time, a bigger kernel volume after the kilning [4], less work labour, better automatisation and less water consumption at higher investment costs.

A Soviet system called Bartnew worked with rotating slanted 1–2° slope long drums. The grains moved 20–60 mm per turn of the drum. In the end the grain had passed every germination stage and was kilned in a vertical kiln with conical chutes [3].

The process could be regulated, 40–70 h for the steeping, 70–200 h for germination and 6–20 h for the kilning [3]. The German Democratic Republic (GDR) continuous malting needed a total of 73–105 h, producing 10,000 t/a. The water was reused for steeping with an addition of 0.2% caustic soda [3].

Moving pile systems by Ostertag, Seeger and Lausmann were turning and moving the malt under spraying or aeration with tempered air. Today tower malting is a semi continuous malt production. The steps on the floors take 9–24 h [4]. The batches are synchronised to the lorry sizes which fetch the malt to the breweries, the production speed is in a weekly rhythm, the silos buffer the amount of the barley harvest, and the demand of malt is by the annual variability of beer production.

4. Continuous brewhouse

The important utilities consumed in the brewhouse are malt grist, hot water for mashing and lautering, steam and water as a cooling liquid. In a batch brewhouse, different batches are processed at the same time and consequently lead to crucial electricity, steam and water consumption peaks.

Milling: usually milling is a continuous process, no matter if wet mill, roller or a hammer mill is used. The malt silo and the milling body are the buffers before and after the milling. Capacities can be reduced, if the mashing process is continuous and not needing the tons per brew in less than 20 min to assure equal treatment of the grains. The comminution degree is dependent on the lautering process following. If continuous the husks do not need to be maintained, like for lautering. This means more yield but less blank worts.

Mashing: infusion mashing is easily to be performed continuously. Plate or tube heat exchangers vary temperature and speed of the mash. Plate heat exchangers have limited applications for products with particles and/or fibres, while tube heat exchangers have the lower energy recovery rates [5]. Mash cannot run against mash like in a plate heat exchanger. An intermedia water circuit is necessary to recuperate energy in the casing pipe. Mixing while heating has to assure equality of the treatment as much as possible.

In 1998, Meura started the development of the continuous brewing concept. A complete pilot plant was installed in 1999. The first operation of the Meurabrew on an industrial scale of 200 hl/h wort (up to 20°P) took place in 2007. A similar order for a plant in Fuzhou, China, was obtained. The entire operation is managed by 45 people, with just 2 men per shift to run the brewing operation from raw material intake to filtered beer during the daytime [1].

Different mash vessels are keeping mash at constant temperature with a specific holding time. A continuous flow passes these vessels. Three parallel filters assure a regular continuous filtration process [1].

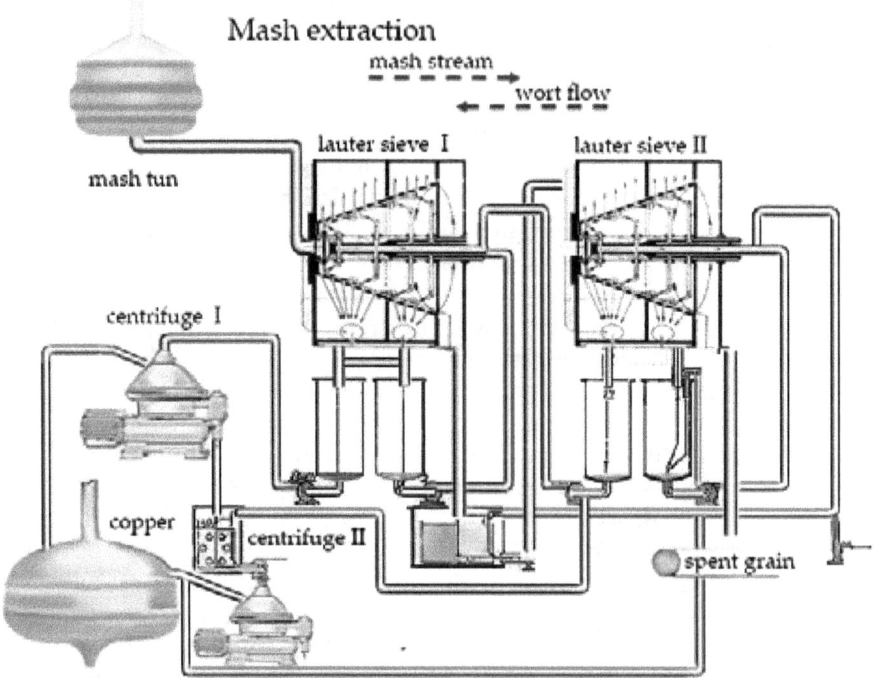

Figure 1.
The continuous step of lautering in a "Pablo Brewhouse" adaptation in 1968 [7].

Lautering: after the mashing process, vacuum rotary filters [6] or decanters may be used to remove the insoluble parts from the wort (**Figure 1**).

A process for the continuous production of wort was described by Harsanyi in 1968 [7]. It was substantially characterised by separating the mash continuously by centrifugal action in at least two stages, sparging the largely dehydrated solids fraction with controlled quantities of water removing the dehydrated solids automatically and subjecting the wort obtained to further clarification before delivering it to the brew kettle. The separation of the liquid mash fraction from the solids is accomplished by means of a special type of centrifuge. The centrifuge has a housing in which a conical, perforated drum rotates. The housing has a first chamber with two compartments and separate liquid cutlets and a second chamber for the removal of solids, the second chamber being arranged at the larger diameter drum end. The mash slurry to be separated is delivered into the centrifuge at its smaller diameter drum end. Disposed within the drum is a hollow shaft having two separate liquid passages to which jet pipes are connected. The shaft rotates at a speed slightly less than that of the perforated drum [7].

Similarly, the continuous lautering system Nessie from Ziemann works, introduced to the public in 2016 [8]. The separation of the mash is carried out via four filter units in cascade arrangement, in which the rotary disc filters perform the separation of wort and spent grains. The sparging of the extract is carried out in parallel using a turbulent counterflow extraction [8]. The time saved is about 160 min (34%) per brew [9]. Worts are less blank and contain more fatty acids, more zinc and less polyphenols. This increases the fermentation speed and the flavour stability [10]. Continuously produced worts have different qualities compared to batchwise-produced worts [3].

Wort boiling: the wort boiling process suits to obtain the following objectives [11]:

- Extraction and isomerisation of hop components
- Hot break formation mainly by coagulation of proteins
- Formation of colourant substances
- Formation of reducing aromatic compounds by Maillard reaction
- Evaporation of undesired volatile aroma compounds
- Decrease of the pH of the wort
- Inspissation of the wort
- Sterilisation of the wort
- Inactivation of the malt enzyme fixation of the wort composition.

This can be done continuous and faster at higher temperatures. Continuous high-temperature wort boiling (C) or high-temperature boiling (HTWB) is an alternative boiling system. The idea is quite old as Dummet described such a system in 1958 and Daris et al. in 1962 [20]. At high temperatures of 130 or 140°C, very satisfying wort analysis data could be obtained although very short boiling times of ca. 5 min were used. The wort was heated up in three steps. In the first two steps, vapour from the flash-off chambers was reused. Considerable energy savings could be obtained due to the short boiling time and energy recuperation. Alternative continuous systems have been developed [11]. A boiling aroma was a limiting factor to the temperature. Most breweries using high-temperature boiling were closed or had to reduce the boiling temperature, as some boiling tastes were not beer typical. Reducing the temperature declined the degree of energy recuperation from the vapours.

Chantrell describes the process of a practical 300 hl HTWB in Great Britain, where HTW was popular in the end of the last century [12]. Wort heating and boiling are influenced by the degree of fouling, which is normal for a three-phase system with solid trub, liquid wort and steam [13]. This leads to flooring especially on the heating zone's surfaces and to differences in the quality of the product in a series of batches. Wort quality changes over a series of brews without intermedi-ate cleaning. The heat transport is decreased, and burnt aromas and trub affect the wort. The thermal load on the wort has to be increased because of the heating profile, which has to be adapted in order to compensate the decreasing heat transfer rate caused by the fouling layer [12, 23].

The wort is collected in the existing kettle at approximately 72°C from the lauter tun. Hop addition and adjustments to colour and gravity are made at this stage. The wort is then pumped through the HTWB to the whirlpool separator. Inside heat exchangers the wort temperature is raised in three successive steps to 140°C. At this temperature, the wort is held for 3 min while passing through an insulated holding tube. The wort then passes into an expansion vessel where the pressure is reduced to a predetermined level. In a second expansion vessel, the pressure is reduced to atmospheric pressure. Energy recovery is achieved with the flash vapours of these two expansion stages. They are used to heat up the first two wort heat exchangers. Only the third heat exchanger requires an external steam supply for the trim heating of the highest temperature. Cleaning of the plant is performed by automatic control and involves three cycles: firstly, the weekly cleaning of the two expansion vessels via spray balls and, secondly, the cleaning of the vapour side of the first and second stage heat exchangers. This cycle is only operated every 2–3 weeks depending on throughput. The main cleaning cycle which follows the wort path through the plant is operated at the end of the 5-day brewing week. Initially, cleaning problems were encountered with the large-diameter holding tubes. These large tubes were difficult to fill, and the low detergent velocity provided no scrubbing action against the protein deposits. The problem has been overcome by the dosing of hydrogen peroxide into these tubes during the caustic cycle. Foaming agents ensure complete cleaning of this section of the plant. The dosage rate for the peroxide is 0.1% by volume of a 30% hydrogen peroxide. For the third stage the heat exchanger is equipped with an automatic self cleaning cycle, using steam to crack the layers usually after one or two thousand hectolitre [12].

At the Meura brewhouse, continuous wort boiling was combined with a hop strainer, if natural hops were used [1]. Decanters or centrifuges were necessary to remove hot break and to avoid yeast slime, especially if continuous fermentation is following. The wort was heated up in-line to boiling temperature. The added hops were homogenised. An adapted agitator assured a sufficient mixing for the trub for-mation. For the chemical/biochemical reaction of turning the S-methylmethionine into dimethyl sulphide, an external agitation must be provided. Clarification is necessarily conducted prior to stripping to avoid fouling the column with hot trub. The wort-settling tank is needed to recover non-oxidised trub from the hot wort in a continuous way. From the clarification unit, the wort is then stripped by a single pass stripping column. The unwanted volatile components are stripped by counterflow clean steam. The wort is pumped continuously from the bottom of the stripping column through the wort coolers. Because fouling is unavoidable, two duplicate built wort coolers have to assure a continuous cooling of the wort; one wort cooler can be cleaned, while the other one is cooling the wort.

Wort cooling can be done within 50–60 min generating a peak consumption during this period. Compared with batch brewhouses, the heat losses and peak of utilities are lower during continuous processing. While batches are pumped from vessel to vessel, air enters the vessels, pipes and valves at each transfer, thus cooling down the facilities. The transfer of batches also enhances the extract, water and energy losses since vessels are never emptied completely [1]. Water evaporates, and sugars, polyphenols and proteins concentrate, forming layers that have to be cleaned before biofilms come up. Continuous systems can be kept in a stable equilibrium for a longer time and need less cleaning, if kept in a hygienic status. This is even more important in the cold section of a brewery.

5. Continuous fermentation

Continuous wastewater treatments with aerobic and anaerobic microorganisms are big continuous fermentations and show the sensitivity of balancing the biological process. Continuous beer fermentation has to fulfil not only the metabolism of substrate. A system of continuous beer fermentation was patented in 1906 by Van Rijn [2]. In 1953 Morton Coutts patented a process known as continuous fermentation at the Waitemata Brewery in New Zealand, which eventually become DB Breweries [14].

Ricketts (1971) referred to continuous beer fermentation systems which date from the end of the nineteenth century. During the 1960s, while introducing large unitanks, interest arose in permanent fermentations. Several systems were devel-oped, and some reached the point of marketability. Some of the anticipated benefits of continuous fermentation were realised, but most breweries, with some notable exceptions, have continued to use the traditional batch approach, using large-capacity vessels [2].

Coutts created a "wort stabilisation process" that clarified the wort and made it more consistent. He separated the main functions of the yeast into two stages, first the yeast growth and then the fermentation. By splitting these two functions, Coutts created a "continuous flow". The brewers had to add raw materials continually to the first stage and draw off a steady gain of finished beer from the second stage, thus allowing the process to run constantly [14].

The rate at which wort is produced must be sufficient to supply the needs of the yeast at all the time. Inevitably, this requires a prefermenter wort collection vessel. Downstream of the fermenter the brewery has to be capable of handling a continuous supply of green beer. Consequences of failure in a continuous fermenting system cause a serious threat to production. Emptying, cleaning, starting a new process and establishment of stable running conditions are long procedures that take at least 2 weeks in time [2].

The reactors took the form of a coil or similar elongated form by using multiple tanks with a continuous wort flow. The open continuous culture system may also consist of a stirred reactor to which medium is introduced by an entry pipe [2]. The rate of medium addition can be altered by a frequency-controlled speed pump, which is controlled by sensors checking the state of fermentation by pH, density, turbidity or gas production. Culture yeast is removed from the reactor via a second pipe which is arranged in the form of a syphon.

In 2013 Müller-Auffermann at the Technische Universität München installed a downward-facing pipe with two reaction zones in each tank. Four tanks could be filled and emptied continuously from the top part of the tanks. The tanks were combined to a reaction cascade. They were equipped inside with a central pipe, with open bottom. The bottom connection of the tank could hence be used to discharge yeast cells and other particles during the process [15]. Inoculation and growth medium were mixed at the point of entry and fed simultaneously and continu-ously into the reactor. Within a discrete "plug" is travelling through the reactor. A minimum of backward and forward mixing had to be assured. So batch growth proceeded. The reactor could be viewed as a continuum of batch cultures. The spatial location is related to culture age. The factors of temperature, inoculation rate and substrate concentration are also influential in a plug flow continuous culture, like in a batch culture. The composition of the culture issued from the reactor is a function of the flowrate. By careful regulation of these parameters, it is possible to establish a steady state at which the product is of a constant and desired composi-tion [2]. Biomass recycling will be a further refinement to be introduced to a plug flow reactor. The biomass is returned to the entry point of the reactor where it is used as inoculum. Used in this way, the reactor requires only to be supplied with fresh substrate [2].

The prolonged nature of continuous fermentation has inherent risks. Extended running times increase the opportunities for microbial contamination. Yeast "variants" may be selected with the concomitant risk of undesirable changes in beer quality. Continuous systems are more sophisticated than many brewery batch fermenters. Skilled personnel must be on-site, night and day, to provide technical support, if a deviation is indicated [2].

In comparison to classical batch fermentation, only one or few fermenting vessels are needed. Furthermore, beer losses are reduced, less pitching yeast is needed, and detergents and sterilants are saved. As long as the process is stable, a consistent beer quality can be expected. Microbiological contaminations or yeast mutation leads to serious consequences [2] especially if no second production line or some beer for blending is available to keep up the delivery capacity.

Continuous fermentation systems, based on immobilised cells, were condemned to failure for several reasons. Engineering problems like excess biomass, problems with CO_2 removal, optimisation of operating conditions, clogging and channelling of the reactor, unbalanced beer flavour, altered cell physiology and cell ageing lead to unrealised cost disadvantages such as high carrier prices at complex and unstable operations [16]. Pilot-plant and full industrial-scale processes showed engineering problems. The carrier material, the reactor design, together with the effect of immobilisation on yeast physiology, and the risk of contamination end up in a hardly predictable flavour profile of the beer produced. Therefore, despite the economic advantages expected, the continuous process has so far been industrially applied only in beer maturation and alcohol-free beer production [16].

The crucial step forward in continuous technology was certainly the development of commercial immobilised yeast reactors. This approach was of sufficient interest to form the subject of an entire European Brewing Convention Symposium "Immobilised Yeast Applications in the Brewing Industry" held in Finland in 1985 [17, 18]. The advantage of immobilised reactors is that very high yeast concentrations are achievable. This allows a very rapid process throughput which is of particular benefit when applied to rapid beer maturation. A single immobilised yeast reactor can eliminate the time-consuming warm conditioning step for diacetyl reduction at the end of a lager beer fermentation [2].

The application of gel occlusion systems in the brewing process, even if associated with many advantages over conventional fermentation technology, has some important drawbacks, particularly diffusional limitations which impact negatively on yeast growth, metabolic activity and beer flavour, Masschelein et al. concluded at EBC Congress in 1984 [18]. Nakanishi et al. recognised that fermentation activity in continuous working fermenters fell gradually during continuous operation of the system. It could be maintained for 2 months by periodic aeration in which 290 mg/g-yeast (dry matter) of oxygen was supplied to the immobilised yeast [17].

Continuous fermentation suits best in breweries making only one style of beer, because its time and capacity consume to stop the process and start up again with a new beer [14]. Immobilised yeast reactors have also found use in new fermentation processes, for example, in the production of low-alcohol or alcohol-free beers [2], where yeast has more clarification tasks than fermenting and propagation. The major strength of the batch system, using several vessels, is that it is able to cope with seasonal or shorter-term fluctuations in demand. It can easily be adapted to vary the spectrum of production of several different beer varieties and qualities. On the other hand, benefits of continuous fermentation are realised when the systems are operating at a stable status for a long period of time with minimum downtime for changes in beer quality [2].

6. Maturation

Some of these yeast metabolism byproducts (vicinal diketones, acetaldehyde, dimethyl sulphide) impart undesirable flavours to the green beer. The main aim of maturation is to reduce the concentration of such unfavourable flavour com-pounds in the green beer, to saturate the final beer with CO_2 and to remove the haze-forming components from beer within 7–30 days [16]. Fumigation with CO_2 under counter-pressure to avoid too much foam may strip the unwanted flavour. The flavour can be removed from the CO_2 with active carbon so that the CO_2 may be collected, compressed and used for further tasks. Continuous clarification is best done with centrifuges or decanters.

7. Filtration

Filtration and stabilisation of the beer are carried out in order to achieve microbial, colloidal and flavour stability so that no visible changes occur for a long time and the beer looks and tastes the same as when it was made [16]. Particular for higher amounts of yeast cells, tangential flow or crossflow filters are good prefilters before flash pasteurisation or membrane filtration. Although batch flushes can extend the continuous filtration of a crossflow filter, fouling layers will clog the membranes. A chemical recovery of the filter modules is necessary; the continuous process has an end. Usually bright beer tanks collect a batch for the final quality control, and they are the buffers for the following filling of the beer in bottles, cans, kegs or road tankers.

8. Continuous bottling

The most expensive and labour-intensive part of the value creation in breweries is the bottling part. Here most breweries produce continuously, as several machines are needed. Depalletisers for new or crated return bottles, washing machines or rinsers, inspection machines fillers, pasteurisers, labelling machines, packers, wrappers, shrink machines and palettisers run more or less continuously. Stops and interruptions have to be buffered by the conveyers, usually able to keep machines running, while the other needs time to repair so that the previous or following machines need not stop. Modern bottling lines have frequency-controlled conveyers and machinery so that the assembly, connected by system bus, can alter their speed to keep up the continuous production.

The big challenge is changing of the products, the beer type, the labels or the shape and size of the bottles. In bottling, when the beer arrives filtered, sterile and stable, mostly physical deviations have to be handled. Product safety has to assure clean, not contaminated bottles. Camera systems or even gamma or X-ray is used to check the bottles, cans or kegs. Rejected containers have to be replaced by the following shipshape containers. Bottle burst leads to splinter showers, where open bottles have to be removed, eventually contaminated by sharp-edged glass.

Mistakes in this process certainly propagate downstream, if not corrected immediately. A dirty bottle becomes a dirty filled and corked bottle, is labelled and packed and—in the worst case—is sold and consumed. Sensors and camera systems should check the system at the highest accuracy possible, as the process goes on and might lead to big amounts of unsafe products at the far end of the beer production and the intersection to the customers and consumers. The more precise the process is performed, the safer the product and the more the consumers' expectations can be met.

9. Inaccuracy of production, measuring, controlling and quality forecast

One problem of continuous fluid dynamics is the dwell time in the system. Flow conditions ought to be simulated in flow models. These have to be simulated and calculated to predict rheological behaviour and chemical or biochemical reactions [13]. In production methods, biologically grown raw materials or process measurements have a certain inaccuracy or mistake, and results cannot be determined. They just can be estimated mathematically.

The average remaining time in a system is not similar to the real remaining time in a system. Molecules or particles entering a system at the same time may have different remaining times [13]. If there are two or more phases, like fluid and solid or gas particles, the continuous phase (usually the fluid of the beer and the foam) may behave totally different from the solids and the gas phase (bubbles) [13]. Coalescence will be influenced by the collision frequency and the behaviour of the substances during the phases [13]. Biochemical processes have lots of influenc-ing factors, like temperature, pH value, viscosity, surface tension, osmotic and hydrostatic pressure, concentration gradients, mechanical influences, electrical effects, zeta potential and many more. Especially microorganisms change their behaviour under different conditions. As evolution does not stop at the brewing vessels, genetic deviations may cause different behaviours of raw materials and microorganisms.

If more than one species is present, synergetic or suppressing effects may end up in biofilms and uncontrollable developments. Working with just one species—mostly *Saccharomyces*—is quite predictable in its final products. As soon as more species come up, things get more and more unpredictable. Hygiene helps to keep processes under control. Dead zones may become lively areas, and biofilms have to be avoided. Hygienic difficulties in cleaning have to be respected, and in fact, continuous systems cannot be cleaned as often, as batch containers if used for a long time.

This causes problems, especially if brands need to have a constant quality to fit to the consumers' expectation which connected to the brand. Lots of homogenisation equipment during continuous production achieve a dense remaining time in the reactors. Energy, shear forces and moving parts are usually combined with abra-sion of wear parts, which means a continuous change of quality of the machinery. Preventive maintenance needs to stop the processes and to start them again.

10. Mathematical, physical, chemical and biological limits by error propagation in continuous production

A lot of factors have an influence on each step of malting and brewing. This ends up in a broad range of quality factors. Especially biologically balanced equilibriums react to changing conditions by complicated, not predictable effects, which can hardly be measured or recognised by sensors (**Table 3**).

Each quality parameter and each sensor, used during the processes, have its specific standard deviation and impreciseness during measurement. This can be respected, when intermedia products are checked for their quality. Corrective arrangements can be used to reach the final quality aim. During continuous production error, propagation may lead to a huge deviation in quality, which is also caused by the given impreciseness of the sensors who should avoid this, especially if lots of sensors are used in following steps to automatically control the continuous production [21].

An imprecise thermometer in the mash process will lead to different amounts of sugars or proteins, which can behave differently in the wort vessel than the wanted product. This may lead to different colours or yeast behaviours. The thermometers, used during fermentation, have a certain deviation as well, which might lead to different metabolisation products. They can be a favourable substance for other yeast or bacteria strains which also create unwanted flavour products. Sensors may detect but also have a deviation which allows unwanted processes. If sensors show a deviation to the quality aims, a continuous system should be able to adjust the process to the predetermined values. If this is not possible, the process has to be stopped. Analysis

Gas phase	Liquid phase (continuous)	Solid phase
Volume Pressure Temperature Gas equilibriums Bubble size Bubble form Molecular size Heat capacity	Temperature Viscosity Surface tension Density Hydrophobic substances Hydrophilic substances pH Heat capacity	Form Size Surface, roughness Density Zeta potential Hydrophobic liquids Hydrophilic liquids Heat capacity
Altered/influenced by		
Temperature	Temperature	Temperature
Pressure	Shear forces	Shear forces
Microorganisms	Microorganisms	Microorganisms
	Enzymes	Enzymes
Time	Time	Time

Table 3.
Influencing physical and microbiological factors during beer production.

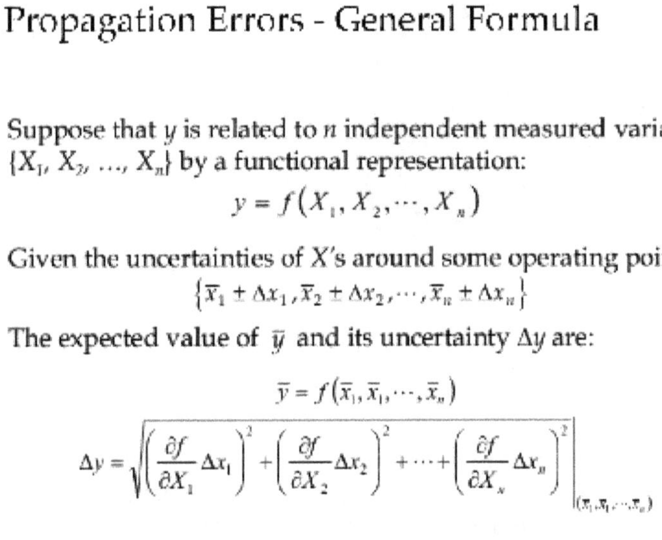

Propagation Errors - General Formula

Suppose that y is related to n independent measured variables $\{X_1, X_2, \ldots, X_n\}$ by a functional representation:

$$y = f(X_1, X_2, \cdots, X_n)$$

Given the uncertainties of X's around some operating points:

$$\{\bar{x}_1 \pm \Delta x_1, \bar{x}_2 \pm \Delta x_2, \cdots, \bar{x}_n \pm \Delta x_n\}$$

The expected value of $\bar{y}$ and its uncertainty Δy are:

$$\bar{y} = f(\bar{x}_1, \bar{x}_1, \cdots, \bar{x}_n)$$

$$\Delta y = \sqrt{\left(\frac{\partial f}{\partial X_1}\Delta x_1\right)^2 + \left(\frac{\partial f}{\partial X_2}\Delta x_2\right)^2 + \cdots + \left(\frac{\partial f}{\partial X_n}\Delta x_n\right)^2}\Bigg|_{(\bar{x}_1, \bar{x}_1, \cdots, \bar{x}_n)}$$

Figure 2.
Mathematical estimation: example for additive error propagation [22].

of the intermedia product may help to decide how the quality targets can be reached by the following process step, in the way batch processes are successfully managed.

The deviations in a naturally given physical, chemical and biological concentra-tions sum up and propagate by each sensor deviation and error used in the process. Calibration might help to a certain extent, but the amount of sensors is growing in continuous processes, so the amount of possible imprecise information is steadily increasing with the amount of measurements and control valves and regulations, which also have a certain deviation. Measurable substances like diacetyl, gravity, conductivity and turbidity are useful, but they are just single parameters in a bunch of varying aroma particles, dependent on microbiological, chemical and physical balances.

In statistics, "propagation of uncertainty" also called "propagation of error" is the effect of variable uncertainties combined. Errors, or more specifically random errors, result in an uncertainty that builds up during long-lasting consecutive

differences in production by treating different phases							
Phases	*mashing*	*lautering*	*boiling*	*fermentation*	*maturation*	*filtration*	*bottling*
solids	grain	spent grain	trub, hops	trub, yeast	trub, yeast	trub	
fluid	mash	wort	wort	beer	beer	beer	beer
gas			steam	O_2, CO_2	CO_2	CO_2	CO_2
production inaccuracies from measurement deviation, influencing the control of the process							
Time [min]	◁	◁	◁	◁	◁	◁	◁
°C	◁	◁	◁	◁	◁	◁	◁
% o.g.	◁	◁	◁	◁		◁	◁
Trub EBC		◁			◁	◁	◁
bar				◁	◁	◁	◁
Aroma				◁	◁	◁	◁

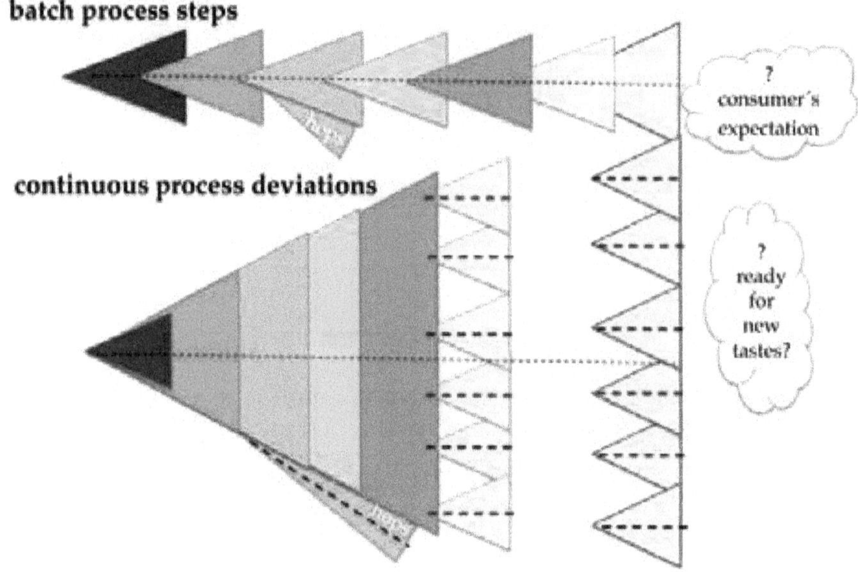

Figure 3.
Deviations in batch and continuous brewing processes.

processes. Measured values are necessary to control the process. All measurements have uncertainties due to limitations (**Figure 2**). Instrument precision and other deviations propagate due to the combination of variables in the function [19].

If biological changes occur in raw materials or microorganisms, errors do not add but multiply, or exponentiate, as several factors change: aromas, natural substances, pH, viscosity, chemical balances, concentrations, compositions, sublimation, dissolving, evaporating, etc. Respecting the rheologically unpredictable behaviour of altering three phases during rheological processes [13], a constant and planned quality can only be expected for a certain time; quality changes and emptying and cleaning of the system become necessary.

Continuously changing products like mash, fermenting of green beer has to come to a stable and defined quality, which fits to the consumers' expectation. This is traditionally achieved by cold stabilisation during maturation, which is also a possibility to check and adjust quality, to blend the beers, and it is a buffer before bottling and racking. Filtration and pasteurisation help to stop the most biological and enzymatic processes. Especially before bottling a quality check has to be conducted, to avoid faulty products being bottled. The product will be kept during distribution and at the consumer's place for a certain time, where it should not change anymore. Finally the consumer's expectation cannot be adjusted to an unpredictable quality (**Figure 3**).

11. Conclusion

Stepwise batch production has been added by continuous brewing processes for more than 100 years. Calculations may predict the results in biological processes to a more accurate degree, if computers obtain more and more data to use. The results in the end are estimations, relying on the number of factors in relation to their influencing parameters. The possibility to measure these factors need to have a known accuracy. This will stay the biggest challenge, as long as natural ingredients and microorganisms play a main role in malting and brewing. Fermenting with more than one yeast strain quality, prediction falls back to the times of spontaneous fermentation, where only few products were of a good taste by accident. The same thing occurs, if contaminations get into the continuous systems, when cleaning always implies a long interruption of the process, especially a long time to adjust the following process up to a stable equilibrium.

Many substances in beer are known, many not. Some substances or physical properties are useful to define the quality of the beer and the intermedia production steps. Few parameters can be used to control the brewing process in the direction of a defined quality aim. All of the raw materials, the determination methods, the regulation equipment and the biological reactions of the microorganisms have deviations, some of them are not measurable or known yet. In a batch process inter-stage products like malt, wort or fermented beer can be checked, and the following processes can be used to adjust the quality goals. Reactions to the deviations in continuous processes should set the system back to the target conditions. Combining several steps to a continuous process with continuous adjustment can be controlled by a self-learning by fuzzy logic or artificial intelligence. As these computers need data from precise sensors to control precise valves, stirrers, pumps, etc., this will lead to a predominantly maintaining, calibrating and scrutinising reaction to the system instead of brewing and creating the quality of the beer.

Nowadays a continuous step is followed by a batch step. The possibility in adjusting the quality with following processes should be given. Also blending needs buffer capacities to equalise the beer to consumer's expectation. For offering different beer types, also blending beverages are necessary to be added. The continuously produced beer might supply a base beer, which is blended before filtration with other batch process beers or water, aromas and lemonades in a nontraditional way.

The brewing process can be performed in steps, continuously and automatically. Perhaps one day continuous processes will be longer stable. The processes may also expire in constant quality results. But brewing is fun, fun that should not be left to the machines.

Author details

Mark Strobl
Hochschule Geisenheim University, Geisenheim, Germany

*Address all correspondence to: mark.strobl@hs-gm.de

References

[1] Bonacchelli B&HF. The Meurabrew the Brewhouse of the Future! [Online]. Available from: http://www.meura.com/uploads/pdf/Meurabrew%20Paper.pdf [Accessed: March 18, 2019]

[2] Buffalo_Brewing_Blog. Buffalo Brewing Blog » Brewery Convention [Online]. 2018. Available from: https://www.buffalobrewingstl.com/brewery-convention/continuous-fermentation.html [Accessed: March 17, 2019]

[3] Kunze W. In: Lebensmittelindustrie MfBIu, editor. Technologie Brauer und Mälzer. 5th ed. Leipzig: VEM Fachbuchverlag Leipzig; 1979

[4] Schuster WN. Die Bierbrauerei Erster Band Die Technologie der Malzbereitung. Stuttgart: Ferdinand Enke Verlag; 1976

[5] Krones. Krones.com. [Online]. 2019. Available from: https://www.krones.com/media/downloads/waermetauscher_en.pdf [Accessed: April 4, 2019]

[6] Narziß L. Back W. Die Bierbrauerei. Band 2: Die Technologie der Würzebereitung, 8th ed. Weinheim, USA: Wiley-VCH Verlag GmbH & Co. KGaA, John Wiley & Sons Inc; 2009

[7] Harsanyi DE. Kontinuierliche Schnellabläuterung nach dem Pablo-System. Brauwelt International. 1968;**108**(Nr. 45/46, 5./7):843-845

[8] Nessie Ziemann Holvrieka. Journal Beer [Online]. 2016. Available from: https://journal.beer/2016/11/14/nessie-revolutionizes-the-brewhouse-novel-mash-filtration-process-by-ziemann-holvrieka/ [Accessed: March 20, 2019]

[9] Klaus Wasmuht DE. Experience is the most Challenging Test. Brewing and Beverage Industry International. In: Sachon VW, editor. Mindelheim: Schloss Mindelburg; 2019

[10] Wasmuht K, Schwill-Miedaner A. With Omnium to Optimum. Rust, Baden: VLB Berlin; 2019

[11] Willaert RG, Baron GV. Wort boiling today—Boiling systems with low thermal stress in combination with volatile stripping. Cerevisia. Belgian Journal of Brewing and Biotechnology. 2001;**26**:217-230

[12] Chantrell NS. Practical experiences in the production of wort by continuous high temperature wort boiling. MBAA Technical Quarterly. 1984;**21**:166-170

[13] Kraume M. Verfahrenstechnik. 2nd ed. Berlin, Heidelberg: Springer Vieweg; 2012

[14] Brooks JR. Continuous Fermentation—Volume 29, Issue 4 [Online]. 2008. Available from: http://allaboutbeer.com/article/continuous-fermentation/ [Accessed: March 17, 2019]

[15] Auffermann-Müller K. More Sustainable Way to Brew Beer: Non-Stop Fermentation Saves Resources May 28, 2013, Technical University Munich. Read more at: https://phys.org/news/2013-05-sustainable-brew-beer-non-stop-fermentation.html#jCp [Online]. 2013. Available from: https://phys.org/news/2013-05-sustainable-brew-beer-non-stop-fermentation.html [Accessed: March 18, 2019]

[16] Brányik T, Vicente AA, Dostálek P, Teixeira JA. Continuous beer fermentation using immobilized yeast cell bioreactor systems. Biotechnology Progress. New York: American Institute of Chemical Engineers (AIChE); 2005; **21**(3):10

[17] Nakanishi K, Onaka TIT, Kubo S. A new immobilized yeast reactor

system for rapid production of beer. In: European Brewery Convention Proceedings of the 20th Congress Helsinki, 1985. Helsinki: European Brewery Convention; 1985. pp. 331-338

[18] Masschelein CA, Carlier A, Ramos-Jeunehomme C. The effect of immobilization on yeast physiology and beer quality in a continuous and discontinuous system. In: Convention EB, editor. European Brewery Convention Proceedings of the 20th Congress Helsinki, 1985. European Brewery Convention; 1985. pp. 339-346

[19] Edu H. Physical Sciences 2 Harvard University, Fall 2007 [Online]. 2007. Available from: http://ipl.physics.harvard.edu/wp-uploads/2013/03/PS3_ Error_Propagation_sp13.pdf [Accessed: April 7, 2019]

[20] Daris AD, Pllock JRA, Gough PE. Journal of Institute of Brewing. Weinheim, USA: John Wiley & Sons Inc.; 1962;**68**:309

[21] Belabbas B. DLR Institute of Communications and Navigation Multi-Sensor Integrity [Online]. Available from: https://www.dlr.de/kn/en/desktopdefault.aspx/tabid-7570/12813_read-32121/gallery-1/216_read-2/

[22] Gehlot T. Slide Share Error Analysis Statistics [Online]. 2014. Available from: https://www.slideshare.net/tarungehlot1/error-analysis-statistics [Accessed: March 22, 2019]

[23] Voigt DJ, Wasmuht K. Fouling during wort boiling—Effects on wort quality. Master Brewers Association of the Americas MBAA TQ. 2006;**43**:207-210

10

Bioprocess Development for Human Mesenchymal Stem Cell Therapy Products

Jan Barekzai, Florian Petry, Jan Zitzmann, Peter Czermak and Denise Salzig

Abstract

Mesenchymal stem cells (MSCs) are advanced therapy medicinal products used in cell therapy applications. Several MSC products have already advanced to phase III clinical testing and market approval. The manufacturing of MSCs must comply with good manufacturing practice (GMP) from phase I in Europe and phase II in the US, but there are several unique challenges when cells are the therapeutic product. Any GMP-compliant process for the production of MSCs must include the expansion of cells *in vitro* to achieve a sufficient therapeutic quantity while maintaining high cell quality and potency. The process must also allow the efficient harvest of anchorage-dependent cells and account for the influence of shear stress and other factors, especially during scale-up. Bioreactors are necessary to produce clinical batches of MSCs, and bioprocess development must therefore consider this specialized environment. For the last 10 years, we have investigated bioprocess development as a means to produce high-quality MSCs. More recently, we have also used bioreactors for the cocultivation of stem cells with other adult cells and for the production of MSC-derived extracellular vesicles. This review discusses the state of the art in bioprocess development for the GMP-compliant manufacture of human MSCs as products for stem cell therapy.

Keywords: bioreactors, quality-by-design, critical process parameters, stem cell potency, standardization

1. Manufacturing cell therapy products

Cell therapy is a growing clinical research and healthcare sector in which living cells are introduced into a patient in an attempt to ameliorate or cure a disease. Stem cell therapy is one of the most promising fields within this sector because the introduced cells have the capacity to differentiate, allowing the repopulation of diseased organs with healthy cells, or to allow even complete organ regenera-tion. This chapter will focus on one specific type of stem cell (MSCs), which are variously defined as mesenchymal stem cells, mesenchymal stromal cells, or (most recently) medicinal signaling cells [1]. These various definitions reflect the controversial origin and functionality of MSCs and uncertainty about their clinical potential [2, 3]. Following encouraging initial results, the unclear or disappointing outcomes of some MSC clinical trials have clouded the picture [4], but the pioneers of this approach still regard MSCs as a promising therapeutic option [5]. One of the key issues in the deployment of MSCs is ensuring they are safe and effective, which requires a well-characterized manufacturing process.

In order to provide enough MSCs for cell therapy, donor cells must be isolated from tissue and then expanded *in vitro* to reach a population of 1–9 × 10^8 cells, which is the typical dose for adult treatment [6]. The success or failure of MSC therapy depends on this *in vitro* expansion process, which was first studied in detail following the failure of the MSC product Prochymal in phase III trials for graft versus host disease (GvHD) [4], whereas a similar product succeeded in phase II. One reason proposed for the contrasting outcomes of each trial was the substantial differences in the MSC expansion step at the manufacturing scale, highlighting the specialized and complex nature of MSCs [4].

1.1 Definition of MSCs and current approved products

MSCs are classified as advanced therapeutic medicinal products (ATMPs) under regulations in Europe and the US. Many countries follow the regulations laid down by the US Food and Drug Administration (FDA), which defines MSCs as cell therapy products, whereas the European Medicines Agency (EMA) defines MSCs as cell-based medicinal products and distinguishes between somatic cell therapy medicinal products (SCTMPs) and tissue engineered products (TEPs) [7]. This means that clinical studies and drug approval are covered by a specific regulatory framework applied at the national or regional level. Manufacturing must therefore be compliant with good manufacturing practice (GMP) regulations that have been tailored for ATMPs, following strict criteria for product specification and release for clinical use. However, the regulatory framework for MSC manufacturing is confounded by ambiguous product definitions reflecting regional differences in the way the regulations are implemented. For example, the EMA requires GMP compliance and manufacturing authorization for phase I material, whereas the FDA does not apply this requirement until phases II and III, and in Canada, GMP compliance is not strictly required at any phase [8]. Even so, various MSC products have been manufactured under these different regulatory jurisdictions and have proceeded through clinical development, in some cases gaining market authorization from the local regulatory agency [9]. Most of these products are allogenic, which means that MSCs from one or more healthy donors are expanded, processed, and stored and then applied to patients as an off-the-shelf product (**Table 1**). In 2016, the allogenic MSC product TEMCELL (developed by Mesoblast) was licensed to JCR Pharmaceuticals, which received market authorization in Japan under a fast-track protocol for patients with steroid-refractory acute GvHD. Mesoblast also conducted a phase III trial with this product in the US, involving 60 patients of the same indication, achieving the primary endpoints (NCT02336230). In 2018, ALOFISEL (Takeda Pharma), an expanded allogenic adipose-derived MSC product, was approved by the EMA to treat complex perianal fistula in patients with Crohn's disease. This was supported by a placebo-controlled trial involving 212 patients [10]. Stempeucel (Stempeutics), an expanded allogenic MSC product, received market authorization from the Drug Controller General of India to treat limb isch-emia in patients with Buerger's disease. However, it is limited to 200 patients on a cost-recovery basis, and a postmarket surveillance study is required. Ninety patients have already received an injection of this MSC product in a phase II trial, achieving a significantly better outcome than standard care [11]. CARTISTEM (Medipost) is an allogenic culture-expanded umbilical cord blood MSC product to treat knee articular cartilage defects in patients with osteoarthritis, grade IV, and following approval for the South Korean market in 2012, its clinical outcomes have remained

	Product 1	Product 2
Exemplary products	ALOFISEL	Queencell
Indication	Crohn's disease, perianal fistula	Regeneration of subcutaneous tissue
Patients per year	23,000 (in EU)*	n.d.
Cell type	Allogenic MSCs	Autologous, patient-specific MSCs
Cell source	Adipose tissue	Adipose tissue
Cells per dose	1.2×10^8 MSCs	7×10^7**
Therapeutic relevant cell properties***	Anti-inflammation, immune modulation	Regeneration, anti-apoptosis
Manufacturing type	Bulk manufacturing	Patient-specific batch
Batch size	Large (min. 100–1000 doses per batch)	Small (1 dose per batch)
Scalability of production	Scale up	Scale out, several batches in parallel
Product storage	Frozen, off-the-shelf	No storage
Stability under storage	Stable >6 month, frozen	Fresh, stable max. 24 hours

**0.003% of all citizens (741 million) in Europe are putative patients.*
***Stromal vascular fraction contains MSCs and other cell types such as preadipocytes, endothelial progenitor cells, pericytes, mast cells, and fibroblast.*
****Following both products have different critical quality attributes (CQAs) and the manufacturing processes have different critical process parameters (CPPs).*
n.d. not determined.

Table 1.
Indication and properties of MSC products impact their manufacturing.

stable over 7 years of follow-up studies [12]. Several autologous MSC products have also been approved in South Korea, meaning that the MSCs are isolated from the patient's own tissue and then manipulated/expanded in a patented process and re-injected into the patient 4–6 weeks later. NEURONATA-R (Corestem) and Cellgram-AMI (Pharmicell) are autologous bone marrow-derived MSCs indicated for amyotrophic lateral sclerosis and acute myocardial infarction, respectively. Two other MSC products derived from adipose-tissue have been approved (Anterogen): a mixture of autologous adipose-derived MSCs with other cells for subcutaneous tissue defects (Queencell) and a pure adipose-derived MSC product for Crohn's fistula treatment (Cupistem) [9]. NEURONATA-R has been designated as an orphan drug by the EMA and FDA.

This brief survey of the market shows that the promise of MSC therapy is materializing, with positive efficacy data in controlled clinical trials followed by regulatory approval for a small number of products.

1.2 The therapeutic properties of MSCs

Although MSCs have been used in cell therapy applications for many years, the fundamental biology of these cells and their precise therapeutic properties are not fully understood. MSCs were initially isolated from bone marrow (bm-MSCs) based on their plastic adherence, but today they are usually isolated from adipose tissue (ad-MSCs) or umbilical cord blood (uc-MSCs), which are more accessible [13]. MSCs are also found in various other adult, fetal, and perinatal tissues [14]. Regardless of their origin, MSCs are heterogeneous and polyclonal cells, with at least three subpopulations defined based on morphology. Type I MSCs are spindle-shaped proliferating cells resembling fibroblasts. Type II MSCs are large, flat, epithelial-like cells, which are more senescent than type I cells and feature visible cytoskeletal structures and granules. Finally, type III MSCs are small round cells with a high capacity for self-renewal [15]. The heterogeneity of MSCs can be considered beneficial in that it ensures that some therapeutically active cells are present, but it reduces the maximum potential efficacy because some of the cells are inactive. However, even monoclonal MSCs become heterogeneous during expansion [16].

Despite the heterogeneity described above, the International Society of Cell Therapy has published a set of minimal criteria that must be met before cells can be defined as MSCs. Such cells must (i) show plastic adherence; (ii) be able to differentiate into cartilage, bone, and fat tissue *in vitro*; and (iii) express the cluster of differentiation (CD) surface markers CD73, CD90, and CD105, but not CD11b, CD14, CD19, CD34, CD45, or HLA-DR [17]. However, this standard set of markers does not distinguish between MSCs and fibroblasts or nonstem mesenchymal cells [18]. Several other markers may be more specific but are only detected in certain MSC isolates or subpopulations. These include stage-specific embryonic antigen-4 (SSEA-4), stem cell antigen-1 (SCA1), nestin, CD44, CD146, CD166, and CD271 [19]. A unique MSC surface marker has yet to be identified.

It is important to note that MSCs cannot be defined merely as a collection of surface markers because this says nothing about their therapeutic effect (**Figure 1**). Initially, the therapeutic potential of MSCs was believed to reflect their ability to migrate into damaged tissues, differentiate *in situ*, and replace damaged or dead cells. However, although MSCs can differentiate *in vitro*, their ability to differentiate *in vivo* has never been confirmed [20]. Current opinion is that MSCs migrate to injury sites and secrete chemoattractants that recruit tissue-specific stem cells,

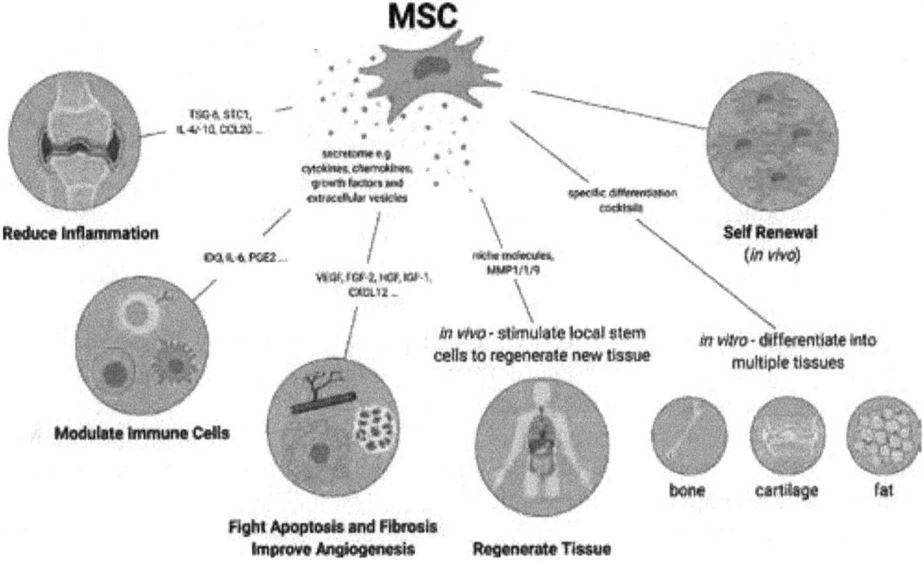

Figure1.
Properties of MSCs and their mode of action. MSCs modulate the host immune systems, e.g., by secreting various trophic factors. Thereby, they reduce inflammation, promote neoangiogenesis, and prevent apoptosis and fibrosis. Further, they stimulate local stem cells to develop new tissue. TSG-6, tumor necrosis factor-inducible gene 6 protein also known as TNF-stimulated gene 6 protein; STC1, stanniocalcin 1; IL-4/6/10, interleukins 4, 6 and 10; CCL20, macrophage inflammatory protein-3; IDO, indoleamine 2,3-dioxygenase; PGE2, prostaglandin E2; VEGF, vascular endothelial growth factor; FGF-2, basic fibroblast growth factor; HGF, hepatocyte growth factor; IGF-1, insulin-like growth factor 1; CXCL12, stromal cell-derived factor 1; MMP1/2/9, matrix metalloproteinase-1/2/9.

which in turn generate new tissues or exert positive immunomodulatory effects [1]. The MSC secretome comprises a pool of cytokines, chemokines, growth factors, and extracellular vesicles (carrying proteins, lipids, and various forms of RNA). This secretome differs widely among MSC isolates and subpopulations and can be used to functionally distinguish between several MSC types (e.g., type I, II, and III cells), revealing that the self-renewable type III cells are therapeutically the most effective [16].

The immunomodulatory properties of MSCs and their secretion of anti-inflammatory molecules and extracellular vesicles are an important therapeutic functionality [14]. MSCs are therefore logical candidates for the treatment of immune disorders, including GvHD, inflammatory bowel disease, multiple sclerosis, rheumatoid arthritis, and diabetes [21]. MSCs also secrete peptides and factors that promote the regeneration of damaged tissue by stimulating cell proliferation and migration, promoting angiogenesis, and suppressing apoptosis and fibrosis [14]. The regenerative capacity of MSCs has been used to treat Alzheimer's disease, bone and cartilage diseases, diabetes, myocardial infarction, and osteoarthritis [22]. Another advantage of MSCs is that they do not form teratomas *in vivo*, which ensures an outstanding clinical safety profile. Human MSCs achieve senescence without evidence of transformation into tumor cells [23].

1.3 The critical quality attributes of MSCs

The biological complexity and heterogeneity of MSCs hamper the translation of laboratory-scale experiments into industrial processes for cost-effective and reliable manufacturing. This can be addressed by developing MSC manufacturing processes that adhere to quality-by-design (QbD) principles [24]. QbD provides a rational framework and integrates scientific knowledge and risk analysis into process development. It is guided by a thorough understanding of the fundamental biology and engineering principles underlying an MSC product and its production process. QbD begins with a description of the desired product quality characteristics, known as the quality target product profile (QTPP). This is used to identify critical qual-ity attributes (CQAs), which are physical, chemical, and biological attributes that define the quality of the product. The QTPP for MSCs describes properties such as identity, purity, and potency, which will be unique for each MSC product and dependent on the therapeutic indication.

1.3.1 Identity

For MSCs, identity often means the cell phenotype, but as discussed above, there is no agreement on a single definition. Identity is often demonstrated by confirm-ing a typical morphology and/or karyotype [25] and by detecting the presence or absence of surface markers. The minimal criteria for MSCs (see above) have led to a misconception that cells meeting these criteria are equivalent in identity and therapeutic functionality. In polyclonal MSC populations, the presence of multiple cell types can be a clinical benefit as stated above [26], and this should be reflected in the identity attributes.

1.3.2 Potency

The functionality and potency of MSCs are closely linked to their therapeutic efficacy and thus the clinical outcome, but potency is used to demonstrate manufacturing consistency for batch release so a measurable property is required. Viability can fulfill the role of a potency indicator because only living cells can act as a therapeutic entity. Potency can also be measured using *in vitro* functional assays that determine MSC activity directly or via an indirect metric that correlates to MSC activity *in vivo*. An assay that measures differentiation potential is only appropriate to describe MSC potency if the therapeutic aim involves engraftment of the cells or tissue formation (notwithstanding the controversy over the assumption that MSCs differentiate *in vivo*, as discussed above). The FDA mandates that potency is measured using quantitative biological assays [27], so the standard approach is to differentiate MSCs *in vitro* by cultivating them in differentiation medium and then testing them after 21 days [17]. Staining for differentiation markers is nonquantitative, so alternative methods such as postdifferentiation RNA or protein analysis [28, 29], or the online monitoring of differentiation by Raman spectroscopy [30], are more suitable.

If the therapeutic effect of MSCs is conferred by the secretome, then the differentiation potential may not be the primary determinant of potency. The profile of secreted factors would be a more appropriate measure, and this could be determined by multiplex enzyme-linked immunosorbent assays (ELISAs) or mass spectroscopy [31]. However, a clear link between the secretome profile and *in vivo* efficacy must be established, so that animal models or cell-based assays can be used to determine the limits of the relevant factors. This is a typical way to move from a complex and highly variable *in vivo* assay to a multiassay approach combining the quantification of viability, target-specific cytotoxicity or cytokine release, surrogate biomarkers (morphological phenotype or released factors that correlate with function), bioactivity (e.g., presentation of surface markers), cell-based assays, and genomic, transcriptomic, and proteomic profiles [32].

1.3.3 Sterility and purity

Impurities are unwanted components from within the process, whereas contaminants come from outside the process. Impurities during MSC manufacturing include unwanted cell types, particles (e.g., residual microcarriers, or plastics and fibers from manufacturing equipment and materials), or components of culture medium. Contaminants include bacteria, fungi, viruses, endotoxins, and mycoplasma. The heterogeneity of MSCs makes it difficult to detect unwanted cell types. MSC preparations should ideally be pure, but fibroblasts are often present as impurities. Cell-specific sorting based on the marker CD166 (which is expressed at higher levels on MSCs) and CD9 (which is expressed at higher levels on fibroblasts) may help to achieve sufficient purity [33]. In other cases, it may be sufficient if most of the cells in the final product (>98%) fulfill the ISCT minimal criteria based on MSC surface markers. All other impurities and contaminants must be measured and the maximum residual levels must be defined to ensure safety and efficacy. A final sterilization step is not possible when the product is living cells, so the entire MSC production process must be carried out under aseptic conditions.

From the QTPP list, CQAs must be identified, which directly influence the safety and efficacy of the MSC product. This means that a risk assessment is carried out to reduce the QTPP list to the most influential attributes based on impact and certainty. According to ICHQ8, a CQA is "*A physical, chemical, biological, or microbiological property or characteristic that should be within an appropriate limit, range, or distribution to ensure the desired product quality.*" Therefore, every process parameter "*whose variability has an impact on a CQA*" is a critical process parameter (CPP) that "*should be monitored or controlled to ensure that the process produces the desired quality.*" There is no precise delimitation of the degree of impact required to define a CPP, so the broad definition of a CPP is generally divided into parameters that have a substantial impact on the CQAs and those with minimal or zero impact. Each process step has multiple CPPs. For example, during the *in vitro* expansion step, CPPs can be directly associated with the MSCs (e.g., cell density and cell age) or raw material attributes (e.g., medium, serum, and growth factors) or operational features of the culture vessel/bioreactor system (e.g., pH, temperature, dissolved oxygen, and agitation). The effect of each CPP on the CQAs must be quantified in a design space. With an appropriate control strategy, the CPPs are kept in their normal operational range, which ensures the production of high-quality MSCs that meet all the required CQAs. Based on the heterogeneity and the complexity of MSCs, each MSC product can have unique CQAs and the corresponding CPPs must be identified case by case.

2. Expansion of human MSCs *in vitro*

Therapeutic applications of MSCs require at least 1×10^8 cells per dose, which is many more than can be isolated by tissue aspiration. All MSC production processes must therefore include an *in vitro* expansion. Having generated or isolated the starting cell population, *in vitro* expansion is followed by harvest, concentration, purification, formulation, fill and finish, storage, and shipping. The manufacturing steps of MSCs are therefore similar to the production of recombinant proteins, but MSCs are more challenging due to the variability of the starting material, the complexity of living cells as a product, an incomplete understanding of their mechanism of action, and the inherent difficulties encountered during product characterization.

2.1 CPPs that affect MSC manufacturing

The properties of MSCs are strongly influenced by the environment because MSCs in nature interact with surrounding cells and tissues, with the extracellular matrix and with various bioactive molecules. Even in an artificial environment like a bioreactor or T-flask, MSCs are very sensitive to their environment, and the most influential factors give rise to CPPs. By identifying CPPs that affect MSC quality, the process can be designed to favor the recovery of MSCs with specific phenotypes of interest, in this case those with the greatest therapeutic efficacy [34, 35]. The CPPs affecting MSC quality are discussed in more detail below.

2.1.1 Cell density and age

During MSC isolation, the seeding density is important because all sources contain different quantities of MSCs. For example, only 1 in 100,000 bone marrow cells is an MSC, whereas in adipose tissue, the ratio is nearer to 1 in 100 [36]. If plastic adherence is selected as a strategy for MSC isolation, the number of adherent cells therefore differs according to the source if a similar number of tissue cells are seeded. Standardization during this step can be achieved by isolating MSCs using a strategy of surface marker sorting, allowing a defined number of cells to be seeded into the culture vessel. The seeding density selected for the *in vitro* expansion step is a CPP. MSCs can be seeded at a very low density (50–100 cells per cm^2) and will proliferate until they achieve confluence. This corresponds to a high expansion factor, but the process takes a long time and requires more rounds of cell division for each seeded cell, so the cells experience significant aging [37]. The aging of MSCs during expansion is a problem, because older cells lose competence to behave as stem cells and have a tendency to enter senescence or even to undergo transfor-mation. The manufacturing of Prochymal provided a clear example of this issue: 10,000 or more doses were manufactured from one donor, and the corresponding expansion stress led to replicative senescence, in which the cells retained a typical

MSC surface marker profile but lost functionality [4]. Aging MSCs are more likely to activate a senescence-associated secretory phenotype and produce pro-inflammatory cytokines such as IL-1, IL-6, and IL-8, which inhibit the regenerative process. The duration of *in vitro* expansion must be considered not only because of senescence, but also due to the phenomenon of clonal impoverishment. MSCs are polyclonal, but prolonged expansion favors the growth of specific cell types or clones. Depending on the expansion time and expansion factor, the cell mixtures may completely differ in phenotype and also in potency. Therefore, although a high expansion factor in a short process time is desirable to achieve high product yields, *in vitro* expansion should never change the properties of MSCs to the extent that it compromises their functionality and potency.

2.1.2 Culture medium

Several basal media have been shown to influence MSC expansion and potency, including Dulbecco's modified Eagle's medium (DMEM), Iscove's modi-fied Dulbecco's medium (IMDM), and MEM alpha (αMEM) [37]. One of the key components of these media is glucose, which is the main carbon source for MSCs. Glucose may be provided at physiological concentrations (1 g/L) or higher (up to 4.5 g/L), the latter variously described as having a negative effect on MSC proliferation and growth factor secretion [38] or no effect at all [39]. Glutamine as a second carbon source is present at concentrations of 2–4 mM and appears essential for MSC growth [40], but its impact on MSC properties is complex, with contradictory results [41–43]. Glutamine is unstable at 37°C and spontaneously degrades to form ammonia. GlutaMAX (dipeptide Ala-Gln) is recommended instead of glutamine to promote MSC expansion [44]. Lactate and ammonia are the most abundant waste products formed by MSCs, and both therefore have the potential to inhibit growth. It therefore follows that glucose, glutamine, lactate, and ammonia levels should be considered as CPPs for the production of MSCs. Several other amino acids may also be relevant, given that the amino acid metabolism of MSCs differs from that of commercial cell lines such as Chinese hamster ovary (CHO) cells [42].

Basal media formulations must be supplemented to achieve MSC expansion. The most important supplement is fetal calf serum (FCS), which is added to a final concentration of 5–20%. FCS strongly influences MSC growth and phenotype, but the specific effectors are unknown because the composition of FCS is variable and lot-dependent [45]. The use of FCS for the manufacture of clinical MSC products is discouraged nowadays, in line with the drive to eliminate all raw materials of animal origin. The complex, uncertain, and variable composition of FCS also makes it difficult to validate for GMP-compliant processes. Finally, the manufacturing process must accommodate steps to eliminate FCS from the final product to avoid potential immunogenicity and allergenicity [46]. FCS can be replaced with human serum and its derivatives, such as human platelet lysate, which promotes MSC growth [47]. However, the same lot-dependent quality issues described above for FCS also apply to human serum [48]. The most acceptable alternative is serum-free or prefer-ably chemically-defined medium, the latter not only serum-free but also lacking any hydrolysates or supplements of unknown composition. MSCs grow well in several commercial serum-free media, including BD Mosaic MSC Serum-free (BD Biosciences), RoosterNourish (Rooster Bio), Mesencult-XF (Stemcell Technologies), StemPro MSC SFM Xeno-Free (Invitrogen), TheraPEAK MSCGM-CD (Lonza), and PPRF-msc6, STK1 and STK2 (Abion) [49]. Growth in chemically-defined medium has also been demonstrated [50]. However, although MSCs showed excellent growth in these serum-free media, they reached senescence earlier, and there were changes in morphology, surface marker profiles, and potency [51]. This does not mean that serum-free and chemically-defined media should be avoided-it is still better to use these media for MSC expansion in order to meet GMP requirements-but further investigations are required to optimize the media composition. The development of serum-free media is mainly driven by companies, which tend not to disclose the precise composition, making it difficult for other researchers to build on the results. In serum-free and defined media, supplemental growth factors such as FGF2 and PDGF are needed to stimulate MSC proliferation, but they also influence MSC potency [18]. Accordingly, chemically-defined media would be preferable for the *in vitro* expansion of MSCs, but growth factor concentrations are important CPPs that affect MSC identity and potency and must be carefully controlled.

2.1.3 Conditions in the culture vessel

MSCs are aerobic cells and any culture vessel must therefore ensure an adequate supply of oxygen. However, the oxygen saturation in standard T-flasks (21% O_2) is far removed from nature (5–7% O_2) [34]. MSCs therefore tend to be oversaturated with oxygen, which can increase the concentration of damaging reactive oxygen species (ROS). Several studies have confirmed that hypoxia enhances MSC proliferation, stabilizes their cell fate, and prevents apoptosis by reducing the levels of caspase-3 [52]. However, rather than imposing hypoxia by preconditioning the cells, it may be better to impose hypoxia during the entire expansion phase, because this mimics their natural niche [53].

In addition to oxygen saturation, temperature and pH are CPPs in every process and can be monitored and controlled very easily. Typically, *in vitro* expansion is carried out at 37°C and neutral pH (7.2–7.4). Expansion at lower temperatures can be advantageous under certain circumstances because this reduces stress (ROS production and frequency of apoptosis) and may yield more potent MSCs. Although the expansion of MSCs has been achieved in the pH range 7.5–8.3 [54], it is unclear how significant variations in pH influence MSC metabolism and whether this affects the secretome. The optimal temperature and pH must be evaluated for each MSC product.

Other CPPs include the parameters grouped under the term hydrodynamics, refer-ring to the potential impact of aeration and agitation. Aeration is required to supply oxygen to the MSCs, but as well as affecting the oxygen saturation, it also generates forces that cause physical stress. In T-flasks, aeration is achieved by the diffusion of oxygen through the surface of the medium, whereas bioreactors must be actively aerated by, e.g., bubbling gas into the liquid. The bursting gas bubbles (cavitation) generate strong forces that can damage cells, although the stress can be reduced by controlling the bubble size [55]. Agitation in bioreactors is generally achieved with impellers, which help to disperse gas (and therefore contribute to aeration) but also maintain a homogenous suspension of cells and nutrients. The creation of a homog-enous environment is advantageous because it avoids gradients of pH, nutrients, or waste products, whose effect on MSCs is unpredictable. Homogenization can also be achieved using pumps or is facilitated by air bubbles. Agitation always generates shear forces, so it is necessary to balance the homogeneity of the cultivation system and the impact of the hydrodynamic forces on the MSCs. Although excessive shear stress is detrimental, hydrodynamic forces can also stimulate MSC growth and increase potency [43]. For these reasons, the mode and rate of aeration and the method and intensity of agitation are CPPs that must be carefully optimized for each process.

2.1.4 Growth surface, cell harvest, and storage

MSCs are anchorage-dependent cells, so the properties of the growth surface also have a significant impact on the process and must be investigated and selected carefully. However, unlike the parameters discussed above, the growth surface does not have to be monitored or controlled during MSC production, so it falls outside the technical definition of a CPP. The expression of certain surface markers by MSCs reflects the stiffness of the growth surface, so it is clear that the surface affects the phenotype [56]. As stated above, the ability to adhere to plastic surfaces is one of the minimal criteria that define MSCs, and tissue-culture plastic is therefore the most commonly-used growth surface. Although all commercial tissue-culture plasticware has a polypropylene base, the surface is often treated differently, and this changes the behavior and properties of the adherent MSCs [37]. MSCs further grow on other surface materials, e.g., glass [57] or dextran [58]. When MSCs are cultivated in serum-free medium, cell growth often requires that the surface is coated with further adhesion-promoting factors, such as fibronectin, vitronectin, or the peptide RGD.

Given that MSCs are anchorage-dependent cells, the harvesting of cells at the end of the *in vitro* expansion step requires an efficient cell detachment method that ideally does not affect functionality or potency. In the laboratory, MSCs can be detached from T-flasks by adding trypsin or other proteases, but this nonspecific proteolysis can affect cell viability and eliminate some MSC surface markers [59]. Proteolytic cleavage is incompatible with the larger-scale processes in bioreactors because longer incubation times are required for the enzymes to work, and even then, the efficiency of cell recovery is low [60]. More importantly, any negative effects of the enzymatic treatment on cell viability and potency are amplified by the longer exposure time, which can inhibit MSC differentiation [61]. These issues can be addressed by adjusting the hydrodynamic conditions to favor cell detachment after limited enzymatic treatment [62]. Alternatively, enzymatic treatment can be circumvented completely by promoting cell detachment using dissolvable growth surfaces [63] or thermosensitive surfaces that release cells following a temperature shift [64, 65]. However, unlike enzymatic treatments, these novel surfaces do not break direct cell-cell bonds and may be unsuitable if single cell is required. The formation of aggregates can be minimized by carefully monitoring the cell density and selecting a harvest point that favors the recovery of single cell, but this must be balanced against the efficiency of expansion given the need to harvest at lower cell densities. The so-called harvest problem, balancing the efficient release of cells against the recovery of cells with desirable properties, has yet to be solved. This highlights the importance of well-defined CPPs at the harvesting stage.

All the approved allogenic MSC products described earlier are cryopreserved, allowing them to be offered as off-the-shelf products that can be stored until quality control and batch release are completed. The use of cryopreserved allogeneic MSCs is the only feasible therapeutic strategy for acute tissue injury syndromes such as stroke, sepsis, or myocardial infarction, because the patient is likely to die before sufficient quantities of autologous MSCs could be prepared. However, cryopreser-vation and thawing have a massive impact on the potency of MSCs [66]. Indeed, even without optimization, fresh MSCs are much more potent than frozen ones [35]. A rule of thumb is to freeze the cells slowly (e.g., 1°C/min) but to thaw them quickly (e.g., direct transfer from storage to a 37°C water bath). The impact of multiple freeze-thaw cycles must be evaluated carefully [67]. The composition of the freezing medium is also important because it often contains dimethyl sulfoxide (DMSO) and FCS as cryoprotectants, the first being cytotoxic and the second undesirable for the reasons already discussed above. Nontoxic alternatives lacking DMSO and FCS have been tested and may be more compatible with MSCs intended for clinical applications [68–70].

In summary, the expansion of MSCs in bioreactors involves multiple CPPs including (i) the source of the initial MSCs before expansion, (ii) the impact of cell density and age, (iii) the effects of the culture medium, (iv) the properties of the bioreactor and aeration/agitation systems, and (iv) the method used for cell harvest and storage. The impact of these CPPs on the quality of MSCs can only be deter-mined by designing robust assays for (i) *in vitro* senescence and genetic stability and (ii) relevant disease-specific mechanism of action and potency. It is clear that there is no one-size-fits-all MSC expansion process and that unique processes must be developed to match different therapeutic objectives. These processes may feature distinct CQAs, meaning that the CPPs may also differ on a case-by-case basis.

2.2 MSC manufacturing for clinical trials

For the 989 interventional clinical trials involving MSCs reported thus far (www.clinicaltrials.gov, search term: mesenchymal stem cell OR mesenchymal stromal cell, 2019/09/27), the MSCs were expanded *in vitro* and in most cases were transfused intravenously at typical doses of $1–2 \times 10^6$ cells per kilogram, never exceeding 12×10^6 cells/kg [3].

The manufacture of protein therapeutics is almost always carried out in bioreactors because they are scalable, controllable via integrated process analytical technology, and most process steps can be automated. This is not the case for MSC products, and a survey of GMP manufacturing at US academic centers has revealed major differences in the various process steps (cell isolation, expansion, and characterization). In the context of cell expansion, 80% of the centers surveyed above used T-flasks or cell factories, whereas only 20% mainly used bioreactors. A broad range of seeding densities was used for cultivation (50–2500 cells/cm^2) and the cultivation time ranged from 1 to 28+ days. The cultivation medium was supplemented with FCS (lot-selected or not) or donor-pooled human platelet lysate (in-house product or commercial product) [71]. All of the centers expanded MSCs under GMP conditions, but with huge variations in the protocol. The production of MSCs in T-flasks is adequate for a small number of patients (30 T-flasks each with a growth surface of 175 cm^2 would be required per patient, assuming each patient is dosed with 416 million cells and the harvesting efficiency is 8×10^4 cells/cm^2 [72]). But for larger clinical trials with >100 patients, the resources required for cell culture would become unsupportable (assuming the conditions stated above, a trial with 140 patients would require 4200 T-flasks filling 32 standard 160-L incubators and 9 full-time personnel to handle the cells). Expansion in T-flasks might also be sufficient for autologous cell therapy, given that only a single patient is involved and it would not be necessary to produce more than 10 doses. However, even for small-scale manufacturing, an automated bioreactor system would offer several advantages over manual cultivation. Given that the entire manufacturing process must be aseptic, closed bioreactors provide much better insurance against contami-nation than an open culture system based on T-flaks. For allogenic MSC products, where up to 1 million doses are produced per batch, bioreactors are the only feasible manufacturing option (**Figure 2**).

2.3 The expansion of MSCs in bioreactors

When an MSC product advances from research to commercial manufacturing, the *in vitro* expansion process must also change. Research is driven by the freedom to test different conditions, but the tests are typically conducted on a small scale. In contrast, commercial products must be manufactured using a standardized process to ensure robustness, and the scale is generally larger. Bioreactors play a key role in large-scale manufacturing because they offer greater traceability due to the control and monitoring of CPPs. The expansion of MSCs in bioreactors allows the

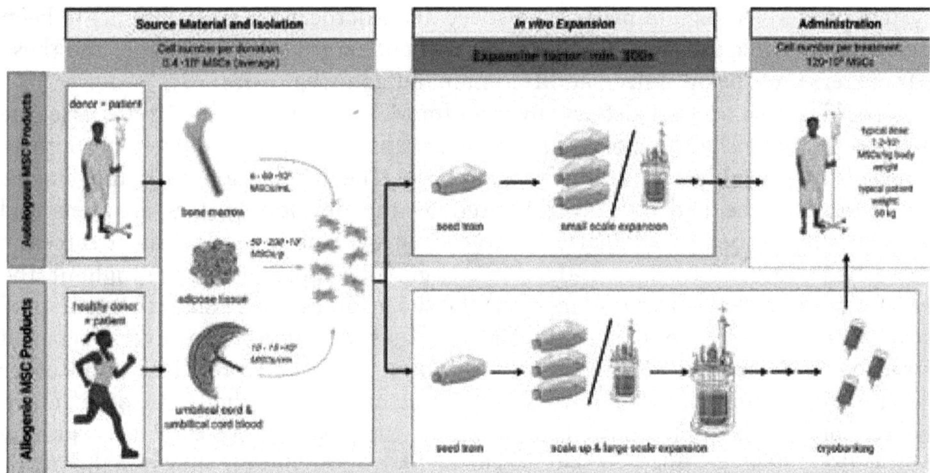

Figure2.
Manufacturing of autologous and allogenic MSC products. Autologous MSC products are isolated from the patient's own tissue, whereas for allogenic MSC products a healthy donor from the same specie donates cells. In the isolation and expansion, there are few differences between the two types of MSC products. Most common sources are bone marrow, adipose tissue, and umbilical cord (blood), all three giving different amount of MSCs. The expansion for both MSC product types differs in scale. Storage is only needed for allogenic MSC products. If we expect that 0.4 × 10^6 MSC are isolated per donation and one dose to treat a single patient is about 120 × 10^6 MSCs, an expansion factor of at least 300-fold is needed. If more doses should be produced from one isolate, e.g., because the patient needs several treatments or in case of allogenic MSC products, the expansion factor dramatically increases.

precise control of the microenvironment, which has a profound influence on cell potency and therapeutic efficacy [73, 74]. For example, the dynamic cultivation of human MSCs in a bioreactor has been shown to induce the secretion of several beneficial growth factors, including BDNF, NGF, VEGF, and IGF-1 [75]. The use of bioreactors also means that the *in vitro* MSC expansion and harvesting steps can be automated, which improves the efficiency of both steps and reduces the amount of hands-on work. The elimination of operator-related errors and contamination risks makes the process more stable, avoiding batch-to-batch variability. Nutrient gradients and abrupt fluctuations in pH caused by manual medium exchange are also avoided. This enables the production of MSCs with consistent identity and potency (CQAs). Many different bioreactor types have been used for the *in vitro* expansion of MSCs, including fixed bed, fluidized bed, and stirred tank reactors, as well as newer innovations such as wave reactors, wall-rotating systems, and vertical wheel reactors [76]. However, most studies have involved only two types of reactor: stirred tank or fixed bed, and these are discussed in more detail below.

Stirred tank reactors are the most widely-used devices for large-scale MSC expan-sion. They are often used with microcarriers, which are small beads that increase the surface area available for cell attachment, although MSCs can also be grown in bioreactors as aggregates or spheroids. The expansion of MSCs growing on microcar-riers is typically a batch-mode manufacturing process because the cells are harvested at a predetermined density. However, fed-batch processes involve a smaller inoculum (100 cells/cm^2, equivalent to five cells per microcarrier) and can thus achieve better economy and a higher expansion factor [77, 78]. There should be minimal (if any) agitation at the beginning of the expansion phase to allow for cell attachment to the microcarriers (if used) or otherwise for the formation of aggregates. However, agitation is required following attachment in order to homogenize the suspension and avoid the formation of large clumps. As discussed above, agitation is an impor-tant CPP and the parameters must be optimized based on the unique combination of system properties (e.g., impeller type/speed and microcarrier size/amount) to keep microcarriers or aggregates in suspension without causing shear damage, and these parameters must be optimized at different manufacturing scales [79].

Fixed bed reactors are also widely used for MSC expansion, and in this case, the cells are grown either on macrocarriers or as capsules (500 μm diameter), both of which form a stable bed at the reactor base. The production of homogeneous condi-tions in the bed can be frustrated by the development of channels and gradients in the bed, particularly in large-scale systems [80]. The shear forces in fixed bed reactors are low ($\sim 0.5 \times 10^{-5}$ N/cm^2) and consistent throughout the reactor with no peaks near the impeller; the shear forces also remain constant at all scales [81]. The *in vitro* expansion of MSCs has been reported in several types of fixed bed reactors [82–85]. One of the major drawbacks of fixed bed and other reactor types compared to stirred tank reactors is the challenge of efficient harvesting. For example, in the reports above, the recovery of viable cells is rarely better than 70%, so this is a key aspect of bioreactor design that remains to be addressed [60, 86].

2.4 Remaining challenges

The earlier sections highlighted several challenges that must be overcome to develop robust processes for the expansion of MSCs in bioreactors, which are summarized briefly below. Furthermore, our current understanding of the CPPs affecting MSC production is rudimentary at best, and more work is required to determine the impact of hydrodynamic factors on the CQAs. Precise online monitoring tools are needed to control CPPs effectively and to measure their influence on cell viability, potency, and secretory profiles. An increase in process understand-ing will facilitate process modeling, to fulfill the requirements of process analytic technology as a prerequisite for GMP manufacturing.

The major challenge for MSC therapy is the development of an *in vitro* expansion process that mimics the natural MSC niche, but nevertheless allows scaled up production for clinical trials without compromising CQAs such as cell functionality and potency. The development of a standardized process is frustrated by the heterogeneity of MSCs, which are isolated from different donors and different tissues, resulting in variable phenotypes and functions. The heterogeneity of primary MSCs can be avoided by working instead with induced pluripotent stem cells (iPSCs), which can differentiate into MSC-like cells with potent therapeutic properties [87]. However, well-controlled *in vitro* expansion processes in bioreactors can also help to reduce the batch-to-batch variation often encountered with MSCs, because parameters such as the seeding density, shear stress during cultivation, and cell density at harvest can be monitored and controlled effectively.

Polyclonal MSCs often show the most potent therapeutic effects, but clonal impoverishment occurs during lengthy expansion phases and this must be avoided if potency is compromised. However, even monoclonal MSCs become heterogeneous over time, generating subpopulations with different morphologies and surface marker profiles. The therapeutic outcome can only be predicted if the MSC pool does not change during expansion, and the well-controlled condi-tions in bioreactors can therefore help to ensure that the cell products remain homogeneous.

Ultimately, even bioreactor-based processes for MSC expansion are constrained by the inbuilt replication limit of MSCs, which leads to senescence after a certain number of generations. Stem cells by definition have an unlimited capacity for self-renewal, but this property is lost *in vitro*. The expansion stress that leads to replicative senescence generates MSCs that maintain their marker profiles but nevertheless lose functionality and therefore therapeutic potency.

The production of MSCs with standardized properties would be facilitated by the development of standardized validated potency assays so that results obtained in different laboratories are truly comparable. The ISCT has taken steps in this direction by publishing standards for the harmonization of potency assays. In a matrix assay approach, they propose to use quantitative RNA analysis for selected gene products, flow cytometry to detect functionally-relevant surface markers, and protein-based assays to map the secretome and determine the immunomodulatory potency of MSCs [88].

3. Additional processes that require MSCs

MSCs are typically the sole product of any MSC cultivation process, but in some applications, the MSCs are used as helper cells to deliver a different product or they are used as a vehicle to produce a specific cellular component. In each case, the CQAs differ significantly from the standard MSC manufacturing process and other CPPs must therefore be considered. We discuss two examples below.

3.1 Production of MSC-derived extracellular vesicles

MSCs are potent therapeutics, but researchers are seeking new ways to achieve the same therapeutic effect without the drawbacks associated with MSC manufacturing, such as the limited availability of potent cells, the complex transfusion process, and the entrapment of MSCs in nontarget organs [89]. As discussed earlier, the therapeutic effect of MSCs reflects the secretion of cytokines, growth factors, and other paracrine signaling molecules, particularly via the release of extracel-lular vesicles that interact directly with target cells and deliver their contents into the cytosol. The advantage of these vesicles over whole MSCs is their much greater stability, which means they can be manufactured, stored, and shipped without losing therapeutic efficacy [90, 91].

The large-scale manufacturing of extracellular vesicles requires the cultivation of MSCs, which secrete these vesicles directly in the culture medium. Scalable production methods are not yet available, and vesicles are currently produced in T-flasks or cell factories without process monitoring. Bioreactors could be used to scale up production, and given there is no need to harvest the MSCs, it would be possible to consider a wider range of bioreactor systems than the relatively narrow selection favored for MSC manufacturing. A fixed bed bioreactor has been used for the continuous production and harvesting of extracellular vesicles, which increased the yield 10-fold compared to T-flasks [92]. Stirred tank reactors with microcarriers might also be suitable, but they have not yet been used for vesicle production [57]. The cells would be exposed to shear forces caused by the impellers and air bubble cavitation, and this may influence vesicle production and potency [93].

The effect of different process parameters on the production of MSC-derived extracellular vesicles has been investigated at the laboratory scale. For example, hMSCs and their vesicles are primed by hypoxic conditions or changes in medium composition, such as the removal of FCS or the addition of priming factors like IFNγ and TNFα [34, 35]. The yield of extracellular vesicles can also be increased by preparing spheroids that mimic *in vivo* conditions, for example by laying down an extracellular matrix and supplying appropriate signaling molecules [89]. Cell density, passage number, and cell origin also affect the vesicle yield. The immortalized cell line hMSC-TERT is more stable than primary MSCs, but the immortalization process has an impact on vesicle production, which must be investigated individually for each cell line because it is not related to the immortalization method [93].

There is currently no standardized large-scale production platform for primed hMSC-derived vesicles, but even if such a platform existed, a corresponding purification process would be required. The laboratory-scale purification of vesicles captured from the culture medium is currently based on a combination of ultracentrifugation, dead-end filtration, precipitation, and size exclusion chromatography, which are difficult to scale up [94, 95]. However, tangential-flow filtration can also be used for large-scale purification, washing, and buffer exchange, and this method should be investigated in more detail for vesicle purification [94, 96]. Extracellular vesicles are even more sensitive to process changes than MSCs, so the influence of multiple cell-dependent, culture, and process parameters on the potency of these vesicles must be determined.

3.2 Cocultivation of MSCs with other cells

The ability of MSCs to restore the activity of dysfunctional cells *in vivo* is the basis of their therapeutic efficacy, but the same interactions can also be exploited *in vitro*. One key example is the interaction between MSCs and pancreatic beta cells, which are widely used for drug screening and cell therapy in the context of diabetes. In both applications, large numbers of functional beta cells are required, but beta cells rapidly lose their functionality when expanded *in vitro*. The loss of beta cell functionality *in vitro* can be prevented by cocultivation with MSCs, which not only stimulate beta cell proliferation but also enhance their glucose-dependent secretion of insulin [97–99].

The major challenge of cocultivation is to balance the demands of two completely different cell types. In large scales, the distribution of cells becomes heterogeneous, which can lead to instability within the bioreactor and lower cell viability. A well-balanced and tightly controlled culture environment is needed to stabilize large-scale cocultures. Because secreted factors are important for the cocultivation of MSCs and beta cells, the hydrodynamic forces in bioreactors, which influence the distribution of secreted molecules, must be considered at an early stage [100]. Furthermore, the optimal cocultivation ratio of the cells must be determined. Established processes can be modified to achieve a new process setup for cocultivation, but it is often beneficial to separate cell expansion from cocultivation (i.e., first expand the pure cultures to generate the cells needed for the coculture and then combine them to improve the function of beta cells in a second process step). For the expansion step, it can be sufficient to improve the growth of beta cells using conditioned medium from the cultivation of MSCs. Alternatively, the expansion and functionalization of beta cells can be combined in one process step [101]. The CPPs for such a complex process can be difficult to identify, but the CQAs of the beta cells are most relevant if the aim of the process is to produce functionalized beta cells for drug screening or cell therapy. Even so, the potency of the MSCs must not be neglected because they are required to stimulate the beta cells. Accordingly, the MSCs must be expanded under controlled and standardized conditions that maximize their beneficial impact on beta cells. In the future, cocultivation bioreactor concepts for MSCs and beta cells must be tested to allow the completely aseptic expansion and cocultivation of both cell types.

4. Conclusions

MSCs are potent therapeutic agents, but their complexity and environmental sensitivity make the GMP-compliant manufacturing of MSC products extremely challenging. Given the range of tissue sources, isolation procedures, and expansion protocols, it is unclear whether MSC products are similar enough across manufacturing sites and whether results can be considered comparable even within the same study. Moreover, the incomplete definition of MSCs makes it difficult to develop objective release criteria. These issues strongly argue for the harmonization and standardization of MSC manufacturing processes, release criteria, and potency assays. The regulatory standards for MSCs are still evolving, and different standards apply in different jurisdictions. MSCs are living cells and cannot be held to the same standards as chemical entities or biopharmaceuticals, both of which can be tested against rigorous and objective quality criteria. The regulations for MSCs should be more flexible, acknowledging that each MSC product is developed for a specific indication, and unique platform technologies, CQAs, and CPPs may therefore be necessary for each manufacturing process. One of the most important platform technologies is the use of bioreactors for cell expansion, because this is the only current strategy that can bring MSC therapy into routine practice. MSCs can also be used as production aids for other products, including beta cells for drug screening or diabetes therapy, and novel biological agents such as extracellular vesicles. In the future, they could even be used for commodity products such as artificial meat. But in all these applications, a robust and scalable manufacturing process will be necessary.

Acknowledgements

We would like to thank the Hessen State Ministry of Higher Education, Research and the Arts for the financial support within the Hessen initiative for scientific and economic excellence (LOEWE-Program, LOEWE Center DRUID (Novel Drug Targets against Poverty-Related and Neglected Tropical Infectious Diseases)). We also received financial support from the Strategic Research Fund of the THM (University of Applied Sciences Mittelhessen). The authors acknowledge Dr. Richard M Twyman for revising the paper.

Author details

Jan Barekzai[1], Florian Petry[1], Jan Zitzmann[1], Peter Czermak[1,2,3] and Denise Salzig

1 Institute of Bioprocess Engineering and Pharmaceutical Technology, University of Applied Sciences Mittelhessen, Giessen, Germany

2 Faculty of Biology and Chemistry, Justus-Liebig-University Giessen, Giessen, Germany

3 Project Group Bioresources, Fraunhofer Institute for Molecular Biology and Applied Ecology (IME), Giessen, Germany

*Address all correspondence to: denise.salzig@lse.thm.de

References

[1] Caplan AI. Mesenchymal stem cells: Time to change the name! Stem Cells Translational Medicine. 2017;**6**:1445-1451. DOI: 10.1002/sctm.17-0051

[2] Sipp D, Robey PG, Turner L. Clear up this stem-cell mess. Nature. 2018;**561**:455-457. DOI: 10.1038/d41586-018-06756-9

[3] Galipeau J, Sensébé L. Mesenchymal stromal cells: Clinical challenges and therapeutic opportunities. Cell Stem Cell. 2018;**22**:824-833. DOI: 10.1016/j.stem.2018.05.004

[4] Galipeau J. The mesenchymal stromal cells dilemma—Does a negative phase III trial of random donor mesenchymal stromal cells in steroid-resistant graft-versus-host disease represent a death knell or a bump in the road? Cytotherapy. 2013;**15**:2-8. DOI: 10.1016/j.jcyt.2012.10.002

[5] Caplan AI. Medicinal signalling cells: They work, so use them. Nature. 2019;**566**:39. DOI: 10.1038/d41586-019-00490-6

[6] Hoffmann D, Leber J, Loewe D, Lothert K, Oppermann T, Zitzmann J, et al. Purification of new biologicals using membrane-based processes. In: Basile A, Charcosset C, editors. Current Trends and Future Developments on (Bio-) Membranes. Elsevier; 2019;**1**:123-150. DOI: 10.1016/B978-0-12-813606-5.00005-1

[7] European Medicines Agency. Reflection Paper on classification of advanced therapy medicinal products

[8] Bedford P, Jy J, Collins L, Keizer S. Considering cell therapy product "good manufacturing practice" status. Frontiers in Medicine. 2018;**5**:118. DOI: 10.3389/fmed.2018.00118

[9] Cuende N, Rasko JEJ, Koh MBC, Dominici M, Ikonomou L. Cell, tissue and gene products with marketing authorization in 2018 worldwide. Cytotherapy. 2018;**20**:1401-1413. DOI: 10.1016/j.jcyt.2018.09.010

[10] Panés J, García-Olmo D, van Assche G, Colombel JF, Reinisch W, Baumgart DC, et al. Long-term efficacy and safety of stem cell therapy (Cx601) for complex perianal fistulas in patients with Crohn's disease. Gastroenterology. 2018;**154**:1334-1342.e4. DOI: 10.1053/j.gastro.2017.12.020

[11] Gupta PK, Krishna M, Chullikana A, Desai S, Murugesan R, Dutta S, et al. Administration of adult human bone marrow-derived, cultured, pooled, allogeneic mesenchymal stromal cells in critical limb ischemia due to Buerger's disease: Phase II study report suggests clinical efficacy. Stem Cells Translational Medicine. 2017;**6**:689-699. DOI: 10.5966/sctm.2016-0237

[12] Park Y-B, Ha C-W, Lee C-H, Yoon YC, Park Y-G. Cartilage regeneration in osteoarthritic patients by a composite of allogeneic umbilical cord blood-derived mesenchymal stem cells and hyaluronate hydrogel: Results from a clinical trial for safety and proof-of-concept with 7 years of extended follow-up. Stem Cells Translational Medicine. 2017;**6**:613-621. DOI: 10.5966/sctm.2016-0157

[13] Murray IR, West CC, Hardy WR, James AW, Park TS, Nguyen A, et al. Natural history of mesenchymal stem cells, from vessel walls to culture vessels. Cellular and Molecular Life Sciences: CMLS. 2014;**71**:1353-1374. DOI: 10.1007/s00018-013-1462-6

[14] Andrzejewska A, Lukomska B, Janowski M. Concise review: Mesenchymal stem cells: From roots

to boost. Stem cells (Dayton Ohio). 2019;**37**:855-864. DOI: 10.1002/stem.3016

[15] Colter DC, Sekiya I, Prockop DJ. Identification of a subpopulation of rapidly self-renewing and multipotential adult stem cells in colonies of human marrow stromal cells. PNAS. 2001;**98**:7841-7845. DOI: 10.1073/pnas.141221698

[16] Rennerfeldt DA, van Vliet KJ. Concise review: When colonies are not clones: Evidence and implications of Intracolony heterogeneity in mesenchymal stem cells. Stem cells (Dayton, Ohio). 2016;**34**:1135-1141. DOI: 10.1002/stem.2296

[17] Dominici M, Le Blanc K, Mueller I, Slaper-Cortenbach I, Marini F, Krause D, et al. Minimal criteria for defining multipotent mesenchymal stromal cells. The International Society for Cellular Therapy position statement. Cytotherapy. 2006;**8**:315-317. DOI: 10.1080/14653240600855905

[18] Martin C, Olmos É, Collignon M-L, de Isla N, Blanchard F, Chevalot I, et al. Revisiting MSC expansion from critical quality attributes to critical culture process parameters. Process Biochemistry. 2017;**59**:231-243. DOI: 10.1016/j.procbio.2016.04.017

[19] Lv F-J, Tuan RS, Cheung KMC, Leung VYL. Concise review: The surface markers and identity of human mesenchymal stem cells. Stem cells (Dayton, Ohio). 2014;**32**:1408-1419. DOI: 10.1002/stem.1681

[20] Ma N, Cheng H, Lu M, Liu Q , Chen X, Yin G, et al. Magnetic resonance imaging with superparamagnetic iron oxide fails to track the long-term fate of mesenchymal stem cells transplanted into heart. Scientific Reports. 2015;**5**:9058 EP. DOI: 10.1038/srep09058

[21] Castro-Manrreza ME, Montesinos JJ. Immunoregulation by mesenchymal stem cells: Biological aspects and clinical applications. Journal of Immunology Research. 2015;**2015**:394917. DOI: 10.1155/2015/394917

[22] Brown C, McKee C, Bakshi S, Walker K, Hakman E, Halassy S, et al. Mesenchymal stem cells: Cell therapy and regeneration potential. Journal of Tissue Engineering and Regenerative Medicine. 2019;**13**:1738-1755. DOI: 10.1002/term.2914

[23] Stenderup K, Justesen J, Clausen C, Kassem M. Aging is associated with decreased maximal life span and accelerated senescence of bone marrow stromal cells. Bone. 2003;**33**:919-926. DOI: 10.1016/j.bone.2003.07.005

[24] Lipsitz YY, Timmins NE, Zandstra PW. Quality cell therapy manufacturing by design. Nature Biotechnology. 2016;**34**:393-400. DOI: 10.1038/nbt.3525

[25] Barkholt L, Flory E, Jekerle V, Lucas-Samuel S, Ahnert P, Bisset L, et al. Risk of tumorigenicity in mesenchymal stromal cell-based therapies—Bridging scientific observations and regulatory viewpoints. Cytotherapy. 2013;**15**:753-759. DOI: 10.1016/j.jcyt.2013.03.005

[26] Mendicino M, Bailey AM, Wonnacott K, Puri RK, Bauer SR. MSC-based product characterization for clinical trials: An FDA perspective. Cell Stem Cell. 2014;**14**:141-145. DOI: 10.1016/j.stem.2014.01.013

[27] FDA/CBER. Guidance for Industry Potency Tests for Cellular and Gene Therapy Products. 2011. Available from: https://www.fda.gov/regulatory-information/search-fda-guidance-documents/potency-tests-cellular-and-gene-therapy-products

[28] Zhang A-X, Yu W-H, Ma B-F, Yu X-B, Mao FF, Liu W, et al. Proteomic identification of differently expressed proteins responsible for osteoblast differentiation from human mesenchymal stem cells. Molecular and Cellular Biochemistry. 2007;**304**:167-179. DOI: 10.1007/s11010-007-9497-3

[29] Aldridge A, Kouroupis D, Churchman S, English A, Ingham E, Jones E. Assay validation for the assessment of adipogenesis of multipotential stromal cells—A direct comparison of four different methods. Cytotherapy. 2013;**15**:89-101. DOI: 10.1016/j.jcyt.2012.07.001

[30] Lazarević JJ, Kukolj T, Bugarski D, Lazarević N, Bugarski B, Popović ZV. Probing primary mesenchymal stem cells differentiation status by micro-Raman spectroscopy. Spectrochimica Acta Part A: Molecular and Biomolecular Spectroscopy. 2019;**213**:384-390. DOI: 10.1016/j.saa.2019.01.069

[31] Cunha B, Aguiar T, Carvalho SB, Silva MM, Gomes RA, Carrondo MJT, et al. Bioprocess integration for human mesenchymal stem cells: From up to downstream processing scale-up to cell proteome characterization. Journal of Biotechnology. 2017;**248**:87-98. DOI: 10.1016/j.jbiotec.2017.01.014

[32] Chinnadurai R, Rajan D, Qayed M, Arafat D, Garcia M, Liu Y, et al. Potency analysis of mesenchymal stromal cells using a combinatorial assay matrix approach. Cell Reports. 2018;**22**:2504-2517. DOI: 10.1016/j.celrep.2018.02.013

[33] Halfon S, Abramov N, Grinblat B, Ginis I. Markers distinguishing mesenchymal stem cells from fibroblasts are downregulated with passaging. Stem Cells and Development. 2011;**20**:53-66. DOI: 10.1089/scd.2010.0040

[34] Noronha NC, Mizukami A, Caliári-Oliveira C, Cominal JG, Rocha JLM, Covas DT, et al. Priming approaches to improve the efficacy of mesenchymal stromal cell-based therapies. Stem Cell Research and Therapy. 2019;**10**:131. DOI: 10.1186/s13287-019-1224-y

[35] Yin JQ , Zhu J, Ankrum JA. Manufacturing of primed mesenchymal stromal cells for therapy. Nature Biomedical Engineering. 2019;**3**:90-104. DOI: 10.1038/s41551-018-0325-8

[36] Tremolada C, Colombo V, Ventura C. Adipose tissue and mesenchymal stem cells: State of the art and Lipogems® technology development. Current stem cell reports. 2016;**2**:304-312. DOI: 10.1007/s40778-016-0053-5

[37] Sotiropoulou PA, Perez SA, Salagianni M, Baxevanis CN, Papamichail M. Characterization of the optimal culture conditions for clinical scale production of human mesenchymal stem cells. Stem cells (Dayton, Ohio). 2006;**24**:462-471. DOI: 10.1634/stemcells.2004-0331

[38] Saki N, Jalalifar MA, Soleimani M, Hajizamani S, Rahim F. Adverse effect of high glucose concentration on stem cell therapy. International Journal of Hematology-Oncology and Stem Cell Research. 2013;**7**:34-40

[39] Weil BR, Abarbanell AM, Herrmann JL, Wang Y, Meldrum DR. High glucose concentration in cell culture medium does not acutely affect human mesenchymal stem cell growth factor production or proliferation. American Journal of Physiology. Regulatory, Integrative and Comparative Physiology. 2009;**296**:R1735-R1743. DOI: 10.1152/ajpregu.90876.2008

[40] Follmar KE, Decroos FC, Prichard HL, Wang HT, Erdmann D, Olbrich KC. Effects of glutamine, glucose, and oxygen concentration on the metabolism

and proliferation of rabbit adipose-derived stem cells. Tissue Engineering. 2006;**12**:3525-3533. DOI: 10.1089/ten.2006.12.3525

[41] Schop D, Janssen FW, van Rijn LDS, Fernandes H, Bloem RM, de Bruijn JD, et al. Growth, metabolism, and growth inhibitors of mesenchymal stem cells. Tissue Engineering Parts A. 2009;**15**:1877-1886. DOI: 10.1089/ten.tea.2008.0345

[42] Higuera GA, Schop D, Spitters TWGM, van Dijkhuizen-Radersma R, Bracke M, de Bruijn JD, et al. Patterns of amino acid metabolism by proliferating human mesenchymal stem cells. Tissue Engineering Parts A. 2012;**18**:654-664. DOI: 10.1089/ten.TEA.2011.0223

[43] Sart S, Agathos SN, Li Y. Process engineering of stem cell metabolism for large scale expansion and differentiation in bioreactors. Biochemical Engineering Journal. 2014;**84**:74-82. DOI: 10.1016/j.bej.2014.01.005

[44] Ikebe C, Suzuki K. Mesenchymal stem cells for regenerative therapy: Optimization of cell preparation protocols. BioMed Research International. 2014;**2014**:951512. DOI: 10.1155/2014/951512

[45] Menard C, Pacelli L, Bassi G, Dulong J, Bifari F, Bezier I, et al. Clinical-grade mesenchymal stromal cells produced under various good manufacturing practice processes differ in their immunomodulatory properties: Standardization of immune quality controls. Stem Cells and Development. 2013;**22**:1789-1801. DOI: 10.1089/scd.2012.0594

[46] Sundin M, Ringdén O, Sundberg B, Nava S, Götherström C, Le Blanc K. No alloantibodies against mesenchymal stromal cells, but presence of anti-fetal calf serum antibodies, after transplantation in allogeneic hematopoietic stem cell recipients. Haematologica. 2007;**92**:1208-1215. DOI: 10.3324/haematol.11446

[47] Smith JR, Pfeifer K, Petry F, Powell N, Delzeit J, Weiss ML. Standardizing umbilical cord mesenchymal stromal cells for translation to clinical use: Selection of GMP-compliant medium and a simplified isolation method. Stem Cells International. 2016;**2016**:6810980. DOI: 10.1155/2016/6810980

[48] Shih DT-B, Burnouf T. Preparation, quality criteria, and properties of human blood platelet lysate supplements for ex vivo stem cell expansion. New Biotechnology. 2015;**32**:199-211. DOI: 10.1016/j.nbt.2014.06.001

[49] Vanda SL, Ngo A, Tzu Ni H. A xeno-free, serum-free expansion medium for ex-vivo expansion and maintenance of major human tissue-derived mesenchymal stromal cells. Translational Biomedicine. 2018;**2**:146. DOI: 10.21767/2172-0479.100146

[50] Salzig D, Leber J, Merkewitz K, Lange MC, Köster N, Czermak P. Attachment, growth, and detachment of human mesenchymal stem cells in a chemically defined medium. Stem Cells International. 2016;**2016**:5246584. DOI: 10.1155/2016/5246584

[51] Gottipamula S, Muttigi MS, Chaansa S, Ashwin KM, Priya N, Kolkundkar U, et al. Large-scale expansion of pre-isolated bone marrow mesenchymal stromal cells in serum-free conditions. Journal of Tissue Engineering and Regenerative Medicine. 2016;**10**:108-119. DOI: 10.1002/term.1713

[52] Hoch AI, Leach JK. Concise review: Optimizing expansion of bone marrow mesenchymal stem/stromal cells for clinical applications. Stem Cells Translational Medicine. 2014;**3**:643-652. DOI: 10.5966/sctm.2013-0196

[53] Henn A, Darou S, Yerden R. Full-time physioxic culture conditions promote MSC proliferation more than hypoxic preconditioning. Cytotherapy. 2019;**21**:S73-S74. DOI: 10.1016/j. jcyt.2019.03.470

[54] Monfoulet L-E, Becquart P, Marchat D, Vandamme K, Bourguignon M, Pacard E, et al. The pH in the microenvironment of human mesenchymal stem cells is a critical factor for optimal osteogenesis in tissue-engineered constructs. Tissue Engineering Parts A. 2014;**20**:1827-1840. DOI: 10.1089/ten.TEA.2013.0500

[55] Czermak P, Pörtner R, Brix A. Special engineering aspects. In: Cell and Tissue Reaction Engineering: With a Contribution by Martin Fussenegger and Wilfried Weber. Berlin, Heidelberg: Springer Berlin Heidelberg; 2009. pp. 83-172. DOI: 10.1007/978-3-540-68182-3_4

[56] Nikukar H, Reid S, Riehle MO, Curtis ASG, Dalby MJ. Control of mesenchymal stem-cell fate by engineering the Nanoenvironment. In: Baharvand H, Aghdami N, editors. Stem Cell Nanoengineering. Hoboken, New Jersey: John Wiley & Sons Inc; 2014. pp. 205-221. DOI: 10.1002/9781118540640.ch12

[57] Elseberg CL, Leber J, Salzig D, Wallrapp C, Kassem M, Kraume M, et al. Microcarrier-based expansion process for hMSCs with high vitality and undifferentiated characteristics. The International journal of artificial organs. 2012;**35**:93-107. DOI: 10.5301/ ijao.5000077

[58] Frauenschuh S, Reichmann E, Ibold Y, Goetz PM, Sittinger M, Ringe J. A microcarrier-based cultivation system for expansion of primary mesenchymal stem cells. Biotechnology Progress. 2007;**23**:187-193. DOI: 10.1021/bp060155w

[59] Tsuji K, Ojima M, Otabe K, Horie M, Koga H, Sekiya I, et al. Effects of different cell-detaching methods on the viability and cell surface antigen expression of synovial mesenchymal stem cells. Cell Transplantation. 2017;**26**:1089-1102. DOI: 10.3727/096368917X694831

[60] Salzig D, Schmiermund A, Grace P, Elseberg C, Weber C, Czermak P. Enzymatic detachment of therapeutic mesenchymal stromal cells grown on glass carriers in a bioreactor. The Open Biomedical Engineering Journal. 2013;**7**:147-158. DOI: 10.2174/1874120701307010147

[61] Goh TK-P, Zhang Z-Y, Chen AK-L, Reuveny S, Choolani M, Chan JKY, et al. Microcarrier culture for efficient expansion and osteogenic differentiation of human fetal mesenchymal stem cells. BioResearch Open Access. 2013;**2**:84-97. DOI: 10.1089/biores.2013.0001

[62] Nienow AW, Hewitt CJ, Heathman TRJ, Glyn VAM, Fonte GN, Hanga MP, et al. Agitation conditions for the culture and detachment of hMSCs from microcarriers in multiple bioreactor platforms. Biochemical Engineering Journal. 2016;**108**:24-29. DOI: 10.1016/ j.bej.2015.08.003

[63] Kalra K, Banerjee B, Weiss K, Morgan C. Developing efficient bioreactor microcarrier cell culture system for large scale production of mesenchymal stem cells (MSCs). Cytotherapy. 2019;**21**:S73. DOI: 10.1016/j.jcyt.2019.03.468

[64] Song K, Yang Y, Wu S, Zhang Y, Feng S, Wang H, et al. In vitro culture and harvest of BMMSCs on the surface of a novel thermosensitive glass microcarrier. Materials Science and Engineering, C: Materials for Biological Applications. 2016;**58**:324-330. DOI: 10.1016/j.msec.2015.08.033

[65] Yang L, Cheng F, Liu T, Lu JR, Song K, Jiang L, et al. Comparison of mesenchymal stem cells released from poly(N-isopropylacrylamide) copolymer film and by trypsinization. Biomedical Materials (Bristol, England). 2012;**7**:35003. DOI: 10.1088/1748-6041/7/3/035003

[66] Moll G, Alm JJ, Davies LC, von Bahr L, Heldring N, Stenbeck-Funke L, et al. Do cryopreserved mesenchymal stromal cells display impaired immunomodulatory and therapeutic properties? Stem cells (Dayton, Ohio). 2014;**32**:2430-2442. DOI: 10.1002/stem.1729

[67] Oja S, Kaartinen T, Ahti M, Korhonen M, Laitinen A, Nystedt J. The utilization of freezing steps in mesenchymal stromal cell (MSC) manufacturing: Potential impact on quality and cell functionality attributes. Frontiers in Immunology. 2019;**10**:1627. DOI: 10.3389/fimmu.2019.01627

[68] Shivakumar SB, Bharti D, Subbarao RB, Jang S-J, Park J-S, Ullah I, et al. DMSO- and serum-free cryopreservation of Wharton's jelly tissue isolated from human umbilical cord. Journal of Cellular Biochemistry. 2016;**117**:2397-2412. DOI: 10.1002/jcb.25563

[69] Ray SS, Pramanik K, Sarangi SK, Jain N. Serum-free non-toxic freezing solution for cryopreservation of human adipose tissue-derived mesenchymal stem cells. Biotechnology Letters. 2016;**38**:1397-1404. DOI: 10.1007/s10529-016-2111-6

[70] Grein TA, Freimark D, Weber C, Hudel K, Wallrapp C, Czermak P. Alternatives to dimethylsulfoxide for serum-free cryopreservation of human mesenchymal stem cells. The International Journal of Artificial Organs. 2010;**33**:370-380. DOI: 10.1177/039139881003300605

[71] Phinney DG, Galipeau J. Manufacturing mesenchymal stromal cells for clinical applications: A survey of good manufacturing practices at U.S. academic centers. Cytotherapy. 2019;**21**:782-792. DOI: 10.1016/j.jcyt.2019.04.003

[72] Olsen TR, Ng KS, Lock LT, Ahsan T, Rowley JA. Peak MSC-are we there yet? Frontiers in Medicine. 2018;**5**:178. DOI: 10.3389/fmed.2018.00178

[73] Sart S, Agathos SN, Li Y, Ma T. Regulation of mesenchymal stem cell 3D microenvironment: From macro to microfluidic bioreactors. Biotechnology Journal. 2016;**11**:43-57. DOI: 10.1002/biot.201500191

[74] Ma T, Tsai A-C, Liu Y. Biomanufacturing of human mesenchymal stem cells in cell therapy: Influence of microenvironment on scalable expansion in bioreactors. Biochemical Engineering Journal. 2016;**108**:44-50. DOI: 10.1016/j.bej.2015.07.014

[75] Teixeira FG, Panchalingam KM, Assunção-Silva R, Serra SC, Mendes-Pinheiro B, Patrício P, et al. Modulation of the mesenchymal stem cell secretome using computer-controlled bioreactors: Impact on neuronal cell proliferation, survival and differentiation. Scientific Reports. 2016;**6**:27791. DOI: 10.1038/srep27791

[76] Elseberg C, Leber J, Weidner T, Czermak P. The challenge of human mesenchymal stromal cell expansion: Current and prospective answers. In: SJT G, editor. New Insights into Cell Culture Technology. InTech; 2017. DOI: 10.5772/66901

[77] Schnitzler AC, Verma A, Kehoe DE, Jing D, Murrell JR, Der KA, et al. Bioprocessing of human mesenchymal stem/stromal cells for therapeutic use: Current technologies and

challenges. Biochemical Engineering Journal. 2016;**108**:3-13. DOI: 10.1016/j.bej.2015.08.014

[78] Hewitt CJ, Lee K, Nienow AW, Thomas RJ, Smith M, Thomas CR. Expansion of human mesenchymal stem cells on microcarriers. Biotechnology Letters. 2011;**33**:2325-2335. DOI: 10.1007/s10529-011-0695-4

[79] Jossen V, Schirmer C, Mostafa Sindi D, Eibl R, Kraume M, Pörtner R, et al. Theoretical and practical issues that are relevant when scaling up hMSC microcarrier production processes. Stem Cells International. 2016;**2016**:4760414. DOI: 10.1155/2016/4760414

[80] Weber C, Pohl S, Poertner R, Pino-Grace P, Freimark D, Wallrapp C, et al. Production process for stem cell based therapeutic implants: Expansion of the production cell line and cultivation of encapsulated cells. Advances in Biochemical Engineering and Biotechnology. 2010;**123**:143-162. DOI: 10.1007/10_2009_25

[81] Weber C, Freimark D, Pörtner R, Pino-Grace P, Pohl S, Wallrapp C, et al. Expansion of human mesenchymal stem cells in a fixed-bed bioreactor system based on non-porous glass carrier—Part A: Inoculation, cultivation, and cell harvest procedures. The International Journal of Artificial Organs. 2010;**33**:512-525

[82] Jossen V, van den Bos C, Eibl R, Eibl D. Manufacturing human mesenchymal stem cells at clinical scale: Process and regulatory challenges. Applied Microbiology and Biotechnology. 2018;**102**:3981-3994. DOI: 10.1007/s00253-018-8912-x

[83] Lechanteur C, Briquet A, Giet O, Delloye O, Baudoux E, Beguin Y. Clinical-scale expansion of mesenchymal stromal cells: A large banking experience. Journal of Translational Medicine. 2016;**14**:145. DOI: 10.1186/s12967-016-0892-y

[84] Mizukami A, de Abreu Neto MS, Moreira F, Fernandes-Platzgummer A, Huang Y-F, Milligan W, et al. A fully-closed and automated hollow fiber bioreactor for clinical-grade manufacturing of human mesenchymal stem/stromal cells. Stem Cell Reviews. 2018;**14**:141-143. DOI: 10.1007/s12015-017-9787-4

[85] Haack-Sørensen M, Follin B, Juhl M, Brorsen SK, Søndergaard RH, Kastrup J, et al. Culture expansion of adipose derived stromal cells. A closed automated quantum cell expansion system compared with manual flask-based culture. Journal of Translational Medicine. 2016;**14**:319. DOI: 10.1186/s12967-016-1080-9

[86] Weber C, Freimark D, Pörtner R, Pino-Grace P, Pohl S, Wallrapp C, et al. Expansion of human mesenchymal stem cells in a fixed-bed bioreactor system based on non-porous glass carrier—Part B: Modeling and scale-up of the system. The International Journal of Artificial Organs. 2010;**33**:782-795

[87] Zhao C, Generation IM. Applications of induced pluripotent stem cell-derived mesenchymal stem cells. Stem Cells International. 2018;**2018**:9601623. DOI: 10.1155/2018/9601623

[88] Galipeau J, Krampera M, Barrett J, Dazzi F, Deans RJ, DeBruijn J, et al. International Society for Cellular Therapy perspective on immune functional assays for mesenchymal stromal cells as potency release criterion for advanced phase clinical trials. Cytotherapy. 2016;**18**:151-159. DOI: 10.1016/j.jcyt.2015.11.008

[89] Cha JM, Shin EK, Sung JH, Moon GJ, Kim EH, Cho YH, et al. Efficient scalable production of therapeutic microvesicles derived

from human mesenchymal stem cells. Scientific Reports. 2018;**8**:1171. DOI: 10.1038/s41598-018-19211-6

[90] Phelps J, Sanati-Nezhad A, Ungrin M, Duncan NA, Sen A. Bioprocessing of mesenchymal stem cells and their derivatives: Toward cell-free therapeutics. Stem Cells International. 2018;**2018**:9415367. DOI: 10.1155/2018/9415367

[91] Abbasi-Malati Z, Roushandeh AM, Kuwahara Y, Roudkenar MH. Mesenchymal stem cells on horizon: A new arsenal of therapeutic agents. Stem Cell Reviews. 2018;**14**:484-499. DOI: 10.1007/s12015-018-9817-x

[92] Whitford W, Ludlow JW, Cadwell JJS. Continuous production of exosomes. Genetic Engineering and Biotechnology News. 2015;**35**:34. DOI: 10.1089/gen.35.16.15

[93] Patel DB, Santoro M, Born LJ, Fisher JP, Jay SM. Towards rationally designed biomanufacturing of therapeutic extracellular vesicles: Impact of the bioproduction microenvironment. Biotechnology Advances. 2018;**36**:2051-2059. DOI: 10.1016/j.biotechadv.2018.09.001

[94] Colao IL, Corteling R, Bracewell D, Wall I. Manufacturing exosomes: A promising therapeutic platform. Trends in Molecular Medicine. 2018;**24**:242-256. DOI: 10.1016/j.molmed.2018.01.006

[95] Kusuma GD, Barabadi M, Tan JL, Morton DAV, Frith JE, Lim R. To protect and to preserve: Novel preservation strategies for extracellular vesicles. Frontiers in Pharmacology. 2018;**9**:1199. DOI: 10.3389/fphar.2018.01199

[96] Tan SS, Yin Y, Lee T, Lai RC, Yeo RWY, Zhang B, et al. Therapeutic MSC exosomes are derived from lipid raft microdomains in the plasma membrane. Journal of Extracellular Vesicles. 2013;**2**:22614. DOI: 10.3402/jev. v2i0.22614

[97] Scuteri A, Donzelli E, Rodriguez-Menendez V, Ravasi M, Monfrini M, Bonandrini B, et al. A double mechanism for the mesenchymal stem cells' positive effect on pancreatic islets. PLoS One. 2014;**9**:e84309. DOI: 10.1371/journal.pone.0084309

[98] Jun Y, Kang AR, Lee JS, Park S-J, Lee DY, Moon S-H, et al. Microchip-based engineering of super-pancreatic islets supported by adipose-derived stem cells. Biomaterials. 2014;**35**:4815-4826. DOI: 10.1016/j.biomaterials.2014.02.045

[99] Gamble A, Pawlick R, Pepper AR, Bruni A, Adesida A, Senior PA, et al. Improved islet recovery and efficacy through co-culture and co-transplantation of islets with human adipose-derived mesenchymal stem cells. PLoS One. 2018;**13**:e0206449. DOI: 10.1371/journal.pone.0206449

[100] Goers L, Freemont P, Polizzi KM. Co-culture systems and technologies: Taking synthetic biology to the next level. Journal of the Royal Society, Interface. 2014;**11**:20140065. DOI: 10.1098/rsif.2014.0065

[101] Petry F, Weidner T, Czermak P, Salzig D. Three-dimensional bioreactor technologies for the cocultivation of human mesenchymal stem/stromal cells and beta cells. Stem Cells International. 2018;**2018**:1-14. DOI: 10.1155/2018/2547098

11

Prospects of Biocatalyst Purification Enroute Fermentation Processes

Michael Bamitale Osho and Sarafadeen Olateju Kareem

Abstract

Biotransformation of broth through fermentation process suffers a major setback when it comes to disintegration of organic substrates by microbial agents for industrial applications. These biocatalysts are in crude/dilute form hence needs to be purified to remove colloidal particles and enzymatic impurities thus enhancing maximum activity. Several contractual procedures of concentrating dilute enzymes and proteins had been reported. Such inorganic materials include ammonium sulphate precipitation; salting, synthetic polyacrylic acid; carboxy-methyl cellulose, tannic acid, edible gum and some organic solvents as precipitants etc. The emergence of organic absorbents such as sodom apple (*Calostropis procera*) extract, activated charcoal and imarsil had resulted in making significant impact in industrial circle. Various concentrations of these organic extracts have been used as purifying agents on different types of enzyme vis: lipase, amylase, protease, cellulase etc. Purification fold and stability of the enzyme crude form attained unprecedented results.

Keywords: Purification, Enzyme, Stability, *Calostropis procera* extract, Activated charcoal, Imarsil

1. Introduction

The fermentation process involves actual growth of the microorganism and product formation under agitation and aeration, optimum environmental conditions to provide uniform and adequate oxygen to the cell for growth and survival. A fermentation process is a biochemical process and, therefore, has requirements of sterility and use of cellular enzymatic reactions instead of chemical reactions aided by inorganic catalysts, sometimes operating at elevated temperature and pressure. Indigenous fermentation process draws the type attention of food scientists for taking the strategies of food security [1]. Microorganisms in fermented foods play major roles in health sector such as production of antimicrobial compounds, antioxidant, and probiotics properties.

The significance of fermentation of different substrates has gained attention in the beginning of the nineteenth century with profound resultant effects. Some products were produced via fermentation viz.; acetone, glycerol, lactic acid, butanol and baker's yeasts. Due to urgent need to treat the World War II army fighters, several metabolites such as amino acids, antibiotics and vitamins were produced via fermentation. Submerged fermentation with larger volumes under aerobic conditions with moderate process control was established during this period. In subsequent years, the fermentation industry has seen constant improvement with leaps and bounds on the production of high-value metabolites, including various antibiotics and growth hormones, using sophisticated bioreactors [2, 3].

Purification of enzyme is often a multifarious process and a number of procedures are normally employed in succession to obtain adequately high purity state. The idea is to use less expensive but simple methods at nascent stages when the volume is large and more expensive and advanced techniques when the volume is relatively small [4]. The aims are to obtain high final degree of purity; enhanced enzyme activity and reproducibility of the products. It obvious that extraction procedures release a number of other cell components like other enzymes, proteins, polysaccharides and nucleic acids apart from the target enzyme into the medium, this often resulted into increasing the viscosity of the solution depending on

their polymeric structure. It is of great significance as to receive knowledge about functional and structural properties of the substance and to predict its applications.

2. Enzyme biosynthesis

Biocatalysis can be defined as the utilization of living materials or molecules to speed up the rate of chemical reactions. The usage of recombinant enzymes, groups of enzymes, naturally occurring enzymes, cell extracts and whole cells, modified or engineered enzymes inclusive. These biological materials have consequential edge over conventional chemical catalysis. Ideally, biochemical reactions occur in aqueous solution, at moderate temperatures and atmospheric pressure which can result in both environmental and economic values as compared with the existing processes at upraised temperatures and pressures, and in organic medium [5]. High cost of enzymes and potential environmental damages resulting from high temperatures and pressures requires substantial energy inputs. Comparatively little known aspect of biocatalysis is the role of protein potency compare to the overall average structure of a protein, mobility/potency is much difficult to estimate experimentally. Like other molecules, proteins are in sustained motion, with stretching, and rotating bonds bending. These motions can bring about to much greater, for example conformational rearrangements following substrate binding or the movement of two domains relative to one another. It is now clear that these motions play important in catalysis and the regulation of enzyme activity [6]. Moreover, recognizing those motions which come up with catalysis will be experimentally challenging. Likewise, modeling them precisely enough to predict their effects in novel arrangement will be tasking.

As important hydrolytic enzymes, amylase and protease represent the two largest groups of industrial enzymes and account for approximately 85% of total enzyme sales all over the globe. At present, more than 3000 different enzymes have been characterized and many of them found their way into biotechnological and industrial applications [7]. One technicality in enzyme technology especially those of starch biosynthetic enzymes is their manipulations to meet the enormous demand of teeming population, safeguard our environment from non-degradable biopolymer and of course satisfy the food industries' need [8]. Some commercially available biocatalysts do not resist industrial processing conditions due to severity of such conditions. Therefore, certain desirable characteristics during isolation and screening of novel enzymes viz.; alkaline stability, halophilicity, and thermostability are foremost to meet the industrial demand. Great deal of attention has been drawn on extremophiles, which are the valuable source of novel biocatalysts [9]. Extracellular enzymes from these halophiles with polymer-degrading ability at low water activity are of significant in many task processes where concentrated salt solutions hinder enzymatic production. The potency of enzymes to maintain its activeness via organic solvents has attracted considerable interest over the past twenty years.

In contrast to in water, numerous advantages of using enzymes in organic solvents or aqueous solutions containing organic solvents have been observed. Generally, enzymes are easily denatured and their activities disappear in the presence of organic solvents. Therefore, enzymes that remain stable in the presence of organic solvents might be useful for biotechnological applications in which such solvents are used [10]. Because salt reduces water activity, a feature in common with organic solvent systems, halophilic enzymes are thought to be valuable tools as biocatalysts in other low-water-activity environments, such as in aqueous/organic and non-aqueous media.

2.1 Purification protocols and applications

There are considerable ways of maintaining enzyme efficiency. Purification technique is a very powerful device has been arduously used to make some economically attainable and high performance enzymes with improved stability [11]. Enzyme purification is imperative for a full apprehension of the description and established process of enzymes. This is usually a multistep process involving biomass separation, concentration, primary isolation, and purification [12]. The contractual methods for the removal of enzymatic debrises or impurities and colloidal particles from fermen-tation broth include ammonium sulphate precipitation which encompassing dialysis for almost 16 hours before product could be recovered and also results into protein denaturation due to conformational changes [13].

Large volume of industrial enzymes are usually not purified. Their recovery is often accomplished by an ultrafiltration step. During enzyme production, desired products are synthesis after several concentration and separation techniques known as downstream processing (DSP). Two factors (time and cost) are the major challenges confronting these conventional techniques and their sustainability and efficiency depend on precise choice of purification methods [14, 15]. Here are some examples of strategies undertaken to improve the performance of enzymes with applications in food industry. Wong et al. [16] investigated strategies employed in starch liquefaction with targeted improvement of thermostability using α-amylase, protein engineering through site-directed mutagenesis and mutant displayed increased half-life between 15 min and 70 min at 100°C evolved. Glucoamylase with specific role as starch saccharifier and targeted improvement of substrate specificity, thermostability and pH optimum was characterized with protein engineering through site-directed mutagenesis alone [17]. Xylose (glucose) isomerase displayed isomerization/epimerization of hexoses, pentoses and tetroses as significance role of pH-activity profile with targeted improvement which resulted in protein engineering through directed evolution and the yield number on D-glucose in wild type was sustained between pH 6.0 and 7.5 and improvably at pH 7.3 as compared with mutant strains enhanced by 30–40%.

The application of polyvinyl alcohol or carbowax for protein and enzyme concentration is being restricted by poor water holding capacity. Moreover, gel filtration technique is also considered arduous and costly to the developing nations [18]. Carboxy-methyl cellulose, tannic acid, edible gum and some organic solvents as precipitants also poses the problem of product recovery [13]. In fishing industry, the use of fast, simple and low cost techniques such as using organic solvents vis ethanol and acetone, successive stages of centrifugation and filtration; and saline solution (ammonium sulfate) were adopted for the separation and partial purification of protein biomolecules obtained from fish by-products beneficiation remains [19–21], aiming to improve the degree of biomolecule purity [21]. The use of ammonium sulphate precipitates during enzyme purification needs protracted separation technique between 12 and 16 h for recovery of product that frequently bring about protein denaturation as gel chromatography is high priced and moderate for developing economies [22]. Application of chromatographic techniques such as gel filtration and ion exchange give rise to purer enzyme fractions, with significant increase in specific activity. These are often used to estimate the molecular mass of the enzyme by comparing protein mixtures of known molecular mass (reference standards) with the unknown.

Pectinase enzyme was precipitated by dissolving it in a 0.1 M, pH 4.2 sodium acetate buffer after mixing with 3 volumes of ice-cold acetone and allowed to stay for 15 min [23]. Chimbekujwo et al. [24] reported the application of SDS-PAGE analysis of purified fungal protease of major protein band with molecular weight of 68 KDa, 13.3 fold and 28% yield. This partially purified enzyme was stable between 30 and 40°C temperature and pH 4–6 which enhanced the activity by Tween-20 and Calcium ions. Moreover, during the production, characterization and anti-can-cer application of extracellular L-glutaminase from the marine bacterial isolate, the enzyme was purified through QFF technique by engaging ethanol precipitation and ion-exchange chromatography and resulted into 2-fold purification with molecular weight 54.8 kDa, specific activity 89.78 U/mg, maximum enzyme activity at 40°C and pH 8.2 and ultimately retained 90% activity for an hour [25].

3. Adoption of organic absorbent materials for purification procedures

The emergence of organic materials such as activated charcoal, *Calostropis procera* latex and imarsil has made significant contributions for industrial applications. Some of these organic absorbents are discussed below:

3.1 Activated charcoal and other carbon particles as purifying agents

Activated charcoal is an adsorbent extensively utilized in the treatment of wastewater and industrial contaminants by reason of its high shifting ability and adaptability for a wide range of pollutants. It is produced from any crucially carbonaceous materials. Coal, cotton waste, tree barks, palm kernel shell, and many agricultural by- products can be made to produce activated carbon and their capacity to remove colors has been investigated. Ferreira and coworkers [26] demonstrated the production, characterization of activated charcoal from castor seed cake through activation with phosphoric acid. Treatment of fino sherry wine with activated charcoal, in combination with other clarifying agents, produces a wine with lower polyphenolic content, good organoleptic characteristics, but its receptive to browning is indistinguishable to that recognized in untreated wine, despite starting from lower levels of color potency. Activated charcoal is used to remove compounds that cause objectionable color, odor and taste in water treatment while its industrial applications require elimination of harmful gases and pesticides and including purification of organic compounds [27]. It is established that 80% of activated charcoal globally produced is used in aqueous-phase adsorp-tion of both organic and inorganic compounds [28]. However, the application of activated charcoal in the decolorization of enzyme-converted glucose syrup had been described; though its application for the purification of microbial biocatalysts has been sparse.

One profound advantage of activated charcoal over conventional purification systems is this swift enzyme purification from composite fermentation broth mixture at a very high purification fold. These conventional procedures of purification of enzyme among others include solvent precipitation; gel filtration and salting out technique. From an industrial application stand point of view, they are quite expensive base on the fact that they are associated with some difficulty of scaling up and plugging leading to viscous and particulate materials when treated with crude enzyme extracts. Additionally, it is not economical for developing nations as the materials disposal or enzyme recovery techniques employed in the separation method might escalate the expenses. As a result, the usage of activated charcoal has been considered as preferred option for enzyme purification method.

In addition to the inexpensiveness of activated charcoal, their efficient surface absorption attributes can be exploited for depolarization of fermented medium for efficacious and efficient recovery and purification of industrial enzymes making the downstream processing in large-scale industrial bioprocesses less economical [29]. López et al. [30] in their investigation on the use of activated charcoal in combination with other fining agents as clarifying agents reported that these carbonized materials acted upon the phenolic compounds thus encouraging their precipitation. In the field of enology, many different substances have been employed as fining agents over time such addition of antioxidants (ascorbic acid, sulfur dioxide and bottling under inert atmosphere. The use of bentonite has been well-proven and reported to have a remarkable effect on the protein content of wine and also hastens the precipitation of the thermolabile protein [31] but has also minimized the polyphenolic content of the wine during production [32].

The structure of activated carbon which is based on the graphite lattice corresponds to a non-graphitizable carbon and macromolecular structure of the precursor residues during heat treatment, and losses small molecules by developing and degradation some cross-linking, so that joining cannot occur. Therefore, cross-linking bring about a fixed design with small vigor, thus producing a permeable system and intercepting the ordering expected during graphitization.

3.2 *Calotropis procera* as purifying agent and its industrial applications

Calotropis procera belongs to the family Asclepidaceae being a native of tropical and subtropical region of Africa, the Middle East, and South and South-East Asia [33]. It is a shrub that produces latex with wide pharmacological profile which is a rich source of biologically active compounds [34]. *C. procera* latex contain several chemical compounds which include calotropagenin glycosides/derivatives [35]; saponins, flavonoids and cardenolides [36, 37]; cardioids such as calotoxin, calotro-pin, uscherin, uscchardin, choline, o-pyrocatechuric acid, glycoside calotropaginin, benzoyllineolone, benzoylisoloneolane, syriogenis and uzariganin etc. [38]. It has been traditionally used for various medicinal purposes such as treatment of animal worms, defense role in plants, acting against herbivorous insects, nematodes and phytopathogenic fungi [39]. Different parts of roots, leaves, flowers and latex from the plant are used in several medicinal preparations [40]. It was also reported to exhibit potent analgesic and weak antipyretic activity in various experimental model, possess antioxidant and anti-hyperglycemic property [41], antihelmintic activity [42, 43]; insecticidal and antifungal proteins and their enzymatic profiles have been characterized [44–46]; and there is an empirical association between antioxidant property and residual peroxidase activity. The milk weed has been established to be efficacious in the chemotherapy of malaria, menorrhagia, fever, leprosy and snake bites. Research works investigated on many biological activi-ties of *C. procera* including osmotin proteins exert antifungal activity [47] and anti-inflammatory potential in rats. *C. procera* latex dispensed to rats revealed pain-killing effects wound healing and toxic [48].

The leaf of *Calotropis procera* is a natural coagulant used traditionally in waste water treatment and it has also been reported that *Calotropis procera* leaf is effective in removal of environmental pollutant, polyphenolic crystal violet dye from aqueous solution of textile effluent [49] which presumed to be ascribed to the presence of peroxidase in *Calotropis procera* leaf that oxidized phenols to phenoxy radicals. Some studies suggest that the insoluble fraction of *C. procera* latex is associated with the noxious effects of this fluid [50]. Contrariwise, some constituents of this fluid cause toxicity in small ruminants [51].

From the investigation carried out by Mafulul *et al.,* [52] in the extraction, partial purification and characterization of peroxidase from *Calotropis procera* leaves, it was revealed that peroxidise purified from *Calotropis procera* leaves in primary purification procedures resulted in 1.613-fold purification of peroxidase from the crude extract. Subsequently, enzyme precipitation using ammonium sulphate with the dialyzed fraction showed 2.04 purification folds. *Calotropis procera* leaves peroxidase maintained above 50% over a temperature range of 20–70 with optimum temperature 50°C.

Furthermore, considering the availability and abundance of *Calotropis procera* fresh leaves in Nigerian distribution coupled with availability of advance purification method, this plant tends to provide a very cheap source of peroxidase for phenolic pollutants' bioremediation for waste treatment especially in oil spill region of Niger Delta. It provides potential alternative peroxidase that can compete with commercially available peroxidases for biotechnological applications.

3.3 Imarsil - an inexpensive synthetic chromatographic absorbent

Imarsil is a novel, inexpensive synthetic chromatographic absorbent and oxidized natural polymer of *Brachystegia nigerica. B. nigerica* is a legume used especially in the eastern states of Nigeria as condiment to thicken soup. Its thickening characteristics have been attributed to the presence of hydrocolloid property or gelling property [53]. Imarsil possesses quick and simple recovery approach more importantly in the clarification of microbial biocatalyst from fermentation broth [11]. Cherry and Fidants, [54] demonstrated the use of carboxy-methyl cellulose, edible gum and tannic acid as precipitants and as well as organic solvents also poses the problem of product recovery. Gel filtration technique is also considered assiduous and expensive in the developing countries.

Several procedures of concentrating protein and dilute enzyme from fermentation cell extracts and media using agricultural residues as coagulants. Furthermore, Kareem *et al.* [55] investigated the use of Imarsil and activated charcoal to purify crude lipase in a two-step purification fold which brought about an increase in specific activity from 5.29 to 20.8 Umg^{-1} with protein reduction of 18.24% in the supernatant and ultimate 3.93-fold purification. The study on crude amylase purification showed that a 40-fold purification was attained with 50% final yield of the total fungi amylase in a 3-step purification technique. The elution pattern of *Rhizopus oligosporus* SK5 amylase on Sephadex G-100 column had peaks at fractions (19–22) and (34–38). This purification fold value is conceived greater than values obtained in previous work [56].

Osho et al. [57] studied on production and optimization of bacterial cellulase using agricultural cellulosic biomass by solid state cultivation where the enzyme was clarified with Whatman No 1 filter paper, partially purified with Imarsil (1% w/v) and incubated at 4°C for 3 h. It was reported that at temperature ranges of 40–90°C, enzyme activity increases in crude and partially clarified states as the relative activity also increased to 50 and 60°C for both forms of cellulase respectively. A decline in activity was noticed as temperature increases for both solutions. However, 90% activity of the partially purified enzyme was retained between 50 and 55°C and activity peaked at 60°C. Partial clarification of enzyme is therefore needed to enhance their stability even at much temperature. Kareem and other coworkers [58] outlined that partial clarification of enzyme using activated charcoal preceding gel filtration will established a high purification fold thus preventing some awkwardness of plugging and scaling up when treating crude extracts that sometimes contain particulate and viscous materials. These studies have further substantiated the use of Imarsil as a coagulating-flocculating agent in purification of crude enzyme extracts.

4. Conclusion

It has been proven that enzymes could be recovered from the fermentation broth by these organic absorbents and flocculating materials, making fermentation procedure less laborious. Following elution process, a highly concentrated and purified enzyme would be obtained at reasonable time. This technique seems to be rapid, cheap and promising in downstream processing of industrial enzymes which leads to an aqueous enzyme concentration. They are also established to be faster and easier to implement than the two or three-step processes of conventional precipitation, dialysis and subsequent chromatography. Thus, these natural coagulating-flocculating materials are of great importance in that they are effective in removal of pollutants and debris from fermented broths without necessarily affects the functional and structural formation of industrial enzymes.

Conflict of interest

The authors declare no conflict of interest.

Author details

Michael Bamitale Osho[1*] and Sarafadeen Olateju Kareem[2]

1 Department of Biological Sciences (Microbiology Unit), College of Natural and Applied Sciences, McPherson University, Seriki Sotayo, Abeokuta, Ogun State, Nigeria

2 Department of Microbiology, Federal University of Agriculture, Abeokuta, Nigeria

*Address all correspondence to: mikebamosho@gmail.com

References

[1] van de Sande T. Socio-economic pitfalls of enhancing indigenous capabilities in household fermentation. Food Control 1997; 303-310. https://doi.org/10.1016/S0956-7135(97)00053-4

[2] Motarjemi Y, Nout MJ. Food fermentation: a safety and nutritional assessment. Joint FAO/WHO Workshop on Assessment of Fermentation as a Household Technology for Improving Food Safety. Bull World Health Organ 1996; 74 (6):553-59. https://apps.who.int/iris/handle/10665/54322

[3] Li S, Li P, Feng F, Luo LX. Microbial diversity and their roles in the vinegar fermentation process. Applied Microbiology and Biotechnology. 2015; 99 (12):4997-5024. http://dx.doi: 10.1007/s00253-015-6659-1

[4] Ramos OL, Malcata FX. Industrial Biotechnology and Commodity Products. In Comprehensive Biotechnology (Third Edition), 2017

[5] Timson DJ. Four Challenges for Better Biocatalysts Review. *Fermentation* 2019, 5, 39; http://dx.doi:10.3390/fermentation5020039

[6] Ahmed R, Saradar M. Immobilization of cellulase on TiO2 nanoparticles by physical and covalent Methods: A comparative study. International Journal of Biochemistry, Biophysics and Molecular Biology. 2014; 51: 314-20.

[7] van den Burg B. Extremophiles as a source for novel enzymes. Current Opinion Microbiology. 2003; 6: 213-218. https://dx.doi:10.1016/s1369-5274(03)00060-2

[8] Verma VC, Agrawal S, Kumar A, Jaiswal JP. Starch content and activities of starch biosynthetic enzymes in wheat, rice and millets. Journal of Pharmacognosy and Phytochemistry 2020; 9(4): 1211-1218 https://doi.org/10.22271/phyto.2020.v9.i4q.11883

[9] Antranikian G, Vorgias CE, Bertoldo C. Extreme environments as a resource for microorganisms and novel biocatalysts. Advanced Biochemistry and Engineering Biotechnology. 2005; 96: 219-262. https://dx.doi: 10.10.1007/b135786

[10] Shafiei P, Karimi K, Taherzadeh MI. Techno-economical study of ethanol and biogas from spruce wood by NMMO-pretreatment and rapid fermentation and digestion. Bioresource Technology. 2011; 102:7879-7886. https://dx.doi:10.1016/j.biortech.2011.05.071

[11] Kareem SO, Adio OQ , Osho MB. Immobilization of Aspergillus niger F7-02 Lipase in Polysaccharide Hydrogel Beads of Irvingia gabonensis Matrix. Enzyme Research Article ID 967056, 2014; 7 pp. http://dx.doi.org/10.1155/2014/967056

[12] Kareem SO, Akpan I, Osho MB. Calotropis procera (Sodom apple)—a potential material for enzyme purification. Bioresource Technology. 2003; 87(1):133-135. https://dx.doi:10.1016/S0906-8524(02)00208.0

[13] Banerjee S, Maiti TK, Roy RN. Production, purification, and characterization of cellulase from Acinetobacter junii GAC 16.2, a novel cellulolytic gut isolate of *Gryllotalpa africana*, and its effects on cotton fiber and sawdust. Annals of Microbiology 2020. 70:28 https://doi.org/10.1186/s13213-020-01569-6

[14] Wang SS, Ning YJ, Wang SN, Zhang J, Zhang GQ , Chen QJ. Purification, characterization, and cloning of an extracellular laccase with

potent dye decolorizing ability from white rot fungus *Cerrena unicolor* GSM-01. International Journal of Biological Macromolecules 2017. 95: 920-927

[15] Antecka A, Blatkiewicz M, Boruta T, Górak A,Ledakowicz S. Comparison of downstream processing methods in purification of highly active laccase. Bioprocess and Biosystems Engineering. 2019. 42:1635-1645. https://doi.org/10.1007/s00449-019-02160-3

[16] Wong DWS, Batt SB, Lee CC, Robertson GH. High-activity barley α-amylase by directed evolution. Protein Journal. 2004; 23(7): 453-460.

[17] Allen MJ, Fang TY, Li, Y. Protein engineering of glucoamylase to increase pH optimum, substrate specificity and thermostability. United States Patents No. 6 2003; 537,792.

[18] Kareem SO, Akpan I. Clarification of amylase extract from moldy bran with Imarsil. Enzyme and Microbial Technology 2003; 33(2-3): 259-261. https://dx.doi:10.1016/S0141-0229(03)00127-3

[19] Khawli FAl, Ferrer E, Berrada H, Barba FJ, Pateiro M, Domínguez R, Lorenzo JM, Gullon P, Kousoulaki K. Innovative green technologies of intensification for valorization of seafood and their by-products. Mar. Drugs 2019. 17; 2-21. https://doi.org/10.3390/md17120689

[20] Oliveira VD, da Cunha MNC, de Assis CRD, Batista JMD, Nascimento TP, dos Santos JF, Lima CD, Marques DDV, Bezerra RD, Porto ALF. Separation and partial purification of collagenolytic protease from peacock bass (*Cichla ocellaris*) using different protocol: Precipitation and partitioning approaches. Biocatalysis and Agricultural Biotechnology 2020. 24. https://doi.org/10.1016/j.bcab.2020.101509

[21] Oliveira VM, Assis CRD, Silva JC, Silva QJ, Bezerra RS, Porto ALF. Recovery of fibrinolytic and collagenolytic enzymes froam fish and shrimp byproducts: potential source for biomedical applications. Bol. Inst. Pesca 2019. 45; 389. https://doi.org/10.20950/1678-2305.2019.45.1.389

[22] Kareem SO, Akpan I. Clarification of amylase extract from moldy bran with Imarsil. Enzyme and Microbial Technology 2003; 33(2-3): 259-261. https://dx.doi:10.1016/S0141-0229(03)00127-3

[23] Sudeep KC, Upadhyaya J, Joshi DJ, Lekhak B, Chaudhary DK, Pant BR, Bajgai TR, Dhital R, Khanal S, Koirala N, Raghavan V. Production, Characterization, and Industrial Application of Pectinase Enzyme Isolated from Fungal Strains. Fermentation 2020. 6 (59); http://dx.doi:10.3390/fermentation6020059

[24] Chimbekujwo KI, Ja'afaru MI, Adeyemo OM. Purification, characterization and optimization conditions of protease produced by *Aspergillus brasiliensis* strain BCW2 Journal of Scientific African 2020. 8; e00398 https://doi/org/10.1016/j.sciaf.2020.e00398

[25] Orabi H, El-Fakharany E, Abdelkhalek E, Sidkey N. Production, optimization, purification, characterization, and anti-cancer application of extracellular L-glutaminase produced from the marine bacterial isolate. Prep Biochemistry and Biotechnology 2020. 50(4); 408-418. https://dx.doi:10.1080/10826068.2019.1703193

[26] Ferreira L.M, de Melo RR, Pimenta AS, de Azevedo TKB, de Souza CB. Adsorption performance of

activated charcoal from castor seed cake prepared by chemical activation with phosphoric acid. *Biomass Conversion Biorefinery* 2020. https://doi.org/10.1007/s13399-020-00660-x

[27] Ochonogor AE. Indigo carmine removal from solutions using activated carbons of Terminalia catapa and Cinnarium schweinfurthi nutshell. Journal of Chemical Society of Nigeria. 2005; 30(2): 88-90

[28] Hayashi J, Yamamoto N, Horikawa T, Muroyama K, Gomes VG. Preparation and characterization of highspecific-surface-area activated carbons from K2CO3-treated waste polyurethane. Journal of Colloid and Interface Science. 2005; 281(2): 437-443.

[29] Kumar CG, Takagi H. Microbial alkaline proteases: from a bioindustrial viewpoint. Biotechnology Advances. 1999; 17(7): 561-594. https://dx.doi:10.1016/S0734-9750(99)00027-0

[30] López S, Castro R, García E, Pazo JSA, Barroso CG. The use of activated charcoal in combination with other fining agents and its influence on the organoleptic properties of sherry wine. European Food Research and Technology. 2001; 212:671-675 http://dx.doi:10.1007/s002170100300

[31] Al Sulaibi MAM, ThiemannC, Thiemann T. Chemical Constituents and Uses of *Calotropis procera* and *Calotropis gigantea* – A Review (Part I – The Plants as Material and Energy Resources) Open Chemistry Journal. 2020; *7*, 1-15 http://dx.doi:10.2174/1874842202007010001

[32] Main GL, Morris JR. Colour of seyval blanc juice and wine as affected by juice fining and bentonite fining during fermentation. American Journal of Enology and Viticulture. 1994; 45:417-422.

[33] Al Sulaibi MAM, ThiemannC, Thiemann T. Chemical Constituents and Uses of *Calotropis procera* and *Calotropis gigantea* – A Review (Part I – The Plants as Material and Energy Resources) Open Chemistry Journal. 2020; *7*, 1-15 http://dx.doi:10.2174/1874842202007010001

[34] Kumar VL, Arya S. Medicinal uses and pharmacological properties of *Calotropis procera*. In: Recent Progress in Medicinal Plants, 11. (Govil E. D., ed.). Studium Press, Houston, TX, USA. 2006; 373-388 pp.

[35] Chundattu SJ, Agrawal VK, Ganesh N. Phytochemical investigation of *Calotropis procera*. Arabian Journal of Chemistry. 2011; https://doi.org/10.1016/j.arabjc.2011.03.011

[36] Yoganandam K, Ganeshan P, NagarajaGanesh B, Raja K. Characterization studies on *Calotropis procera* fibers and their performance as reinforcements in epoxy matrix. Journal of Natural Fibers. 2019. http://dx.doi.org/10.1080/15440478.2019.1588831

[37] Patil **SP.** Calotropis gigantea *assisted green synthesis of nanomaterials and their applications: a review.* Beni-Suef University Journal of Basic and Applied Sciences 2020. 9:14 https://doi.org/10.1186/s43088-020-0036-6

[38] Tu LX, Duan WZ, Xiao WL, Fu CX, Wang AQ, Zheng Y. *Calotropis gigantea* fiber derived carbon fiber enable fast and efficient absorption of oils and organic solvents. Separation Purification Technique 2018. *192*; 30-35. http://dx.doi.org/10.1016/j.seppur.2017.10.005

[39] Xiao, W.; Wang, N.; Niu, B.; Fu, C.; Zhou, L.; Zheng, Y. Polyethylene sulfone assisted shape construction of *Calotropis gigantea* fiber for preparing a sustainable and reusable oil sorbent. Cellulose, 2019. *26;* 3923-3933. http://dx.doi.org/10.1007/s10570-019-02356-6

[40] Cao, E.; Duan, W.; Yi, L.; Wang, A.; Zheng, Y. Poly(mphenylenediamine) functionalized *Calotropis gigantea*n an

fiber for coupled adsorption reduction for Cr(VI). Journal of Molecular Liquids 2017. *240*; 225-232. http://dx.doi.org/10.1016/j.molliq.2017.05.087

[41] Cao E, Duan W, Wang A, Zheng Y. Oriented growth of poly (mphenylenediamine) on *Calotropis gigantea* fiber for rapid adsorption of ciprofloxacin. Chemosphere 2017. *171*; 223-230. http://dx.doi.org/10.1016/j.chemosphere.2016.12.087

[42] Yi LS, Liang GW, Xiao WL, Duan WZ, Wang AQ, Zheng Y. Rapid nitrogen-rich modification of *Calotropis gigantea* fiber for highly efficient removal fluoroquinolone antibiotics. Journal of Molecular Liquids. 2018. *256*; 408-415. http://dx.doi.org/10.1016/j.molliq.2018.02.060

[43] Kaur R, Kaur H. *Calotropis procera* an effective adsorbent for removal of Congo red dye: isotherm and kinetc modelling. Model. Earth System Environment 2017. *3* http://dx.doi.org/10.1007/s40808-017-0274-3

[44] Alshahrani NDSTS, Aref IM, Nasser RA. Allelopathic potential of *Calotropis procera* and Eucalyptus species on germination and growth of some timber trees. *Allelopathy Journal* **2017**. *40*; 81-94. http://dx.doi.org/10.26651/2017-40-1-1068

[45] Cavalcante GS, Morais SM, Andre WPP, Ribeiro WLC, Rodrigues ALM, Lira FCML. Chemical composition and in vitro activity of *Calotropis procera* (Ait.) latex on Haemonchus contortus. Veterinary Parasitology. 2016; 226:22-25. http://dx.doi.org/10.1016/j.vetpar.2016.06.012

[46] Cavalcante GS, Morais SM, André WPP, Araújo-Filho JV, Muniz CR, Rocha LO. Chemical constituents of *Calotropis procera* latex and ultrastructural effects on Haemonchus contortus. Brazilian Journal of Veterinary Parasitology. 2020; 29(2): e001320. https://doi.org/10.1590/S1984-29612020045

[47] Doshi H, Satodiya H, Thakur MC, Parabia F, Khan A. Phytochemical screening and biological activity of *Calotropis procera* (Ait). R.Br. (Asclepiadaceae) against selected bacteria and Anopheles stephansi Larvae. International Journal of Plant Research. 2011; 1(1): 29-33. https://doi:10.5923/j.plant.20110101.05

[48] Shaker KH, Morsy N, Zinecker H, Imhoff JF, Schneider B. Secondary metabolites from *Calotropis procera* (Aiton). Phytochemistry Letters. 2010; 3(4): 212-216.

[49] Ali H, Muhammad SK. Biosorption of crystal violet from water on leafe biomass of *Calotropis procera*. Journal of Environmental Science and Technology. 2008; 1(3):143-150. http://dx.doi.org/10.3923/jest.2008.143.150

[50] Kumar VL, Shivkar YM. In vivo and in vitro effect of latex of *Calotropis procera* on gastrointestinal smooth muscles. Journal of Ethnopharmacology. 2004; 93(2-3): 377-379. http://dx.doi.org/10.1016/j.jep.2004.04.013

[51] El Sheikh HA, Ali BH, Homeida AM, Hassan T, Idris OF, Hapke HJ. The activities of drug-metabolizing enzymes in goats treated orally with the latex of *Calotropis procera* and the influence of dieldrin pretreatment. Journal of Comp Pathology. 1991; 104(3): 257-268. http://dx.doi.org/10.1016/S0021-9975(08)80038-2

[52] Mafulul SG, Joel EB, Barde LA, Lenka JL, Ameh AA, Phililus MG. Extraction, Partial Purification and Characterization of Peroxidase from *Calotropis procera* Leaves. Journal of Advances in Biology & Biotechnology. 2018; 18(1): 1-10. http://dx.doi:10.9734/JABB/2018/41709

[53] Odum DC. Proximate composition and Sugars of the hydrocolloid extract of Brachystegia nigerica. Nigerian Journal of Microbiology. 2000; 14:77-82.

[54] Cherry JR, Fidants AL. Directed evolution of industrial enzymes: An update. Current Opinion Biotechnology. 2003; 4: 438-443. http://dx.doi.org/10.1016/s0958-1669(03)00099-5

[55] Kareem SO, Akpan I, Popoola TOS, Sanni LO. Purification and characterization of thermostable glucoamylase from Rhizopus oligosporus SK5 mutant obtained through UV radiation and chemical mutagenesis. Biokemistri. 2014; 26 (1) 19-24.

[56] Selvakumar PA, Shakumary L, Helen A, Pandey A. Purification and Characterization of glucoamylase produced by Aspergillus niger in solid state fermentation. Letters in Applied Microbiology. 1996; 23: 403-406. http://dx.doi.org/10.1111/j.1472-765x

[57] Osho MB, Nwagala PN, Ojo EE. Production and Optimization of Bacterial Cellulase Using Agricultural Cellulosic Biomass by Solid State Cultivation. Current Biotechnology. 2017; 6: 349-355. http://dx.doi.org/10.2174/2211550106666170109151156

[58] Kareem SO, Adebayo OS, Balogun SA, Adeogun AI, Akinde SB. Purification and Characterization of Lipase from Aspergillus flavus PW2961 using Magnetic Nanoparticles. Nigerian Journal of Biotechnology. 2017; 32:77-82 http://dx.doi.org/10.4314/njb.v32i1.11.

12

Integrated Biorefinery Approach to Lignocellulosic and Algal Biomass Fermentation Processes

Felix Offei

Abstract

Lignocellulosic and algal biomass have been suggested as relatively sustainable alternatives to sugar and starch-based biomass for various fermentation technologies. However, challenges in pretreatment, high production costs and high waste generation remains a drawback to their commercial application. Processing cellulosic and algal biomass using the biorefinery approach has been recommended as an efficient and cost-effective pathway since it involves the recovery of several prod-ucts from a single biomass using sequential or simultaneous processes. This review explored the developments, prospects and perspectives on the use of this pathway to add more value and increase the techno-economic viability of cellulosic and algal fermentation processes. The composition of lignocellulosic and algal biomass, the conventional ethanol production processes and their related sustainability issues are also discussed in this chapter. Developments in this approach to lignocellulosic and algal biomass has shown that valuable products at high recovery efficiencies can be obtained. Products such as ethanol, xylitol, lipids, organic acids, chitin, hydrogen and various polymers can be recovered from lignocellulosic biomass while ethanol, biogas, biodiesel, hydrocolloids, hydrogen and carotenoids can be recovered from algae. Product recovery efficiencies and biomass utilisation have been so high that zero waste is nearly attainable. These developments indicate that indeed the appli-cation of fermentation technologies to cellulosic and algal biomass have tremendous commercial value when used in the integrated biorefinery approach.
Keywords: fermentation, integrated biorefinery, lignocellulosic biomass, algae, ethanol

1. Introduction

Concerns over the depletion and environmental effects of greenhouse gas (GHG) emissions from the use of fossil fuels has led to the extensive search for alternative, renewable and sustainable fuels. Currently, the highest contributor to GHG emissions is the transportation sector through fuel combustion. Biomass is currently the only abundant renewable energy source for the direct production of fuel. Typical fuels currently produced from biomass include bioethanol, biogas, biodiesel, bio-butanol, syngas and bio-oil. Bioethanol is currently the largest alternative fuel produced globally at 106 billion litres per annum [1].

Sugar and starch-based biomass have been the primary choice of raw mate-rial for the production of food and fuel grade ethanol for various commercial applications. They however face enormous competing interests often illustrated with the food-vs.-fuel debate [2]. Lignocellulosic and algal biomass have been suggested as relatively sustainable alternatives. They have been hundreds of extensive research on the factors that influence their efficiency as substrates for ethanol production. The major drawbacks noted in these studies during their application include: the need for pretreatment processes, higher production costs and high waste generation [3]. A processing approach that has potential to maximise the profitability and minimise waste generation from the use of cellulosic and algal biomass as feedstock is the integrated biorefinery approach. The integrated biorefinery concept refers to the use of single or multiple technologies to produce several high value products from a single or multiple biomass [4]. This approach to biomass processing is considered more efficient, economical and sustainable.

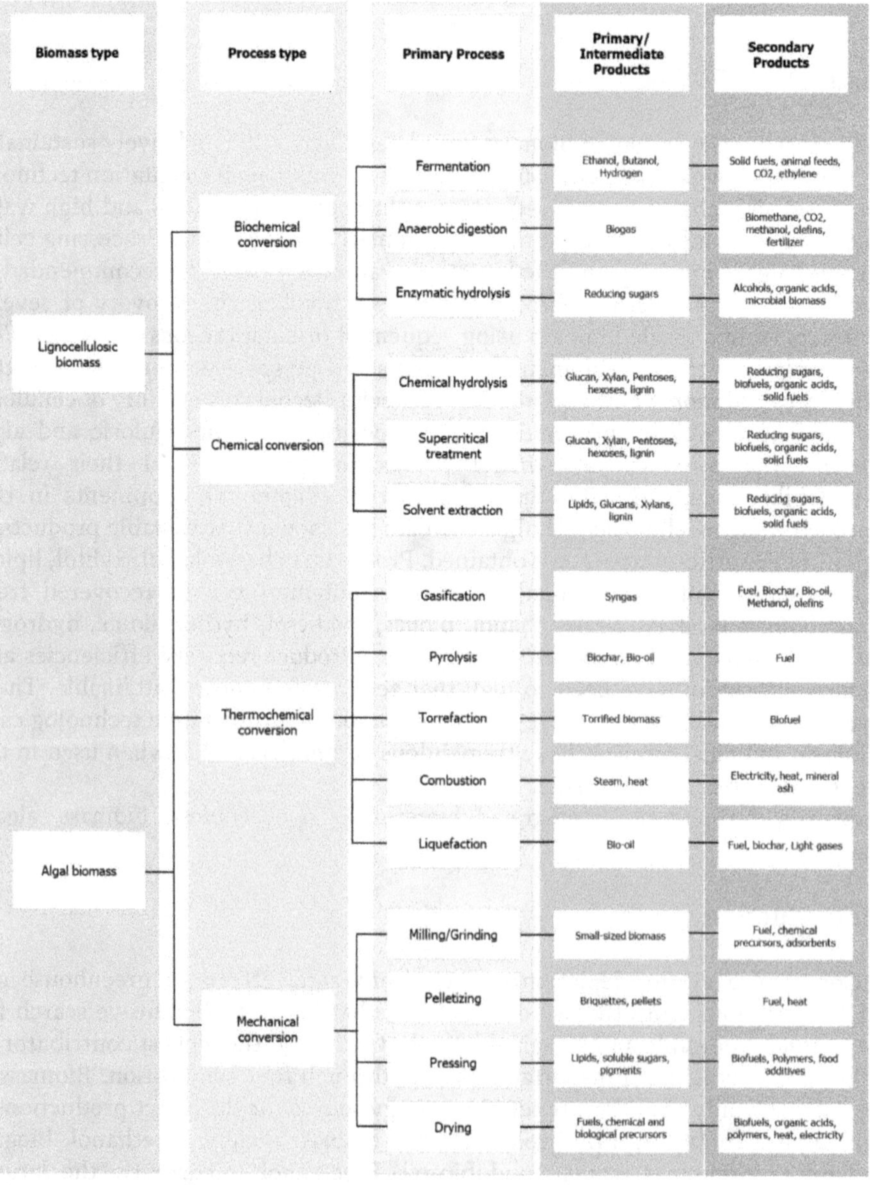

Figure 1.
Typical biorefinery conceptual scheme.

Biorefineries generally integrate various biomass conversion technologies to produce fuels, power, heat and other value-added products from biomass. These refineries have evolved over the last two decades in several phases. Phase I biorefineries convert a single raw material to a single product. Phase II converts a single raw material using multiple processing tools to obtain a broad range of products. Phase III biorefineries, commonly referred to as integrated biorefineries use a wide range of raw materials and technologies simultaneously or sequentially to produce a wide range of valuable products [5]. Some integrated biorefineries use various feedstock and technologies to produce biofuels as main products along with co-products such as platform chemicals, heat and power [5].

The International Energy Agency sums up the description of the biorefinery concept as "the sustainable processing of biomass into a spectrum of marketable products and energy" [6]. It expands the concept to include a wide range of technologies that separate biomass resources into their basic polymeric units such carbohydrates, proteins, lipids and even elementals which can be converted to valuable products including fuels, heat and chemicals. Biorefinery as an entity is described as a facility or network of facilities where various processing technologies are integrated to obtain multiple products from a single or several types of biomass [6]. Bioethanol is currently the leading energy product recovered from biomass using the biorefinery approach.

Sugar and starch-based biomass have been the primary choice of material for the production of food and fuel grade ethanol for various commercial applications but has an enormous competing interest often illustrated with the food-vs.-fuel debate. Lignocellulosic and algal biomass have been suggested as relatively sustainable alternatives. However, difficulties in pretreatment, high waste generation and high processing costs remains a drawback to their commercial application. Processing cellulosic and algal biomass using the biorefinery approach has been recommended as an efficient and cost-effective pathway since several valuable products can be recovered using sequential or simultaneous processes as illustrated in **Figure 1** [4]. This review explored the developments made in the use of this pathway to add more value and increase the techno-economic viability of cellulosic and algal fermenta-tion processes. The composition of lignocellulosic and algal biomass, the conven-tional ethanol production processes and their related sustainability issues are also discussed in this chapter.

2. Lignocellulosic biomass for biorefinery applications

Lignocellulosic biomass typically refers to plant materials composed primarily of cellulose, hemicellulose and lignin. This type of biomass usually includes forest materials, agricultural residues, wood processing residues and non-edible plant materials usually referred to as energy crops (**Table 1**). In the context of biofuel production, lignocellulosic biomass are referred to as second generation biomass which is used to differentiate them from sugar and starch based biomass (1st generation biomass) and algal biomass (3rd generation biomass). They are typically composed of 40–50% cellulose, 25–30% hemicellulose and 15–20% lignin [30]. The effective use of these three primary components would significantly determine the economic viability of cellulosic ethanol production.

Cellulose refers to the linear polymer made up of glucose monomer units bonded together by β-1,4 glycosidic bonds. Hemicellulose refers to branched heteropolymers of xylose, glucose, galactose, mannose, arabinose and some uronic acids. Lignin is primarily made up of three major phenolic components, namely p-coumaryl alcohol, coniferyl alcohol and sinapyl alcohol [20]. The ratio of these

Biomass type	Biomass	Cellulose (%)	Hemicellulose (%)	Lignin (%)	Reference
Agricultural residues	Sugarcane bagasse	49	29.6	27.2	[7]
	Barley straw	37.5	37.1	16.9	[8]
	Rice husk	33.4	30.0	18.3	[9]
	Corn cob	44	36.4	18.0	[10]
	Corn stover	36.5	31.3	13.6	[11]
	Rye straw	42.1	24.4	22.9	[12]
	Rapeseed straw	37.0	24.2	18.0	[13]
	Wheat straw	40.0	33.8	26.8	[14]
	Rice straw	36.6	22.0	14.9	[15]
	Sunflower stalk	33.8	24.3	19.9	[16]
	Sorghum bagasse	45.3	26.3	16.5	[17]
	Barley hull	34.0	36.0	19.0	[18]
	Banana peels	13.0	15.0	14.0	[19]
	Cotton stalk	31.0	11.0	30.0	[20]
	Coffee pulp	36.9	47.5	19.1	[20]
	Wheat bran	14.8	39.2	12.5	[20]
	Sugarcane tops	35.0	32.0	14.0	[21]
	Jute fibres	45.0	18.0	21.0	[20]
	Oat straw	31.0	20.0	10.0	[20]
	Soya stalks	34.5	24.8	9.8	[22]
Municipal and industrial wastes	Newspapers	60.3	16.4	12.4	[23]
	Paper sludge	60.8	14.2	8.4	[24]
	Brewer's spent grain	21.0	32.8	25.6	[25]
Woods and grasses	Softwood stems	44.5	21.9	27.7	[6]
	Switchgrass	35.4	26.5	18.2	[6]
	Bamboo	50.0	20.0	23.0	[20]
	Eucalyptus	51.0	18.0	29.0	[20]
	Hardwood stems	55.0	40.0	25.0	[26]
	Pine	49.0	13.0	23.0	[20]
	Poplar wood	51.0	25.0	10.0	[20]
	Olive tree	25.2	15.8	19.1	[27]
	Water hyacinth	22.1	50.1	5.4	[28]
	Spruce	43.8	20.8	28.3	[29]
	Oak	45.2	24.5	21.0	[29]

Table 1.
Composition of typical lignocellulosic biomass used in biorefinery applications.

components varies between various plant tissues as shown in **Table 1**. The cellulose units are packed into microfibrils which are attached to each other by hemicelluloses and amorphous polymers of different sugars as well as other polymers such as pectin covered by lignin. The units of individual microfibrils in crystalline cellulose are packed so tightly that neither enzymes nor water molecules can enter the complex framework [20]. This high molecular weight and ordered tertiary structure of natural cellulose makes it insoluble in water. However, some parts of the microfi-brils have a less ordered, non-crystalline structure referred to as amorphous regions [31]. The crystalline regions of cellulose are more resistant to biodegradation than the amorphous parts while cellulose with low degree of polymerisation will be more susceptible to cellulolytic enzymes. The composition of typical lignocellulosic bio-mass that have been considered for various biorefinery applications are presented in **Table 1**.

3. Algal biomass for biorefinery applications

Marine biomass accounts for over 50% of primary biomass produced globally but has been the least harnessed for various applications [32]. It is mainly grouped into two, namely macroalgae (commonly known as seaweeds) and microalgae. However, cyanobacteria is conventionally regarded as a form of algae often called blue-green algae [33]. Both groups have been used in the production of various biofuels. Microalgae has been explored predominantly as substrate for bio-oils and biodiesel while macroalgae has been used mainly in bioethanol and biogas production [32].

Marine algae are plant-like multicellular organisms that live attached to hard substrata such as rocks in coastal areas [34]. Their basic structure consists of a thallus, which forms the body of the organism and a holdfast, a structure on its base which allows it to be attached to hard surfaces such as rocks near the shoreline of coastal areas. Brown seaweeds are the largest in size, growing up to 4 m in length for some species. Green and red seaweeds are smaller ranging from a few centimetres in some species to a meter in others [35]. According to the FAO [36], 8.2 and 15.8 million tons of brown and red seaweed respectively were produced in the year 2013. This was valued at USD 1.3 billion and 4.1 billion for the brown and red seaweeds respectively. For the green seaweed 14,800 tons valued at USD 15.7 million was produced globally in the year 2013 [36]. The enormous difference in the production values of the brown and red from the green seaweed can be attributed to the valu-able hydrocolloids such as alginate, carrageenan and agar found only in the red and brown seaweeds.

The structural differences found between land-based plants and algae gives algal biomass an advantage of a higher yield per hectare. In comparison to land-based plants, seaweeds have an average yield per hectare per year of 730,000 kg while sugarcane, sugar beet, maize and wheat have 68,260; 47,070; 4,815 and 2,800 kg respectively [37]. The high yields from macroalgae in general is attributed to the low energy required in the formation of its supporting tissue during growth. Seaweeds can also absorb nutrients across its entire surface and can be cultivated three dimensionally in water [37].

Seaweeds are composed of carbohydrates, proteins, lipids and minerals which ranges from 30 to 60%, 10–40%, 0.2–3% and 10–40%, respectively [38]. Besides their unique and varying composition, seaweeds have been grouped into three, based on their pigmentation. They are rhodophyceae (red seaweeds), phaeophyceae (brown seaweeds) and chlorophyceae (green seaweeds) based on their pigments r-phycoerythrin, chlorophyll and xanthophyll, respectively [39].

Biomass type	Species	Carbohydrate	Protein	Lipid	Ash	Ref.
Macroalgae (seaweed)	*Chaetomorpha linum*	54	—	—	22	[35]
	Caulerpa lentillifera	38.7	10.4	1.1	37.2	[40]
	C. linum	29.8	8.6	2.6	30.5	[41]
	Codium fragile	58.7	15.3	0.9	25.1	[42]
	Ulva fasciata	31.3	14.4	1.5	28.0	[43]
	Ulva lactuca	54.3	20.6	6.2	18.9	[44]
	Ulva pertusa	52.3	25.1	0.1	22.5	[38]
	Ulva rigida	53	23.4	1.2	21.7	[45]
	Chondrus pinnulatus	64.4	22.5	0.2	12.9	[46]
	Cryptonemia crenulata	47	—	—	19	[47]
	Kappaphycus alvarezzi	60.7	17.4	0.8	21.1	[48]
	K. alvarezzi	55			23	[47]
	Eucheuma cottonii	26.5	9.8	1.1	46.2	[40]
	Gelidium amansii	66.0	20.5	0.2	13.3	[42]
	Gigartina tenella	42.2	27.4	0.9	24.5	[46]
	Hypnea charoides	57.3	18.4	1.5	22.8	[26]
	Hypnea musciformis	39	—	—	22	[47]
	H. musciformis	37	—	—	30	[47]
	Hydropuntia dentata	31.2	10.3	3.2	38.7	[43]
	Lomentaria hakodatensis	40.4	29	0.7	29.9	[46]
	L. digitata	64.2	3.1	1.0	11.9	[49]
	Laminaria japonica	51.9	14.8	1.8	31.5	[44]
	Sargassum fulvellum	39.6	13	1.4	46	[44]
	Sargassum polycystum	33.5	5.4	0.3	42.4	[40]
	Sargassum vulgare	32.6	10.3	1.0	27.2	[43]
	Saccharina latissima	16.8	10.1	0.5	34.6	[49]

Biomass type	Species	Carbohydrate	Protein	Lipid	Ash	Ref.
Microalgae	*Scenedesmus acutus*	39.0	8.0	41.0	2.0	[50]
	Scenedesmus obliquus	25.0	48.8	22.5	12.9	[51]
	Pseudochoricystis ellipsoidea	19.3	27.5	45.4	2.3	[51]
	Chlorogloeopsis fritschii	37.8	41.8	8.2	4.6	[51]
	Chlorella vulgaris	16.7	41.0	10.0	13.4	[52]
	Chlorella emersonii	37.9	9.0	29.3	2.8	[51]
	Chlorella zofingiensis	11.5	11.2	56.7	4.8	[51]
	Spirulina sp.	15.1	50.1	12.3	7.6	[51]
	Nannochloropsis sp.	37.3	32.2	25.0	5.5	[53]
	Schizochytrium limacinum	25.3	12.4	56.7	5.6	[53]
	Chlorella vulgaria	43.4	28.2	17.9	10.5	[53]
	Scenedesmus sp.	35.4	24.6	10.5	29.5	[53]
	Chlamydomonas reinhardtii	35.5	34.2	24.2	6.1	[53]
	Dunaliella tertiolecta	21.7	61.3	2.9	13.5	[54]
	Botryococcus braunii	2.4	39.6	33.0	7.5	[52]
	Spirulina platensis	11.0	42.3	11.0	7.1	[52]
	Chaetoceros muelleri	34.2	16.3	43.4	—	[51]

Table 2.
Composition of typical algal biomass used in biorefinery applications.

Algal biomass composition has been found to vary based on several factors such as the season, availability of nutrients, water salinity and availability of sunlight (**Table 2**) [55]. The algal component of primary importance to bioethanol production is the carbohydrates (polysaccharides), since they currently form the only fraction that can be fermented to ethanol. Generally, some algae are composed of large fractions of complex sulphated polysaccharides which are uniquely different in each group serving as their cellular storage and structural support tissue [56]. The composition of typical algal biomass that have been considered for various biorefinery applications are presented in **Table 2**.

4. Processes for bioethanol production

The conversion of cellulosic and algal biomass to bioethanol usually involves four major processes excluding biomass selection. They include biomass pretreatment, hydrolysis of pretreated biomass, fermentation of biomass hydrolysates and ethanol recovery from the fermentation broth using distillation and dehydration processes [46]. The various efficiencies of each process will influence the final ethanol yield therefore each process condition and catalyst used is carefully selected and in most cases optimised to maximise the process efficiencies.

One of the most influential processes in bioethanol production from cellulosic and algal biomass is pretreatment. This process is used to render biomass susceptible to further breakdown by separating the cellulose, hemicellulose and lignin fractions. The selection of an efficient and cost effective biomass pretreatment method has been a major hurdle in cellulosic bioethanol production and its commercialisation for several decades. Different pretreatment mechanisms have been developed with varying degrees of efficiency [20]. All these methods have been developed with a common aim of finding a good balance between efficiency, cost, environmental effects and energy use. So far, all the methods developed have come with intrinsic advantages and disadvantages. Some common disadvantages experienced include: degradation of sugars, formation of inhibitors, high energy requirements, catalyst requirements, difficulties in catalyst recovery, challenges in waste treatment and high overall costs [20]. One or more these drawbacks are experienced in the various pretreatment processes currently developed. Nonetheless, a careful comparison and risk analysis could be used to distinguish and select one from the other. The biomass specificity for particular pretreatments could be explored to see the variations in the interactions between various cellulosic and algal biomass and various pretreatment methods as a solution.

The hydrolysis process in bioethanol production is one of the most limiting stages in the entire production process since it is the stage where the sugars to be converted to ethanol is obtained. Hydrolysis simply refers to cleavage or division through the addition of water molecules. In the context of complex sugars (polysac-charides), it involves the use of a water molecule by a catalyst to break the glycosidic linkages within their polymeric form (di-, tri-, oligo- or polysaccharide) to their monomeric form (monosaccharides or reducing sugars). During the cleavage of sugars, a hydrogen atom (H^+) is gained by one part of the polymeric structure whiles the other gains a hydroxyl group (OH^-). Thus, the separation continues until all polymeric units are reduced to their individual monomeric form [46].

The hydrolysis of cellulosic biomass for bioethanol production involves the breakdown of polymeric units such as cellulose and hemicellulose whiles the hydrolysis of algal biomass (particularly macroalgae) involves the breakdown of polymeric units such as laminarin, ulvan, alginate, carrageenan, mannitol, agar and cellulose. The simple sugars (monosaccharides) recovered from both agal and cellulosic biomass include glucose, galactose, rhamnose, mannose, fucose, xylose and arabinose for fermentation to ethanol [57]. The common methods that have been used in cellulosic and algal biomass hydrolysis includes dilute acid thermal [58], dilute alkaline thermal [59], enzymatic [3] and thermal [58] hydrolysis. All other hydrolysis methods are usually derivatives of these and are usually broadly grouped under physical, chemical, thermal and biological hydrolysis. Two or more of these methods are often combined to improve the efficiency of monomeric sugar recovery.

Enzymatic hydrolysis, particularly the use of cellulases in both cellulosic and algal biomass hydrolysis, has been promoted extensively over all other forms of hydrolysis. This is because enzymes are considered more environmentally friendly in their application and generate no inhibitors as is the case with chemical catalysts. Three major cellulase activity systems have been identified to be involved in cellulosic hydrolysis. The enzymes involved in these systems include endoglucanases, exoglucanases (cellodextrinases) and β-glucosidases [60]. Cellulase synthesis is predominant among fungi such as *Trichoderma reesei*, *Aspergillus niger*, and *Humicola insolens*; and bacteria such as *Bacillus subtilis*, *Streptomyces drodowiczi*, and *Bacillus pumilusand* [20]. Studies in enzymatic hydrolysis have focused on process optimization, improving cellulase activities, optimisation of reaction conditions, enzyme-to-substrate ratios and enzyme recovery and reuse strategies. The ideal final enzyme or enzyme cocktail should have high hydrolytic efficiencies on the preferred biomass, operate at mildly acidic or alkaline pH, be resilient to process stresses and be cost-effective [30].

The fermentation process in bioethanol production is the stage within which the reducing sugars obtained after hydrolysis are converted to ethanol by an organ-ism. This process is always dependent on the overall ethanol production pathway selected. Currently, the ethanol pathways that have been used in cellulosic and algal biomass processing include: separate hydrolysis and fermentation (SHF), simultaneous saccharification and fermentation (SSF), simultaneous saccharification and co-fermentation (SSCF) and consolidated biomass processing (CBP) [46]. SHF is the most common and most well-developed approach which allows the use of the optimal conditions for both the hydrolysis and fermentation processes [61]. It offers the flexibility of choosing various hydrolysis processes, a feature which cannot be found in the use of the SSF approach. The SSF process involves the co-application of the enzyme for saccharification and the organism for fermentation to the pretreated biomass in the reactor under similar conditions of operation. This process is considered more cost-effective than SHF but comparisons on its process efficiency relative to SHF is currently inconclusive [46].

5. Integrated biorefinery applications to lignocellulosic biomass

Processing of cellulosic biomass using the biorefinery approach has often had its roots in the processing of first generation biomass. Typical first generation biomass such as corn, sugarcane and cassava (mostly in Africa and Asia) are still the most preferred feedstock in commercial fermentation processes. The biorefinery way of processing corn by microbial fermentation often yields ethanol, citric acid, lactic acid or lysine as the main product depending the primary product goal of the biorefinery [62]. Conventionally, the starch fraction of the corn is processed to dextrose via enzymatic pathways before microbial fermentation to the desired product. Corn fibre, gluten meal and corn steep liquor are the usual by-products obtained in a corn biorefinery which are of enormous value. Corn fibre which is lignocellulosic in nature can be further hydrolysed to obtain glucose, xylose and other monomeric sugars which can be further fermented to products such as ethanol, xylitol and acetate [63]. Gluten meal from corn which is very high in proteins can be used as feed for livestock and poultry or as substrates for various pharmaceutical products and commercial polymers [64]. Corn steep liquor which is also high in proteins is often used as a nitrogen source in various fermentation processes [65].

Sugarcane biorefineries are usually very interesting due to the unique composition of sugarcane which is usually 11–16% sucrose, 70–75% water and 10–16% fibre [63]. Sugarcane processing begins with the extraction of cane juice which imme-diately leads to the generation of solid residue in the form of sugarcane bagasse. Sugarcane bagasse is very high in fibre and is considered a lignocellulosic biomass. This bagasse can be valorised in a relatively more complex pathway to ethanol and other chemicals using microbial fermentation technologies or simply used as fuel in boilers for the generation of steam and electricity. The latter is the predominant process application of bagasse in industry currently. Sugarcane alone as a single biomass can be processed to obtain first generation ethanol from the cane juice and second generation ethanol from the bagasse. Additionally, sugar processing plants which use sugarcane obtain molasses as a sucrose-rich by-product which can also be used as substrate for ethanol production [63].

The potential co-production of ethanol and xylitol from sugarcane bagasse was examined in a study by Unrean and Ketsub [66]. In the study, cellulose and hemicellose fractions of the sugarcane bagasse were separated using sulphuric acid and enzymatic hydrolysis processes. The pretreated cellulose was used as substrate for the recovery of ethanol using *Saccharyomyces cerevisiae* as the fermenting organism while hemicellulose hydrolysate was used as substrate for the recovery of xylitol with *Candida tropicalis* as the fermenting organism. The product recoveries reported from the use of the bagasse was 0.44 g/g total glucose for ethanol and 0.50 g/g total xylose for xylitol. An economic analysis within the same study revealed a 2.3 fold increase in profitability for the integrated ethanol and xylitol production process over standalone cellulosic ethanol production [66].

Cellulosic pulp and paper mill waste in the form of primary sludge was examined in a study as a substrate for the production of bioethanol and biolipids [67]. In the integrated study, bioethanol and biolipids were both obtained from the hydrolysates of the primary sludge at yields of 9% and 37.8%, respectively. *S. cerevisiae* was used as the fermenting organism for bioethanol while the oleaginous yeast *Cutaneotrichosporon oleaginosum* was used as the organism for the biolipids production. A unique addition to the biorefinery process was the use of the unhydrolysed primary sludge as a cement additive or fibre reinforcement material in comparison with conventional Portland cement. The comparison of the compression load between the two materials indicated that the unhydrolysed paper mill material had 102% higher compressive strength than the Portland cement [67]. This unique application of fermentation based and non-fermentation based processes to harness the use of the pulp and paper residual biomass in a zero waste approach can be explored for other lignocellulosic biomass.

Dairy manure, a nitrogen rich cellulosic biomass has been examined as a substrate for the co-production of fumaric acid and chitin [68]. Fumaric acid is commonly used in food flavouring and preservation while chitin is a natural biopolymer with applications in the water treatment and pharmaceutical industries. In the study by Liao et al. [68], *Rhizopus oryzae* ATCC 20344 was applied as fermenting organism in a one-pot fermentation process to obtain fumaric acid in the liquid medium of the broth while chitin was found in the resulting fungal biomass formed in the broth. A maximum fumaric acid yield of 31% and a chitin yield of 0.21 g/g fungal biomass (from 11.5 g/l fungal biomass concentration) was obtained [68].

Wheat straw and corn stover have been studied as substrates for the co-production of hydrogen and ethanol [69]. In the study, genetically engineered *Escherichia coli* were applied in a dark fermentation process as means to maximise the simultaneous production of the two products. The engineered strained of *E. coli* produced a 30% increase in the co-production yield of hydrogen and ethanol. The yields obtained were 323 ml H_2/g total reducing sugars (TRS) and 3.5 g ethanol/g TRS for wheat straw and; 337.1 ml H_2/g TRS and 2.9 g ethanol/g TRS for corn stover [69].

Lignin utilisation has been a very important of part of the goal to maximise the use of lignocellulosic biomass in a biorefinery context. Considerably large quanti-ties of lignin-rich by-products are generated from the conversion of cellulosic biomass to biofuels and various organic compounds. The efficient use of the lignin can improve the overall economics of the commercial use of lignocellulosic biomass [70]. A wide range of polymeric materials which can be used as precursors for even more valuable products have successfully been derived from lignin. They include polyesters, epoxy and phenolic resins, hydrogels, graft polymers, vanillin and poly-amides. Vanillin in particular is an important compound used as flavouring agent in the food and pharmaceutical industries. It has also been considered as a precursor to hydrogels, polyester epoxide and polyethylene. Direct lignin recovery from ligno-cellulosic biomass can be effected using the Kraft process, lignosulfonates process, organosolv process, steam explosion or using ligninolytic enzymes such as lignin peroxidase, manganese peroxidase and laccase [70].

6. Integrated biorefinery applications to algal biomass

Several studies have used the integrated biorefinery approach to maximise the use of algal biomass and improve both their economic and process sustainability. This approach was used in the processing of the green seaweed, *C. linum* to co-produce bioethanol and biogas in a single study [41]. A bioethanol yield of 0.41 g/g reducing sugar (0.093 g/g pretreated seaweed) was obtained after the pretreatment, enzymatic hydrolysis and fermentation of the seaweed biomass. The enzymatic hydrolysis was done with a crude enzyme from *Aspergillus awamori* at 45°C and pH 5 for 30 hours while the fermentation was done with *S. cerevisiae* at 28°C for 48 hours while shaking at 150 rpm. The fermentation broth was then distilled to recover the ethanol while the residue referred to as vinasse was used as the feed for anaerobic digestion. The anaerobic digestion of the vinasse which was done at 38°C in a 0.5 l digester for 30 days yielded 0.26 l/g VS of biomethane [41]. The final waste generated was 0.3 g/g biomass which represents a substrate utilisation of up to 70%. This approach did indeed enhance the use of the substrate.

Ashokkumar *et al.* [71] also made a similar attempt with the biorefinery approach. They considered the integrated conversion of the brown seaweed *Padina tetrastromatica* to both biodiesel and bioethanol. The crude lipids content was first extracted from the biomass using various solvents to obtain a yield of 8.15% w/w biomass. This was processed further through transesterification (the process of exchanging the organic group R' of an ester with the organic group R' of an alcohol) to obtain a final biodiesel yield of 78 mg/g biomass. The residual biomass after lipids extraction was hydrolysed and fermented using baker's yeast to obtain a bio-ethanol yield of 161 mg/g residual biomass [71]. This study demonstrated that the integration of biodiesel and bioethanol production processes on a single seaweed biomass can efficiently harness both the lipid and carbohydrate fraction which could form up to 70% of the entire biomass.

A unique application of the biorefinery approach was used by Xu *et al.* [72]. In their study, mannitol was first removed from the brown seaweed *L. japonica* leaving behind an alginate rich suspension. The alginate suspension was used as substrate for volatile fatty acid (VFA) production via fermentation. The VFAs produced were recombined with the mannitol to produce lipids through fermentation with the oleaginous yeast, *Cryptococcus curvatus*. During the alginate fermentation process several by-products were obtained including; acetate, succinate, lactate, formate, propionate, butyrate and ethanol. A maximum lipids yield of 48.3% was achieved. The lipids obtained were very high in oleic acid (48.7%), palmitic acid (18.2%) and linoleic acid (17.5%) which indicates a fatty acids composition similar to vegetable oil [72]. The lipids can therefore be used for a myriad of applications including culinary processes and biodiesel production.

Dong *et al.* [50] were able to effectively hydrolyze the microalgae *S. acutus* to obtain reducing sugars while making the lipids more easily extractable. An ethanol concentration of 22.7 g/l was obtained from the algae while the recovery of lipids was in the range of 82–87% of total lipids after ethanol removal [50]. There was no adverse effect observed on lipids recovery due to either the acid pretreatment or the fermentation of soluble sugar processes which preceded the lipids extrac-tion. The fatty acid methyl esters concentration was also found to be high for the lipids recovered which makes it a good substrate for biodiesel production. Lee *et al.* [73] also recovered similar products of lipids and ethanol from the microalgae, *D. tertiolecta*. In their study, 48 g lipids were extracted from 220 g of the microalgae while the residual biomass after lipid extraction was found to have a carbohydrates content of 51.9%. Upon fermentation with *S. cerevisiae*, 0.14 g ethanol/g residual biomass (0.44 g ethanol/g glucose) was obtained from the residual biomass [73]. The successful demonstration of potential biodiesel and bioethanol co-production from microalgae indicatives high potential improvements in the economic feasibility of microalgal biorefineries.

The red macroalgae, *Gracilaria verrucosa* was used a substrate for the co-production of agar (a hydrocolloid) and ethanol [74]. In the study, 33% agar was extracted from the biomass while the residual pulp was enzymatically hydrolysed to obtain 0.87 g reducing sugars/g cellulose. The hydrolysate obtained from the pulp was fermented with *S. cerevisiae* to produce ethanol with a yield of 0.43 g/g reducing sugars. A mass balance assessment in the study indicated that for every 1000 kg of dried algal biomass, 280 kg of agar can be obtained together with 38 kg of ethanol. Additionally, 20 and 25 kg of lipid and protein, respectively can be obtained from the residual pulp after agar extraction [74]. In a similar approach, the hydrocolloid, carrageenan was first extracted from the seaweed *E. cottonii* before the application of the residual pulp in ethanol production [3]. The carrageenan extraction led to an increase in the cellulose fraction to 64% in the residual seaweed pulp. Ethanol yields of 0.25–0.27 g/g residual seaweed pulp were obtained using *S. cerevisiae* as the fermenting organism [3].

Co-production of biosolar hydrogen and biogas was explored on the microalgae *C. reinhardtii* as a means to evaluate the integrated biorefinery approach to processing the biomass [75]. Hydrogen was first produced using the sulphur deprivation method. This method involves the cultivation of algal cells in a sulphur-containing medium until the cells reach the stationary growth phase. Cell pellets are then harvested and re-suspended in sulphur-free medium followed by incubation in light at 600 μmol/m^2/s under room temperature. The production of hydrogen prior to anaerobic digestion of the microalage resulted in a 123% increase in biogas generation from an initial 587 ml biogas/g volatile solids with 66% CH_4 content [75].

In another biorefinery process, the microalgae *Nannochloropsis* sp. was used as substrate for the recovery of three different valuable products [76]. Supercritical CO_2 was used to extract 45 g lipids/100 g dry biomass and 70% of pigments which were mainly carotenoids. The residual microalgal biomass after extraction was used as an efficient substrate to produce hydrogen at a yield of 60.6 ml/g dry biomass through dark fermentation with *Enterobacter aerogenes* [76]. Harnessing these valuable products from a single biomass shows high economic prospects for microalgal biorefineries.

7. Additional prospects for biorefinery applications

Prospects in other biomass conversion pathways such as thermochemical, mechanical and chemical cannot be completely ignored and in some cases entirely replaced with the biochemical processes proposed (**Figure 1**). Thermochemical processes such gasification which involves the application of heat to biomass at high temperatures (> 700°C) in the presence of low oxygen concentrations can be used to obtain syngas (mixture of methane, hydrogen, carbon dioxide and carbon monoxide) [77]. Syngas can be used as a standalone fuel or a platform chemical for the production of alcohols and organic acids. Alternatively, biomass can be subjected to a pyrolysis process which involves the use of temperatures between 300 and 600°C in the absence of oxygen to convert the biomass to a liquid bio-oil with biochar and light gases as by-products [78]. Such thermochemical processes could be considered as downstream processes after lignocellulosic and algal biomass fermentation where large non-cellulose fractions are generated as side-streams. A variant thermochemical process is hydrothermal treatment or upgradation. It involves the use of high temperature (200–600°C) and pressure (5–40 MPa) liquids often in the form of supercritical water to produce various liquid fuels [33].

Mechanical processes which do not typically change the composition of biomass but tend to reduce sizes or separate impurities or other components are usually applied in most biorefinery processes. It is particularly popular when handling and pre-treating lignocellulosic biomass [79]. However, there are mechanical processes that are considered complete standalone processes which generates their own useful products. A typical example is briquetting. Briquettes are often in the form of relatively evenly sized pellets produced by the compression of carbon-rich biomass. They are known to burn longer and produce a lower net greenhouse gas emissions which promotes their use as good substitutes to coal, charcoal and raw firewood [80]. Such a process could be used as a downstream process after lignocellulosic and algal biomass fermentation to minimise waste generation and add more value to residual materials.

8. Sustainability and circular economy perspectives of cellulosic and algal biorefineries

Circular economies principally emphasise the development of economic systems that eliminate waste and continuously utilise resources. In the context of biomass resources, an alternative term often used is Circular bioeconomy. Biomass is emerging as the primary renewable resource to tackle several challenges especially with regards to greenhouse gas emissions and depleting fossil fuels [6]. Therefore several technologies and multi-technology integration systems are being promulgated as the backbone for a Circular bioeconomy. The European Union describes this Circular bioeconomy as one that encompasses the formation of various renewable biological resources and their conversion to several high-value bio-based products such as food, feed, chemicals, and energy [81]. At the heart of this economic model is the biorefinery concept which has been elaborately described in this review. The biorefinery concept's role especially for algal and lignocellulosic biomass processing is to optimise the conversion of these biomass to achieve the goals principally set for the circular bioeconomy [82]. Lignocellulosic biomass utilisation will be key to the success of the bioeconomy because they are the primary components of most biological wastes generated especially from crop production and processing. The unique benefits derived from the use of algal biomass in particular includes no arable land requirements, high biomass productivity and no reliance on fresh water and fertiliser sources [2]. This makes it an equally important resource for the circular bioeconomy.

The circular bioeconomy and the circular economy in a broader context have direct positive ripple effects on the social, economic and environmental concerns associated with current economic development models. These three aspects of any development process form the pillars of sustainability. It is therefore nearly impossible to dissociate the circular economy from sustainability. The role of lignocellulosic and algal biorefineries in sustainable development can be found directly in a number of the Sustainability Development Goals (SDGs) proposed by the United Nations. They include: *Zero hunger* (Goal 2) through the provision of affordable feed for livestock farming; *Clean water and sanitation* (Goal 6) through the utilisa-tion of algal blooms which forms a major health hazard for coastal communities; *Affordable and clean energy* (Goal 7) through the conversion of cellulosic and algal biomass to biofuels; *Decent work and economic growth* (Goal 8) through the creation of small and medium scale biorefinery businesses and employment opportunities; *Industry, innovation and infrastructure* (Goal 9) through the creation of new and innovative co-product pathways using the biorefinery approach; *Sustainable cities and communities* (Goal 11) through energy recovery from the biodegradable frac-tions of municipal solid wastes; *Responsible Consumption and Production* (Goal 12) through the multi-product recovery from the same biomass leading to a reduction in waste fractions and; *Climate Action* (Goal 13) through the reduction in greenhouse gas emissions from crop production residue decay and direct combustion [83].

A reduction or absence of waste streams especially for agro residual biomass which is promoted by Goal 12 of the SDGs is a direct attribute of the zero waste concept. This concept refers to the design and management of products and processes in a systematic form to avoid and eliminate waste, and to recover all resources from the waste stream [84]. Resource recovery from waste streams is the primary point of intersection between the integrated biorefinery concept and the zero waste concept. The utilisation of cellulosic agro residues such stalks from various cereals reduces the apparent greenhouse gas emissions from their decay or direct combus-tion. This forms a simple yet effective climate change mitigation measure for both developed and developing countries.

9. Conclusions

The studies described in this chapter have highlighted the considerable benefits from the use of integrated processing technologies on lignocellulosic and algal biomass. The most obvious feature is the increased use of the substrate and the minimization of waste generated. The less obvious feature is the improvements in the economic sustainability of commercial cellulosic and algal biorefineries. These studies show that the potential range of products including fuels, chemicals and polymers that current and future biorefineries could produce is currently very extensive. Research and development efforts are adding almost daily to products and co-products of known fermentation-based biorefinery pathways. The most important consideration which has pushed research even further is the importance attached to the sustainability of processes in recent years. Sustainability is now an equally important consideration in addition to economic feasibility, product yield, process efficiency and selectivity. This is due to the importance of developing climate smart yet cost-effective technologies and processes which will protect and preserve ecosystems for present and future generations. The integrated biorefinery approach has therefore become indispensable to productive and sustainable biomass processing.

Acknowledgements

The author is grateful to Moses Mensah, Robert Aryeetey and George vanDyck for their continuous technical support.

Conflict of interest

The author declares no conflict of interest.

Author details

Felix Offei
Department of Marine Engineering, Regional Maritime University, Accra, Ghana

*Address all correspondence to: felix.offei@rmu.edu.gh

References

[1] Sawin J, Rutovitz J, Sverrisson F. Advancing the global renewable energy transition. Tech. rep; 2018.

[2] Kraan S. Mass-cultivation of carbohydrate rich macroalgae, a possible solution for sustainable biofuel production. Mitigation and Adaptation Strategies for Global Change. 2013 Jan;18(1):27-46.

[3] Tan IS, Lee KT. Enzymatic hydrolysis and fermentation of seaweed solid wastes for bioethanol production: An optimization study. Energy. 2014 Dec 15;78:53-62.

[4] Gavrilescu M. Biorefinery systems: an overview. Bioenergy research: advances and applications. 2014 Jan 1:219-241.

[5] Pande M, Bhaskarwar AN. Biomass conversion to energy. In Biomass conversion 2012 (pp. 1-90). Springer, Berlin, Heidelberg.

[6] Cherubini F. The biorefinery concept: using biomass instead of oil for producing energy and chemicals. Energy conversion and management. 2010 Jul 1;51(7):1412-1421.

[7] Maeda RN, Serpa VI, Rocha VA, Mesquita RA, Santa Anna LM, De Castro AM, Driemeier CE, Pereira Jr N, Polikarpov I. Enzymatic hydrolysis of pretreated sugar cane bagasse using Penicillium funiculosum and Trichoderma harzianum cellulases. Process Biochemistry. 2011 May 1;46(5):1196-1201.

[8] García-Aparicio MP, Oliva JM, Manzanares P, Ballesteros M, Ballesteros I, González A, Negro MJ. Second-generation ethanol production from steam exploded barley straw by Kluyveromyces marxianus CECT 10875. Fuel. 2011 Apr 1;90(4):1624-1630

[9] Abbas A, Ansumali S. Global potential of rice husk as a renewable feedstock for ethanol biofuel production. BioEnergy Research. 2010 Dec;3(4):328-334.

[10] Wang L, Yang M, Fan X, Zhu X, Xu T, Yuan Q. An environmentally friendly and efficient method for xylitol bioconversion with high-temperature-steaming corncob hydrolysate by adapted Candida tropicalis. Process Biochemistry. 2011 Aug 1;46(8):1619-1626

[11] Liu CZ, Cheng XY. Improved hydrogen production via thermophilic fermentation of corn stover by microwave-assisted acid pretreatment. International Journal of Hydrogen Energy. 2010 Sep 1;35(17):8945-8952.

[12] Gullón B, Yáñez R, Alonso JL, Parajó JC. Production of oligosaccharides and sugars from rye straw: a kinetic approach. Bioresource technology. 2010 Sep 1;101(17):6676-6684.

[13] Lu X, Zhang Y, Angelidaki I. Optimization of H2SO4-catalyzed hydrothermal pretreatment of rapeseed straw for bioconversion to ethanol: focusing on pretreatment at high solids content. Bioresource technology. 2009 Jun 1;100(12):3048-3053.

[14] Talebnia F, Karakashev D, Angelidaki I. Production of bioethanol from wheat straw: an overview on pretreatment, hydrolysis and fermentation. Bioresource technology. 2010 Jul 1;101(13):4744-4753.

[15] Yadav KS, Naseeruddin S, Prashanthi GS, Sateesh L, Rao LV. Bioethanol fermentation of concentrated rice straw hydrolysate using co-culture of Saccharomyces cerevisiae and Pichia stipitis. Bioresource technology. 2011 Jun 1;102(11):6473-6478.

[16] Caparrós S, Ariza J, López F, Nacimiento JA, Garrote G, Jiménez L. Hydrothermal treatment and ethanol pulping of sunflower stalks. Bioresource technology. 2008 Mar 1;99(5): 1368-1372.

[17] Goshadrou A, Karimi K, Taherzadeh MJ. Bioethanol production from sweet sorghum bagasse by Mucor hiemalis. Industrial Crops and Products. 2011 Jul 1;34(1):1219-1225.

[18] Panichelli L, Gnansounou E. Estimating greenhouse gas emissions from indirect land-use change in biofuels production: concepts and exploratory analysis for soybean-based biodiesel. 2008.

[19] Monsalve JF, Medina de Perez VI, Ruiz Colorado AA. Ethanol production of banana shell and cassava starch. Dyna. 2006 Nov;73(150):21-27.

[20] Menon V, Rao M. Trends in bioconversion of lignocellulose: biofuels, platform chemicals & biorefinery concept. Progress in energy and combustion science. 2012 Aug 1;38(4):522-550.

[21] Jeon YJ, Xun Z, Rogers PL. Comparative evaluations of cellulosic raw materials for second generation bioethanol production. Letters in applied microbiology. 2010 Nov;51(5):518-524.

[22] nee'Nigam PS, Gupta N, Anthwal A. Pre-treatment of agro-industrial residues. InBiotechnology for agro-industrial residues utilisation 2009 (pp. 13-33). Springer, Dordrecht.

[23] Lee DH, Cho EY, Kim CJ, Kim SB. Pretreatment of waste newspaper using ethylene glycol for bioethanol production. Biotechnology and Bioprocess Engineering. 2010 Dec 1;15(6):1094-1101.

[24] Peng L, Chen Y. Conversion of paper sludge to ethanol by separate hydrolysis and fermentation (SHF) using Saccharomyces cerevisiae. Biomass and Bioenergy. 2011 Apr 1;35(4):1600-1606.

[25] Pires EJ, Ruiz HA, Teixeira JA, Vicente AA. A new approach on brewer's spent grains treatment and potential use as lignocellulosic yeast cells carriers. Journal of agricultural and food chemistry. 2012 Jun 13;60(23):5994-5999.

[26] Malherbe S, Cloete TE. Lignocellulose biodegradation: fundamentals and applications. Reviews in Environmental Science and Biotechnology. 2002 Jun;1(2): 105-114.

[27] Cara C, Ruiz E, Oliva JM, Sáez F, Castro E. Conversion of olive tree biomass into fermentable sugars by dilute acid pretreatment and enzymatic saccharification. Bioresource technology. 2008 Apr 1;99(6): 1869-1876.

[28] Aswathy US, Sukumaran RK, Devi GL, Rajasree KP, Singhania RR, Pandey A. Bio-ethanol from water hyacinth biomass: an evaluation of enzymatic saccharification strategy. Bioresource technology. 2010 Feb 1;101(3):925-930.

[29] Shafiei M, Karimi K, Taherzadeh MJ. Pretreatment of spruce and oak by N-methylmorpholine-N-oxide (NMMO) for efficient conversion of their cellulose to ethanol. Bioresource Technology. 2010 Jul 1;101(13): 4914-4918.

[30] Knauf M, Moniruzzaman M. Lignocellulosic biomass processing: a perspective. International sugar journal. 2004;106(1263):147-150.

[31] Arantes V, Saddler JN. Access to cellulose limits the efficiency of enzymatic hydrolysis: the role of amorphogenesis. Biotechnology for biofuels. 2010 Dec;3(1):1-1.

[32] Adams JM, Ross AB, Anastasakis K, Hodgson EM, Gallagher JA, Jones JM, Donnison IS. Seasonal variation in the chemical composition of the bioenergy feedstock Laminaria digitata for thermochemical conversion. Bioresource technology. 2011 Jan 1;102(1):226-234.

[33] Singh R, Bhaskar T, Bala gurumurthy B. Hydrothermal upgradation of algae into value-added hydrocarbons. In Biofuels from Algae 2014 Jan 1 (pp. 235-260). Elsevier.

[34] Ab Kadir MI, Ahmad WW, Ahmad MR, Misnon MI, Ruznan WS, Jabbar HA, Ngalib K, Ismail A. Utilization of eco-colourant from green seaweed on textile dyeing. In Proceedings of the International Colloquium in Textile Engineering, Fashion, Apparel and Design 2014 (ICTEFAD 2014) 2014 (pp. 79-83). Springer, Singapore.

[35] Schultz-Jensen N, Thygesen A, Leipold F, Thomsen ST, Roslander C, Lilholt H, Bjerre AB. Pretreatment of the macroalgae Chaetomorpha linum for the production of bioethanol–Comparison of five pretreatment technologies. Bioresource Technology. 2013 Jul 1;140:36-42.

[36] FAO, Year book of Fishery Statistics. Available from: ftp://ftp.fao.org/FI/STAT/summary/default.htm. 2013. Accessed 28 January 2016

[37] Adams JM, Gallagher JA, Donnison IS. Fermentation study on Saccharina latissima for bioethanol production considering variable pre-treatments. Journal of applied Phycology. 2009 Oct;21(5):569-574.

[38] Lee SY, Chang JH, Lee SB. Chemical composition, saccharification yield, and the potential of the green seaweed Ulva pertusa. Biotechnology and bioprocess engineering. 2014 Nov 1;19(6): 1022-1033.

[39] Borines MG, De Leon RL, McHenry MP. Bioethanol production from farming non-food macroalgae in Pacific island nations: Chemical constituents, bioethanol yields, and prospective species in the Philippines. Renewable and Sustainable Energy Reviews. 2011 Dec 1;15(9):4432-4435.

[40] Matanjun P, Mohamed S, Mustapha NM, Muhammad K. Nutrient content of tropical edible seaweeds, Eucheuma cottonii, Caulerpa lentillifera and Sargassum polycystum. Journal of Applied Phycology. 2009 Feb;21(1):75-80.

[41] Yahmed NB, Jmel MA, Alaya MB, Bouallagui H, Marzouki MN, Smaali I. A biorefinery concept using the green macroalgae Chaetomorpha linum for the coproduction of bioethanol and biogas. Energy Conversion and Management. 2016 Jul 1;119:257-265.

[42] Hong IK, Jeon H, Lee SB. Comparison of red, brown and green seaweeds on enzymatic saccharification process. Journal of Industrial and Engineering Chemistry. 2014 Sep 25;20(5):2687-2691.

[43] Offei F, Mensah M, Kemausuor F. Cellulase and acid-catalysed hydrolysis of Ulva fasciata, Hydropuntia dentata and Sargassum vulgare for bioethanol production. SN Applied Sciences. 2019 Nov;1(11):1-3.

[44] Kim NJ, Li H, Jung K, Chang HN, Lee PC. Ethanol production from marine algal hydrolysates using Escherichia coli KO11. Bioresource technology. 2011 Aug 1;102(16): 7466-7469.

[45] El Harchi M, Kachkach FF, El Mtili N. Optimization of thermal acid hydrolysis for bioethanol production from Ulva rigida with yeast Pachysolen tannophilus. South African Journal of Botany. 2018 Mar 1;115:161-169.

[46] Offei F, Mensah M, Thygesen A, Kemausuor F. Seaweed bioethanol production: A process selection review on hydrolysis and fermentation. Fermentation. 2018 Dec;4(4):99.

[47] Rhein-Knudsen N, Ale MT, Ajalloueian F, Yu L, Meyer AS. Rheological properties of agar and carrageenan from Ghanaian red seaweeds. Food Hydrocolloids. 2017 Feb 1;63:50-58.

[48] Fayaz M, Namitha KK, Murthy KC, Swamy MM, Sarada R, Khanam S, Subbarao PV, Ravishankar GA. Chemical composition, iron bioavailability, and antioxidant activity of Kappaphycus alvarezzi (Doty). Journal of agricultural and food chemistry. 2005 Feb 9;53(3):792-797.

[49] Manns D, Deutschle AL, Saake B, Meyer AS. Methodology for quantitative determination of the carbohydrate composition of brown seaweeds (Laminariaceae). Rsc Advances. 2014;4(49):25736-25746.

[50] Dong T, Knoshaug EP, Davis R, Laurens LM, Van Wychen S, Pienkos PT, Nagle N. Combined algal processing: A novel integrated biorefinery process to produce algal biofuels and bioproducts. Algal Research. 2016 Nov 1;19:316-323.

[51] Biller P, Ross AB. Pyrolysis GC–MS as a novel analysis technique to determine the biochemical composition of microalgae. Algal Research. 2014 Oct 1;6:91-97.

[52] Sydney EB, Sturm W, de Carvalho JC, Thomaz-Soccol V, Larroche C, Pandey A, Soccol CR. Potential carbon dioxide fixation by industrially important microalgae. Bioresource technology. 2010 Aug 1;101(15):5892-5896.

[53] Yao L, Gerde JA, Lee SL, Wang T, Harrata KA. Microalgae lipid characterization. Journal of agricultural and food chemistry. 2015 Feb 18;63(6):1773-1787.

[54] Shuping Z, Yulong W, Mingde Y, Kaleem I, Chun L, Tong J. Production and characterization of bio-oil from hydrothermal liquefaction of microalgae Dunaliella tertiolecta cake. Energy. 2010 Dec 1;35(12):5406-5411.

[55] Dahiya A. Algae biomass cultivation for advanced biofuel production. InBioenergy 2015 Jan 1 (pp. 219-238). Academic Press.

[56] Chirapart A, Praiboon J, Puangsombat P, Pattanapon C, Nunraksa N. Chemical composition and ethanol production potential of Thai seaweed species. Journal of applied phycology. 2014 Apr;26(2):979-986.

[57] Lee KT, Ofori-Boateng C. Biofuels: production technologies, global profile, and market potentials. In Sustainability of biofuel production from oil palm biomass 2013 (pp. 31-74). Springer, Singapore.

[58] Yazdani P, Zamani A, Karimi K, Taherzadeh MJ. Characterization of Nizimuddinia zanardini macroalgae biomass composition and its potential for biofuel production. Bioresource technology. 2015 Jan 1;176:196-202.

[59] Trivedi N, Gupta V, Reddy CR, Jha B. Enzymatic hydrolysis and production of bioethanol from common macrophytic green alga Ulva fasciata Delile. Bioresource technology. 2013 Dec 1;150:106-112.

[60] Lynd LR, Weimer PJ, Van Zyl WH, Pretorius IS. Microbial cellulose utilization: fundamentals and biotechnology. Microbiology and molecular biology reviews. 2002 Sep 1;66(3):506-577.

[61] Kim HM, Wi SG, Jung S, Song Y, Bae HJ. Efficient approach for bioethanol production from red

seaweed Gelidium amansii. Bioresource technology. 2015 Jan 1;175:128-134.

[62] Bevan MW, Franssen MC. Investing in green and white biotech. Nature biotechnology. 2006 Jul;24(7):765-767.

[63] Hossain GS, Liu L, Du GC. Industrial bioprocesses and the biorefinery concept. In Current Developments in Biotechnology and Bioengineering 2017 Jan 1 (pp. 3-27). Elsevier.

[64] Shukla R, Cheryan M. Zein: the industrial protein from corn. Industrial crops and products. 2001 May 1;13(3):171-192.

[65] Agarwal L, Dutt K, Meghwanshi GK, Saxena RK. Anaerobic fermentative production of lactic acid using cheese whey and corn steep liquor. Biotechnology letters. 2008 Apr;30(4):631-635.

[66] Unrean P, Ketsub N. Integrated lignocellulosic bioprocess for co-production of ethanol and xylitol from sugarcane bagasse. Industrial crops and products. 2018 Nov 1;123:238-246.

[67] Zambare VP, Christopher LP. Integrated biorefinery approach to utilization of pulp and paper mill sludge for value-added products. Journal of Cleaner Production. 2020 Nov 20;274:122791.

[68] Liao W, Liu Y, Frear C, Chen S. Co-production of fumaric acid and chitin from a nitrogen-rich lignocellulosic material–dairy manure–using a pelletized filamentous fungus Rhizopus oryzae ATCC 20344. Bioresource technology. 2008 Sep 1;99(13):5859-5866.

[69] Lopez-Hidalgo AM, Sánchez A, De León-Rodríguez A. Simultaneous production of bioethanol and biohydrogen by Escherichia coli WDHL using wheat straw hydrolysate as substrate. Fuel. 2017 Jan 15;188:19-27.

[70] Ahmad E, Pant KK. Lignin conversion: a key to the concept of lignocellulosic biomass-based integrated biorefinery. In Waste biorefinery 2018 Jan 1 (pp. 409-444). Elsevier.

[71] Ashokkumar V, Salim MR, Salam Z, Sivakumar P, Chong CT, Elumalai S, Suresh V, Ani FN. Production of liquid biofuels (biodiesel and bioethanol) from brown marine macroalgae Padina tetrastromatica. Energy conversion and Management. 2017 Mar 1;135:351-361.

[72] Xu X, Kim JY, Oh YR, Park JM. Production of biodiesel from carbon sources of macroalgae, Laminaria japonica. Bioresource technology. 2014 Oct 1;169:455-461.

[73] Lee OK, Kim AL, Seong DH, Lee CG, Jung YT, Lee JW, Lee EY. Chemo-enzymatic saccharification and bioethanol fermentation of lipid-extracted residual biomass of the microalga, Dunaliella tertiolecta. Bioresource technology.2013 Mar1;132:197-201.

[74] Kumar S, Gupta R, Kumar G, Sahoo D, Kuhad RC. Bioethanol production from Gracilaria verrucosa, a red alga, in a biorefinery approach. Bioresource technology. 2013 May 1;135:150-156.

[75] Mussgnug JH, Klassen V, Schlüter A, Kruse O. Microalgae as substrates for fermentative biogas production in a combined biorefinery concept. Journal of biotechnology. 2010 Oct 1;150(1):51-56.

[76] Nobre BP, Villalobos F, Barragan BE, Oliveira AC, Batista AP, Marques PA, Mendes RL, Sovová H, Palavra AF, Gouveia L. A biorefinery from Nannochloropsis sp. microalga–extraction of oils and pigments. Production of biohydrogen from the

leftover biomass. Bioresource technology. 2013 May 1;135:128-136.

[77] Spath PL, Dayton DC. Technical and economic assessment of synthesis gas to fuels and chemicals with emphasis on the potential for biomass-derived syngas. Preliminary Screening. 2003. http://handle.dtic.mil/100.2/ADA436529.

[78] Yan G, Yao W, Fei W, Yong J. Research progress in biomass flash pyrolysis technology for liquids production. Chemical Industry and Engineering Progress. 2001 Jan 1(8):13-17.

[79] Senneca O. Kinetics of pyrolysis, combustion and gasification of three biomass fuels. Fuel processing technology. 2007 Jan 1;88(1):87-97.

[80] Azasi VD, Offei F, Kemausuor F, Akpalu L. Bioenergy from crop residues: A regional analysis for heat and electricity applications in Ghana. Biomass and Bioenergy. 2020 Sep 1;140:105640.

[81] European Commission. A Sustainable Bioeconomy for Europe: strengthening the connection between economy, society and the environment, Brussels, Belgium. 2018.

[82] Ubando AT, Felix CB, Chen WH. Biorefineries in circular bioeconomy: A comprehensive review. Bioresource technology. 2020 Mar 1;299:122585.

[83] United Nations Development Programme. Transforming our world: The 2030 agenda for sustainable development. UN Summit, New York, 2015.

[84] Zaman AU, Lehmann S. The zero waste index: a performance measurement tool for waste management systems in a 'zero waste city'. Journal of Cleaner Production. 2013 Jul 1;50:123-132.

PERMISSIONS

All chapters in this book were first published by InTech Open; hereby published with permission under the Creative Commons Attribution License or equivalent. Every chapter published in this book has been scrutinized by our experts. Their significance has been extensively debated. The topics covered herein carry significant findings which will fuel the growth of the discipline. They may even be implemented as practical applications or may be referred to as a beginning point for another development.

The contributors of this book come from diverse backgrounds, making this book a truly international effort. This book will bring forth new frontiers with its revolutionizing research information and detailed analysis of the nascent developments around the world.

We would like to thank all the contributing authors for lending their expertise to make the book truly unique. They have played a crucial role in the development of this book. Without their invaluable contributions this book wouldn't have been possible. They have made vital efforts to compile up to date information on the varied aspects of this subject to make this book a valuable addition to the collection of many professionals and students.

This book was conceptualized with the vision of imparting up-to-date information and advanced data in this field. To ensure the same, a matchless editorial board was set up. Every individual on the board went through rigorous rounds of assessment to prove their worth. After which they invested a large part of their time researching and compiling the most relevant data for our readers.

The editorial board has been involved in producing this book since its inception. They have spent rigorous hours researching and exploring the diverse topics which have resulted in the successful publishing of this book. They have passed on their knowledge of decades through this book. To expedite this challenging task, the publisher supported the team at every step. A small team of assistant editors was also appointed to further simplify the editing procedure and attain best results for the readers.

Apart from the editorial board, the designing team has also invested a significant amount of their time in understanding the subject and creating the most relevant covers. They scrutinized every image to scout for the most suitable representation of the subject and create an appropriate cover for the book.

The publishing team has been an ardent support to the editorial, designing and production team. Their endless efforts to recruit the best for this project, has resulted in the accomplishment of this book. They are a veteran in the field of academics and their pool of knowledge is as vast as their experience in printing. Their expertise and guidance has proved useful at every step. Their uncompromising quality standards have made this book an exceptional effort. Their encouragement from time to time has been an inspiration for everyone.

The publisher and the editorial board hope that this book will prove to be a valuable piece of knowledge for researchers, students, practitioners and scholars across the globe.

LIST OF CONTRIBUTORS

María Jesús Callejo, Wendu Tesfaye, María Carmen González and Antonio Morata
Universidad Politécnica de Madrid, Spain

Octavio García-Depraect, Daryl Rafael Osuna-Laveaga and Elizabeth León-Becerril
Department of Environmental Technology, Centro de Investigación y Asistencia en Tecnología y Diseño del Estado de Jalisco, A.C., Guadalajara, Jalisco, México

Yonglan Tian and Huayong Zhang
Research Center for Engineering Ecology and Nonlinear Science, North China Electric Power University, Beijing, China

Edmond Sanganyado
Marine Biology Institute, Shantou University, Shantou, Guangdong, China

Pascal Drouin and Renato J. Schmidt
Lallemand Animal Nutrition, Milwaukee, Wisconsin, United States

Lucas J. Mari
Lallemand Animal Nutrition, Aparecida de Goiânia, Goiás, Brazil

Mark Strobl
Hochschule Geisenheim University, Geisenheim, Germany

Spyridon Achinas and Gerrit Jan Willem Euverink
Engineering and Technology institute Groningen, University of Groningen, Groningen, Netherlands

Keukeu Kaniawati Rosada
Department of Biology, Padjadjaran University, Sumedang, Indonesia

Mohamed Hawashi, Tri Widjaja and Setiyo Gunawan
Department of Chemical Engineering, Institut Teknologi Sepuluh Nopember (ITS), Surabaya, Indonesia

Jerry O. Ugwuanyi and Augustina N. Okpara
Department of Microbiology, University of Nigeria, Nsukka, Nigeria

Jan Barekzai, Florian Petry, Jan Zitzmann and Denise Salzig
Institute of Bioprocess Engineering and Pharmaceutical Technology, University of Applied Sciences Mittelhessen, Giessen, Germany

Peter Czermak
Institute of Bioprocess Engineering and Pharmaceutical Technology University of Applied Sciences Mittelhessen, Giessen, Germany
Faculty of Biology and Chemistry, Justus-Liebig-University Giessen, Giessen, Germany
Project Group Bioresources, Fraunhofer Institute for Molecular Biology and Applied Ecology (IME), Giessen, Germany

Michael Bamitale Osho
Department of Biological Sciences (Microbiology Unit), College of Natural and Applied Sciences, McPherson University, Seriki Sotayo, Abeokuta, Ogun State, Nigeria

Sarafadeen Olateju Kareem
Department of Microbiology, Federal University of Agriculture, Abeokuta, Nigeria

Felix Offei
Department of Marine Engineering, Regional Maritime University

Index

Printed in the USA
CPSIA information can be obtained
at www.ICGtesting.com
JSHW051354091023
49903JS00006B/143